AN ENGINEERED WORLD

Engineering Studies

Edited by Gary Lee Downey and Matthew Wisnioski

Matthew Wisnioski, *Engineers for Change: Competing Visions of Technology in 1960s America*

Amy Sue Bix, *Girls Coming to Tech! A History of American Engineering Education for Women*

Jessica Smith, *Extracting Accountability: Engineers and Corporate Social Responsibility*

Elizabeth Reddy, *¡Alerta! Engineering on Shaky Ground*

Konstantinos Chatzis, *Forecasting Travel in Urban America: The Socio-Technical Life of an Engineering Modeling World*

Ann Johnson and Johannes Lenhard, *Predictions: How Mathematical Tools Conquer and Divide*

Edward Beatty and Israel G. Solares, *An Engineered World: The Role of Engineers in Global Modernity*

AN ENGINEERED WORLD

THE ROLE OF ENGINEERS IN GLOBAL MODERNITY

EDITED BY EDWARD BEATTY AND ISRAEL G. SOLARES

The MIT Press
Cambridge, Massachusetts
London, England

The MIT Press
Massachusetts Institute of Technology
77 Massachusetts Avenue, Cambridge, MA 02139
mitpress.mit.edu

The MIT Press would like to thank the anonymous peer reviewers who provided comments on drafts of this book. The generous work of academic experts is essential for establishing the authority and quality of our publications. We acknowledge with gratitude the contributions of these otherwise uncredited readers.

This book was set in Stone Serif and Stone Sans by Westchester Publishing Services. Printed and bound in the United States of America.

Library of Congress Cataloging-in-Publication Data is available.

ISBN: 978-0-262-55335-3

EU Authorised Representative: Easy Access System Europe, Mustamäe tee 50, 10621 Tallinn, Estonia | Email: gpsr.requests@easproject.com

CONTENTS

SERIES FOREWORD

We live in highly engineered worlds. Engineers play crucial roles in the normative direction of localized knowledge and social orders. The *Engineering Studies Series* highlights the growing need to understand the situated commitments and practices of engineers and engineering. It asks: What is engineering for? What are engineers for?

Drawing from a diverse arena of research, teaching, and outreach, engineering studies raise awareness of how engineers imagine themselves in service to humanity and how their service ideals impact the defining and solving of problems with multiple ends and variable consequences. It does so by examining relationships among technical and nontechnical dimensions and how these relationships change over time and from place to place. Its researchers often are critical participants in the practices they study.

The *Engineering Studies Series* publishes research in historical, social, cultural, political, philosophical, rhetorical, and organizational studies of engineers and engineering, paying particular attention to normative directionality in engineering epistemologies, practices, identities, and outcomes. Areas of concern include engineering formation, engineering work, engineering design, equity in engineering (gender, racial, ethnic, class, geopolitical), and engineering service to society.

The *Engineering Studies Series* thus pursues three related missions: (1) advance understanding of engineers, engineering, and outcomes of engineering work; (2) help build and serve communities of researchers and learners in engineering studies; and (3) link scholarly work in engineering studies to broader discussions and debates about engineering education, research, practice, policy, and representation.

Gary Lee Downey and Matt Wisnioski, Editors

ACKNOWLEDGMENTS

We first discussed what a global history of engineering might look like in the fall of 2018 at the Kellogg Institute for International Studies at the University of Notre Dame, where Solares held a Fulbright García Robles Fellowship and Beatty is a faculty fellow. Our conversations often focused on the apparent puzzle of engineers, nearly omnipresent in our earlier, individual research on multinational corporations and Mexican technology history, respectively, but nearly invisible in the historiography on economic and business history. Those initial conversations quickly led to further reading and preliminary data gathering on the professional careers of thousands of individuals who graduated from US engineering schools, which soon led to gathering data on graduates from engineering programs in Mexico, Germany, the UK, and France. We applied for and received three-year funding from the Science and Technology Studies program of the National Science Foundation (grant #204026), allowing us to continue and deepen this project in the years since.

Creating a network of scholars who work on the history of engineering in diverse global contexts and who think about an integrated (or interactive) global context was a central objective of ours from the beginning. This edited volume constitutes one significant step toward that goal. Funding from the National Science Foundation and a grant from the Notre Dame International office allowed us to host two successive workshops to develop the overarching themes and individual chapters for this volume. The Kellogg Institute at Notre Dame provided generous and critical support for each event, with our special appreciation for the fine work and support of Therese Hanlon.

We extend a deep appreciation to all those who participated in the workshop discussions, including (beyond those who contribute chapters here) Gabriela Recio, Ángela Vergara, and Nikhil Menon. Our chapter authors have been generous, incisive, and timely (!) through the entire process. We are grateful for the support and perceptive critique offered by Gary Downey and Matt Wisnioski, the editors of the Engineering Studies series at the MIT Press, from three anonymous reviewers of the manuscript, and the important support and advice from Justin Kehoe and others at the MIT Press.

1 INTRODUCTION: TOWARD AN ENGINEERED WORLD

EDWARD BEATTY AND ISRAEL G. SOLARES

In little over a generation or two, the modern profession of engineering rose out of longstanding but typically more informal practices. Engineering, in its broadest sense, encompasses humans' strategic efforts to manipulate the environment and construct the means to do so more effectively. Engineers are generally a subcategory of "technicians," ranging from skilled labor to mechanics to those with increasingly specialized knowledge about "how things work."[1] But the transformation of the profession in the late nineteenth and early twentieth centuries was primarily founded on new or reformed institutions: university-based training in a relatively standard curriculum with common theory and practice elements; associations that promoted professional ethos, identity, and standards; publications that certified and disseminated new engineering knowledge in the form of multiedition texts and serial journals; and employment within large organizations, both private and public. Together, these institutions created patterns of professional practice and identity that have characterized engineering to the present day.[2]

In this transformation, largely accomplished in the half-century or so after 1870, engineers and engineering moved from the longstanding domains of military and civil works, typically hired by the state, to a much closer association with private industry and, perhaps most of all, to science. By the last decades of the century, university-trained and professionalized engineers had increasingly become the critical intermediaries between a world understood in scientific terms and the application of scientific knowledge to solve instrumental problems. Work that had been called the "useful arts" became "applied science," and to claim engineering expertise increasingly required access to the emerging institutions of the field. By the early twentieth century, engineers had become society's privileged experts in technology and technical knowledge.[3]

This volume examines the dramatic and global expansion of modern professional engineering between roughly 1870 and 1950 or so. Over just a few decades, the number of people who called themselves "engineers" (or who were recognized as such by others) grew from a small and eclectic collection of individuals to one of the most numerous, mobile, and influential professional groups of the twentieth century, central actors in an increasingly science-based world. Tens of thousands of university-trained engineers and other professionalized technical experts, a few famous but most anonymous, worked to further the technical, organizational, and political development of both capitalist and socialist systems in the twentieth century. This was the case in Western Europe and the United States, and it was also true in imperial domains and independent countries around the world, where the institutions of modern engineering were established at roughly the same time. Across the globe, university-trained professional engineers became advance agents of modernity, imperial power, and corporate authority during this era, if not its designers and directors. The world we inherit and inhabit today has been significantly shaped by the expertise, authority, and ubiquity of engineers and the role they came to occupy in organizations and societies.

Histories of engineers and engineering are typically told within national settings, focused on the distinctive elements of British, French, German, American, Indian, Mexican, or Chinese engineering, for example.[4] We argue that modern, professional engineering emerged as a fundamentally global phenomenon without disputing the importance of distinct national cultures and nation-building projects. Global movements, exchange, and influence were as important as national traditions and interests. The case studies in this volume illustrate the global mobility and interactions of engineers and engineering institutions, knowledge, and cultures. No other profession in the nineteenth and early twentieth centuries was as globally mobile. This mobility was not just transnational but was globally dispersed, resulting from activities of diffusion and exchange. National histories of engineering cannot be understood apart from this global story. Furthermore, engineers played a central role in the design and management of distinct versions of globality in the late nineteenth and early twentieth centuries: imperial expansion to the east and south, cosmopolitan corporate power reaching across the globe, and increasing nationalistic control of industries and industrial supply chains in a globally competitive world.[5] The lines within and between these realms were always indistinct and overlapping.

Despite their importance, however, engineers remain understudied relative to their historical role. As one review of the literature notes, studies of engineers and engineering have too often been relegated to "intellectual marginality and conceptual insignificance," with some notable exceptions.[6] Histories of engineers have remained specialized, generally obscure, and frequently constrained within institutional or national contexts. Engineers are often invisible to the users of their work, and despite the fully engineered state of the world in which we live, they are too often seen merely as interchangeable pieces of technical expertise. Scholars' attention has more often been drawn to entrepreneurs, business tycoons or corporate managers, skilled craftsmen, or common laborers. Engineers often fall somewhere between management and labor, bearing elements of each but claimed by nor identified as either. Histories of expert knowledge focus overwhelmingly on scientists (engineers' older and more popular cousins), and histories of engineering have typically been situated within institutional or national traditions rather than in a global context and overwhelmingly focused on engineering in the West.[7]

In contrast, the research gathered in this volume offers a global perspective on modern engineering and engineers' work. Engineers, after all, labored at the very center of global capitalist and communist systems.[8] They played a critical role in designing and constructing the transportation, communication, and military infrastructure for the global movement of goods, were employed as the agents (and often managers) of capital, and carried scientific and technical knowledge and expertise with them as they moved and worked across the world. Their technical expertise, bought and sold in the employment market, often translated into technocratic authority within corporations, nation-states, and imperial systems. It is no coincidence that engineering as a modern profession became critical to private- and public-sector organizations, planning, and leadership during the so-called first globalization, circa 1870–1930.[9] The chapters in this volume suggest that there was arguably no occupation in the world more "global" (and perhaps more influential) in the early twentieth century than engineering.

This book sketches a global history of engineering circa 1870–1950 through the lens of eight case studies of engineers working in distinct but interconnected spaces around the world.[10] Each chapter in this book examines engineers working in distinct fields, from mining and metallurgy to civil and mechanical engineers and meteorologists who drew on their engineering education. Their work was situated in East Asia, India, South Africa, Latin America, London, and the United States, but all interacted with engineering knowledge and interests transnationally. Despite the variety among these case histories, they exhibit strong common elements.

Three broad themes emerge from these chapters that begin to provide the outlines of a global history of the profession. First, they highlight interconnected communities and networks that engineers embodied and created, with particular attention to the emergence of engineering institutions and practices: the university, professional associations, technical publications, and organizational employment within private or public bureaucracies. Second, the chapters illustrate the rapid expansion of modern engineering across time and space in an integrated and global fashion. In each chapter, we see the interplay between different temporal and spatial scales. They illustrate the iterative establishment of engineering institutions in countries around the world, each drawing transnationally on earlier models and, in turn, informing others to come. They show the constant mobility and exchange between local efforts and global models, ranging across local, regional, national, imperial, corporate, and global spaces. In other words, the chapters give a dynamic sense of how engineering became globally integrated. Third, the chapters trace the transmission of norms and rules through global influences and interactions, emphasizing how modern engineering is as much a cultural, political, and social practice as it is a technical occupation. As engineering institutions and practices replicated and reinforced the dominant middle-class, gendered, and racialized origins of the profession, the mobility of engineers and their institutional models carried these cultural norms around the world and into new areas of application, including public-sector policy-making and economic planning. Each of these themes has previously been identified and examined by scholars within national contexts. We aim for a truly global perspective. This introductory chapter explores these three themes in broad strokes before presenting an overview of the eight case studies that follow.[11]

INSTITUTIONS AND THE MODERN PRACTICE OF ENGINEERING

Engineering—humans' efforts to manipulate their environment to their own advantage—can be traced to the dawn of the species. Toolmaking, early efforts to control instead of simply follow the flora and fauna that fed hunting and gathering peoples, and the purposeful use of water and fire all embody the same impulse as any feat of modern engineering. Efforts to engineer a human-centered environment expanded with the rise of urban population centers. Polities nurtured and rewarded the expertise necessary to design and build the physical infrastructure of ancient cities like Copán, Varanasi, Rome, and Beijing. Through the ages, those we might readily identify as "engineers" undertook the planning, design, and labor management necessary to construct impressive works of hydrology, agricultural engineering, building and monument construction, road design, and military technologies and fortifications. This was the case across the ancient world, from the Mediterranean to East Asia, and the practice of high-level, project-based engineering expertise continued through the Middle Ages and into the early modern era.[12]

The institutions that define engineering in the modern world began to appear in early modern Europe, taking the form of university and training programs as well as organizations to represent and defend the interests of those doing engineering work. These early developments were most advanced in France, where formalized engineering instruction was organized in schools to prepare young men for work in the administration of the country and its colonies.[13] Universities and guild organizations provided institutional foundations for a few, but around the world, most who were involved in engineering work before the modern era acquired their expertise through forms of apprenticeship and direct experience. They often found work or patronage from authorities who sought the technical expertise necessary to achieve their military and civil objectives of state-building or conquest in local or imperial spheres, but engineers' work was not always constrained by the boundaries of sovereign power.

Technical experts have long moved across political boundaries to apply their knowledge, earn a living, or respond to geopolitical interests and incentives. Indeed, the origins of patent laws in the fifteenth century lay in efforts by some European sovereigns to incentivize technical experts from outside their jurisdictions to bring their expertise and exploit it locally. The history of technical expertise in early modern Europe, the North American colonies, and the early United States is a history of mobile technicians: sent, lured, hired, or seeking opportunities in which they would utilize, acquire, share, or exchange knowledge and expertise. Indeed, every country within what we glibly distinguish as "the West" benefited from sources of technical expertise, knowledge, and institutional models that came from elsewhere. This was the case in the movement of people and ideas among countries like England, France, Prussia, the Spanish Empire, and Saxony, for example, and also for regions like Scandinavia, the early national United States, and Latin America. Engineering has long had transborder and transnational characteristics.[14]

Despite early roots in France and Germany, the profession we define as modern engineering did not emerge more broadly until the latter half of the nineteenth century, and it did so in a distinctly expansionary, imperial, and globalizing context. Although many of the characteristics of the modern engineering profession can be traced to early modern antecedents, it was only in the nineteenth century that the practice and profession of engineering took its current form with the complete package of institutions: systematized, increasingly specialized, and relatively standard curricular training in universities; in the professional norms established by voluntary associations and societies that operated at national and subnational levels; in print publications that conveyed technical knowledge in regularly published journals and texts; and in the employment of professional engineers to do the work of design, planning, and management within large organizations—increasingly, private-sector firms as well as agencies of the state. These four institutional contexts became identified with the profession of engineering, and participation in all of them emerged as a common feature among the professionals who claimed—or were ascribed—an engineering identity.[15]

Other paths to attaining technical expertise remained, including more specialized or vocational-technical training and especially experiential, on-the-shop-floor learning. Shop culture survived in many places but increasingly around the edges of a modern profession defined by new institutional arrangements. The growing influence and authority of these institutions was driven by the demand for technical expertise generated by industrialization, imperial expansion, corporate growth, and the increasingly global movement of capital, goods, people, and ideas in the decades after about 1870.[16] As a consequence, engineering transformed from the diverse and more locally isolated traditions of previous generations into one of the most numerous and truly global professions of the twentieth century. Modern engineering was in a sense reborn within the newly established university systems of this era, in new professional associations, and in organizational employment. These new institutions, in turn, were shaped by the global mobility of engineers, engineering knowledge, and institutional models in every country.

New institutions created venues and modalities for the movement of engineers and engineering knowledge. Young men (nearly all men) left their hometowns to seek university training.[17] For most, this meant relocation to attend a state- or provincial-level school or training center; for others, it meant travel across state or provincial lines to attend a regional or national program. Before the dramatic expansion of US engineering schools in the 1870s, hundreds of young men from the United States and around the world sought engineering training in European universities such as those in Freiberg or Paris.[18] As US universities began to surpass their European predecessors in scale if not in reputation in the last decades of the nineteenth century, young men from Europe and around the world reversed the to-Europe flow and sought training in North America in a pattern that would continue and expand through the twentieth century.[19] Meanwhile, institutions for engineering and technical training had long existed in Latin America and Asia, and late nineteenth-century efforts to expand and reform existing institutions and create new ones drew students from across regional and national lines. From West Point (1802) and Georgia Tech (1885) to the Tecnológico de Monterrey in Mexico (1943) and the Indian Institutes of Technology system (1950), new schools with curricular and training programs, as well as new professional associations, were modeled after older European institutions. In the United States, Greece, Portugal, and Japan (for example), the designers and teaching staff of early technical training programs came from Europe. Patterns of institutional borrowing, adoption, and adaptation of school models played out around the world between the mid-nineteenth and the mid-twentieth centuries as thousands of young men pursued technical training and thereby access to new forms of economic opportunity and social prestige via the expert authority conferred by an engineering degree. Once trained, engineering graduates sought employment that, for some, might take them to distant work sites and often across provincial and national borders. As one recent study notes, "Engineering is a profession where the global and the historical are intertwined more than any other."[20]

In the second half of the nineteenth century, sovereignties around the world, both public and private, dramatically expanded their administrative capacity to manage state-building and capital-enhancing projects at home and to extend their reach abroad. State- and nation-building projects included founding universities to produce the technical expertise that enabled public administration to govern territory, build infrastructure, foment economic activity, and extend imperial authority to colonial and semicolonial settings.[21] In the private sector, business entities that pursued raw material production, manufacturing, and distribution activities sought to expand their internal administrative capacities to manage vertical and horizontal integration and new technological systems in firms of increasing scale, complexity, and often geographic reach. This was especially but not only the case for multinational enterprises.[22] In both realms—public and private, state and corporate—university-trained engineers proved extraordinarily mobile and quickly became indispensable to planning, design, and management within these organizations.[23]

The creation of nationally based engineering schools, professional associations, and technical publications drew from existing global networks of knowledge, extended along pathways forged by trade, imperial initiatives, and capital investments. In turn, new national engineering institutions served to facilitate and dramatically expand those networks. These institutions provided engineers with their professional formation before they moved to dispersed workplaces, following national and imperial interests as well as increasingly global flows of investment and trade capital. But engineers did not always exclusively identify their own professional interests with those of their countries of origin and their employers. They frequently moved across public and private organizational boundaries, as well as along them, pushed by their own wanderlust and pulled by market opportunities. Individual engineers sought jobs and sold their talents to firms and agencies in ways that did not always follow the logic of corporate, national, or imperial allegiances but were instead defined by their own career-building aspirations. This was certainly the case for a relatively small number of prominent consulting engineers, and it was also true for thousands of more anonymous technical professionals. Moreover, the professional organizations of engineers sometimes took on an extra-national character, such as in the formation of binational and international associations, as well as in international memberships in national organizations. In other words, global engineering networks—and pathways for the flow of technical knowledge—were not simply defined by organizational lines of authority but were shaped by a global market for technical expertise. Engineers trained in one national setting often found employment within the national or imperial orbit of another, mixing the ambitions and interests of both in a kind of "inter-imperialism."[24]

Technical knowledge was embodied in the human capital of individual engineers but also in a rapidly expanding volume of print materials, from technical publications to telegraph messages. Corporate, national, and imperial domains often shaped but did not confine the distribution of and access to these modalities for information exchange. The movement of people, goods, capital, and ideas depended on the infrastructure designed and built by engineers working for private and public organizations: steamships, ports, railroads, mail service, and telegraph lines, as well as by the less visible infrastructure of political order and imperial authority that made communications infrastructure and physical mobility possible. Railroads (and ports and telegraph lines) constituted the major civil engineering projects of the nineteenth century, at once the product of engineering expertise and the means to extend engineers' spatial mobility. Global mail delivery carried an increasing and unending stream of print publications to homes, offices, and worksites worldwide: trade journals, textbooks, and other forms of print-based knowledge. Trade periodicals in the mining sector, for example, provided one forum for the relatively rapid, analog back-and-forth sharing of field trials and experimental results for various mineral refining processes.[25] Centers of printing, publication, and translation of technical works proliferated during this era in Europe, the Americas, and Asia.

Modern engineering emerged from nationally based institutions but at the same time was shaped by the transgression of organizational boundaries. Global mobility lay at the center of this historical experience: the mobility of engineers across national and local boundaries; the exchange of information through professional organizations and technical publications; the circulation of institutional models and technological designs between individuals, firms, and geographies; and the operation of technical and management systems as solutions to local problems. We will return to these issues in the concluding chapter of the volume. The result was a profession that exhibited global influences and embodied two strong forces. One constituted the outward, imperialist expansion of engineering talent, knowledge, and institutional models from Western Europe and the United States. The other constituted more locally based efforts, played out in diverse places worldwide, to develop technical capacities through new education, training, and support institutions. Although historians have often seen these two trends as sequential—with local institutions outside the West as an imposed or defensive response to the diffusion of Western expertise—they were actually contemporaneous and entwined, more co-created than sequential or causal. Interaction and exchange between global and local influences created opportunities for the mobility of technical knowledge.[26]

ENGINEERING ACROSS TIME AND SPACE

Any attempt to construct a *global* history of professional engineering carries the challenges of synthesizing temporal and spatial scales: periodization, boundaries, and storylines. The trajectories of engineering's growth between roughly 1870 and 1950 vary widely within and between nations in ways that challenge any common periodization. Imposing a simple chronological timeline on the first half-century or so of modern engineering, across a global dimension, risks privileging the temporal pattern of one place over others and running roughshod over local variations. We cannot, for example, tell a global history that is essentially constrained to the engineering experience of the United States and Western Europe and the subsequent diffusion of those engineers around the world, a "West and the Rest" story. This is especially the case for a global history that aims to capture, in some fashion, the origins, growth, and especially the interactions and relationships of modern engineering in communities, countries, and regions worldwide. Virtually everywhere, this era witnessed the establishment of new institutions, a dramatic increase in numbers of professionalized engineers, and their rising influence in design, planning, and management work within private and public organizations. If not exactly simultaneous, national engineering institutions emerged in close temporal proximity and as a result of a common context of global interactions and national imperatives.

Similarly, no single spatial scale can account for the different effects that modern engineering has exerted on the emerging industrial world over the past century and a half. Local, national, imperial, corporate, and global scales shaped engineering lives and practices and interacted in dynamic relationships. Technical work within any local space implicated global networks that traversed geographical, cultural, and social distances. These scales were not always hierarchical, sequential, or neatly embedded in one another, from the small to the large, the early to the late, or the particular to the universal. Moreover, movement between these scales was never frictionless, and none were impermeable to external pressures. Engineering practices and cultures in countries around the world combined multiple influences, including colonial impositions or commercial displacements even as they drew heavily on existing, local practices and institutions. Engineers worked in and across all these dimensions. In short, the local or national and the imperial or global were at times collaborative and synthetic, at times in tension and conflict. But they were always characterized by interaction, as engineers, engineering institutions, and technical knowledge moved back and forth across spatial boundaries.[27]

Finally, histories of engineering during this period often risk telling a progressive, positivist story of dramatic expansion and service to humanity. They often highlight the ways that the work of engineers has contributed to the technological wonders and increased standard of living that some of us enjoy in the late twentieth and early twenty-first centuries. There are good reasons to be wary of such a linear, teleological, and often hagiographic account. Failure, unintended consequences, negative externalities, and the reification of exclusionary social norms have marked engineering work as much as welfare-enhancing (or profit-enhancing) technical innovations. The balance sheet of beneficial and detrimental outcomes has not yet been fully accounted or is at least highly contested, despite our collective and absolute dependence on the marvels of modern engineering. Whether Panglossian or Pandoran, we live in an engineered, Anthropocenic world.[28] With these caveats in mind, we can trace several patterns in the development of modern engineering across temporal and spatial scales.

Everywhere around the world, the institutions and practices of modern engineering arose out of interactions between universal visions and local practices and settings. This was true in the first efforts to institutionalize engineering education and professional associations in the United States, at West Point or Georgia Tech, for instance, as it was in China and Chile. On the one hand, engineering knowledge existed in a cosmopolitan, universal space, founded on a scientific understanding of nature and on the scientific method of problem-solving. University-based engineering curricula were increasingly designed around the basic elements of physics, chemistry, thermodynamics, and materials science, depending on the field. New university programs aimed to produce civil, mechanical, electrical, or chemical engineers who would possess a transparent body of knowledge and a set of skills on which employers could rely. By the turn of the twentieth century, engineering expertise had become a practice that sought legitimacy and status based on empirical, rational, and scientific claims.[29]

On the other hand, the work of engineers was always rooted in particular places and local problems: in bridge or product or plant design, at the worksite, on the shop floor, or at the drafting table and the rock face. Their work was functional, instrumental, and oriented toward specific projects, in specific places, and in the interest of specific organizations. Immobility was a constant and essential characteristic of engineering. Engineers' place-based work rooted them in local and national cultures and tied them to local and national interests. At times, however, engineers' rootedness was disrupted by opportunities and assignments created as the forces of empire and capital pulled engineers across borders. The late nineteenth and early twentieth centuries were such a time, marked by a disruptive acceleration of global pressures and opportunities, ending with decreased global mobilities after the 1920s. The interaction between the particular and the universal, between pragmatic problem-solving and cosmopolitan science, between corporate and professional interests, between the local and the global and the immobile and mobile have long been present in the work of engineers who worked mentally if not physically across these spatial scales.[30]

Most engineers from the 1870s onward found employment within bureaucratic units and hierarchies of corporations or states and were directed toward projects that served the immediate, localized interests of those organizations. In a globalizing era, rapidly rising investment in industrial enterprise, raw material supply, and transport facilities created the demand for more specialized technical expertise. These demands were heightened by intense competition among the imperial interests of industrial powers, which pushed organizations both public and private to expand their scale and administrative capacity. At the same time, private firms sought to adopt new sources of motive power, specialized metallurgical and chemical processes, and ever-larger scale systems of production (and attendant cost-cutting measures), all of which demanded sophisticated coordination of flows, processes, labor, and production units. As a consequence, firms sought technically trained experts to advise, install, and manage the work. Industrialists, along with their allies in government, were often the primary movers in creating new institutions for technical training, and engineers became the designers, managers, and directors of corporate enterprises.[31]

If the interplay between universal science and local problem-solving within organizations lay at the center of engineering work, the early history of professional engineering centered on its explosive growth in the United States and Western Europe. Within just a few decades, the production of university-educated professional engineers grew from nearly zero to tens of thousands in the half-century or so after 1870. In the United States, the number of engineers grew a remarkable twenty-five times faster than the national population between 1870 and 1950.[32] Following the Morrill Land Grant Act of 1862—adopted by Congress in the absence of the southern states and designed to support institutions dedicated to "the agricultural and mechanical arts"—engineering degree programs rose quickly from just a handful to 70 by 1872 and at least 110 by 1920, with new degrees in electrical and chemical engineering. During the same period, industry and educational leaders established national and many regional engineering associations for mechanical, mining, chemical, and electrical engineers (joining the older traditions of military and civil engineering) as engineering became increasingly specialized in distinct fields of knowledge and practice, each with its own institutions and subcultures. By 1920, over 50,000 students were enrolled in US engineering degree programs, and membership in the main national associations exceeded 55,000.[33]

Across the Atlantic, several German states boasted arguably the strongest engineering institutions in Europe by 1900, with deeper historical origins than elsewhere. New investments in technical and engineering training by provincial governments after 1850 created what one historian calls "the first educational-industrial complex in the Western World."[34] By the 1920s, technical training programs (especially the Technische Hochshulen) were educating over 20,000 students each year, with a quarter million employed in 1937.[35] French investment in technical education grew more slowly, though also from strong early traditions. The *polytechnique* schools remained more focused on the technical trades than formal engineering and the *grandes écoles* more on the needs of the state than of industry, although both dated from early nineteenth-century initiatives. These schools focused primarily on producing engineers for French public service and colonial postings. In Britain, the longstanding tradition of apprenticeship and shop-floor training remained dominant well into the mid-twentieth century. New university programs in Britain were established but grew slowly, producing by the 1920s just a fraction of the number of graduates in Germany and the United States. Across

Western Europe, new, field-based professional associations joined university training to establish the institutional foundations of the profession, each with thousands of members by 1910. As in the United States, new European investments in engineering after the 1870s sought to produce a plentiful supply of technically trained manpower in a context of rapid industrialization, state-building, and imperial competition. All this gave professional engineering in the industrializing North Atlantic economies a particular momentum and practitioners who would increasingly wield their expertise and authority on a global scale.[36]

Newly trained and professionalized engineers in the United States and Western Europe quickly exceeded local demand and spilled over across the world, part and parcel of imperial expansion and economic globalization during these same decades. Most engineers in the late nineteenth and early twentieth centuries found work within bureaucracies of companies and states with intrinsically pragmatic, utilitarian, and site-specific aims. As a consequence, engineers quickly became one of the most globally mobile of occupations, building roads, railroads, ports, telegraph lines, dams and bridges, and military installations to achieve imperial (and, in reaction, anti-imperial) goals; designing systems for the extraction, refining, and transport of minerals; solving problems in large-scale, export-oriented agricultural production; and (among many other tasks) providing the technical support for military expeditions and other forms of imposed authority. Civil engineers moved across borders to build the infrastructure of transportation and communication; mining and metallurgical engineers (and allied experts) were perhaps the most geographically mobile of all engineering fields, pursuing opportunities in a globally integrated industry. Agronomists and agricultural engineers typically worked in localized spaces, but many moved on exploratory expeditions and to develop export opportunities in tropical regions. Electrical and chemical engineers proved relatively less mobile internationally.[37]

Engineers' physical mobility across spatial scales had a major impact on their communities of origin, as well as on their globally dispersed workplaces and communities of employment. Indeed, engineers' global mobility helped define "the West" as a single space, characterized by technological innovation of a particular kind and set against "the Rest" or "the Global South," loosely considered as most of Asia, Africa, Latin America, and subordinated social groups in North America and Europe. Between 1870 and the 1930s, much of the Global South was overwhelmed by an influx of trade, capital, and engineers and engineering projects from the United States and Western Europe. An older (and stubbornly persistent) historiography has often imposed a distinctly polarized temporal and spatial lens on the process of economic and technological development. It has often used the language of Western modernity, progress, and technological diffusion, contrasting with accounts of "backwardness," "lagging," and a mimetic and dependent response in the rest of the world. More recently, scholars have tended toward the language of "translation," "exchange," or "coproduction," sometimes identifying forms of "Indian," "Asian," or "Islamic" modernity (for example) that drew from Western influences and models but built on local capacities and cultures and responded to local circumstances.[38]

Indeed, in colonial and postcolonial settings with strong, local engineering traditions (such as parts of Latin America and Asia), the arrival of Western projects and Western-trained engineers often squeezed out existing opportunities for local engineering talent or established enclaves isolated from local practice. Yet, the arrival of Western-trained engineers could also lead to substantial learning. The two outcomes were not necessarily mutually exclusive. Foreign engineers carried with them knowledge and experience at the global frontier of technological advancement. They brought information, even if they were often not able or willing to share full knowledge. In many places, foreign engineers worked alongside local technicians, though not always on equal status, creating, at times, the face-to-face conditions for learning "spillover" of knowledge. Nevertheless, limits to exchange and interaction were often substantial, typically based in culture and politics, and stubborn obstacles to technology transfer constrained mobilities.[39] The presence and pressures of engineers from abroad could also inspire or assist local interests in establishing or expanding technical training institutions, as in a number of nineteenth-century engineering schools in India and Japan, for example.[40]

In many colonies, countries, and communities worldwide, engineering capabilities and institutions had substantial local roots well before the late-century arrival of Western-trained engineers. Famously, Muhammed Ali sought to develop modern industry in Egypt and established an Egyptian polytechnical school in 1836. Mexico boasted the first technical university in the Western hemisphere, established in 1792. But nearly everywhere, new efforts to establish engineering institutions were undertaken in the latter half of the nineteenth century and continued into the first half of the twentieth. These initiatives sometimes built on earlier engineering practice and cultures but were also in conversation with new models from the North Atlantic. All were undertaken amid evolving economic challenges and opportunities in this era of imperial expansion and globalization. The timing is not coincidental. Regional initiatives were motivated by several factors: efforts to mimic and catch up to the early industrializers, take advantage of new commercial opportunities in a globalizing economy, and proactively resist Western imperial pressures, both military and economic. From Japan to Mexico and Brazil, as elsewhere, engineering expertise and "the most modern" technology were viewed as the sin qua non of regional and national strength or sometimes simply of survival.[41]

Early initiatives to modernize and expand local engineering education began in India (1847), Japan (1871), and China (1895), as well as in many nations in Latin America, including Chile (1853), Mexico (1867), Argentina (1870), Peru (1872), Venezuela (1881), and Brazil (1893), for example.[42] In some places, colonial authorities established new institutions; in others, it was national governments. Often, these initiatives were hampered by imperial pressures, weak state capacity, or civil conflict well into the twentieth century. Yet in their motivation and their orientation, these efforts were little different from those undertaken almost contemporaneously in the West. New institutions everywhere built on local technical training institutions while also drawing on the global circulation of new scientific and institutional models. They played out on various timetables from the mid-nineteenth into the mid-twentieth century, responded to the interests of industrializing economies, and were part and parcel of aspirations for economic and political sovereignty via nationalist state-building projects but set within a globalizing context.[43] All sought to increase technological and therefore economic self-sufficiency.

Early efforts to attract or locally train engineers and other technical experts were typically not "national" in character but responded to regional (i.e., subnational) interests. Efforts to build technical capacity were nearly always linked to industrialization, promoted by industrial investors themselves and their political supporters (or vice versa), investing in the human capital infrastructure they believed necessary for industrial enterprises.[44] Only later did local efforts take on a national character with the weakening of imperial influence. In India, this was linked early to the Swadeshi movement and manifested in the creation of the Bengal Technical Institute in 1906 but did not fully flower until after independence in the 1950s. In China, substantial new initiatives were pushed by the Nationalist government in the

1930s, tripling the number of students enrolled in engineering programs at Jiaotong, National Central, and Qinghua universities (for example), creating a foundation for further expansion under Communist rule in the 1950s.[45] Increasing pressures of international competition and nationalist interests shaped and reshaped technical capacity-building initiatives in both the West and the Global South, in both capitalist and communist societies. The institutional foundations for technological capacities on a national scale emerged only when political movements were able to direct sufficient state resources to manage, develop, and extend capacity-building projects that drew on local interests as well as on global flows of knowledge, institutional models, and people.[46]

In many places around the world, engineering and technological capacities expanded along a roughly similar sequence.[47] Early initiatives typically involved entrepreneurs or the state importing and employing engineers from Europe or the United States. At the same time, local families and firms might send young men abroad to study at European or US universities or to work there to gain practical experience, sometimes subsidized by local or national governments. These initiatives were often followed by local efforts to establish institutes, training centers, or research stations, sometimes in direct affiliation with support from local business enterprises. In regions with strong pro-industrial interests and exposure to foreign models, these activities might escalate to lobbying for the creation of full-fledged university-based engineering programs, which constituted the most significant institutional commitment by private and public actors to develop engineering capacities. University programs might be followed by initiatives to establish specialized research centers. In some places, these arose from partnerships between the private and public sectors; in others, as state-led rather than corporate-based research and development units.

By the mid-twentieth century, regional and national efforts to build technical capacity were joined by external support from foreign foundations or international organizations, such as the Rockefeller and Ford foundations, or United Nations agencies, where geopolitical motives mixed, sometimes uncomfortably, with local economic or nationalist interests. The result at each step was a gradual "nationalization" of technical expertise and engineering capacity into the middle decades of the twentieth century: Indianization, Mexicanization, or Sinoization, among others.[48] By the 1940s or so, most of the larger, politically independent countries of the world had established the domestic institutions for a nationalized engineering capacity (albeit with varying results).[49] Nationalization of technical capacity thus appeared as a twofold phenomenon in the early and middle decades of the twentieth century. It involved efforts to expand engineering capabilities within countries across Latin America, Asia, the Middle East, and Africa. But these efforts worked in uneasy relation with constraints to national capacity building imposed by the mere presence of (and sometimes preference for) foreign engineers, the extractive orientation of foreign corporate interests, the work of transnational research foundations, and the geopolitical interests of global powers and multinational institutions.

Can we tell an integrated global history of modern, professionalized engineering institutions and practices that is not centered on the rise of the West? We have noted the extraordinary expansion of engineering capacity in Western Europe and the United States, the circulation of engineering practices through imperial and corporate paths, and the movement of tens of thousands of engineers trained in the West across the globe. But many factors undermine a straightforward diffusionist narrative: mobility was multidirectional, local engineering traditions could be substantial, interactions and learning occurred at local sites, and out of these interactions arose new systems of knowledge, training, and practice, as illustrated by the great variety of engineering institutions and capacities that emerged in this era around the world. Despite common themes, there remain multiple histories. Western influence during this era of imperial and corporate expansion was substantial but not determinative. Modern, professionalized engineering quickly became a truly global presence, mobilized by Western imperialism and corporate activities but also built on local practices and locally based efforts of capacity building around the world.

SOCIAL IDENTITY AND SOCIAL ENGINEERING

Who were the actors at the center of this story, and how did their social position and cultural identity shape their work? Who wielded the visible hand of technical management or the iron fist of imperial expansion? Modern engineering developed within a broader historical context of expertise in the "practical arts" but also within a social context that shaped engineers' worldview and activities. Its institutions were both the product of, and agents for, societal interests and cultural norms. The technical objective of engineering was never purely a transformation of nature but also the regulation of human practices. Engineers' role in the design, construction, and administration of technology was, ultimately, also an exercise of power. Locally established engineering practices in many corners of the world had to confront the dominant norms of Western engineering as those spread globally along the paths of imperial and corporate interests, as well as through local efforts to adopt and adapt the institutions of Western engineering.

In the United States, the emerging institutions of professional engineering mirrored and reinforced the social norms of a masculine, white, middle-class, and largely Protestant professional culture.[50] This orientation served to broaden and democratize access to expert knowledge for many white males in the late nineteenth and twentieth centuries; thousands of middle-class youth gained educational opportunities in new university programs. But at the same time, it worked to exclude nearly all others along gender, ethnic, and racial lines. Socially demarcated boundaries of exclusion helped engineers and the institutions that produced and supported them enhance their status relative to other social and occupational groups.

Women were active users and designers of technology in the nineteenth and early twentieth centuries but were largely excluded from social understandings of engineering. Women were nearly invisible in engineering programs until the 1960s in the United States, as were ethnic and racial minority populations well into the second half of the twentieth century. Hundreds of women pursued engineering degrees in US and European universities before the 1960s, but these numbers were tiny compared to tens of thousands of their male counterparts, at under 3 percent of the total. Women were typically excluded from engineering spaces such as professional associations, meetings, and industrial research labs.[51] Their peers, the media, and historians have long portrayed female engineers as exceptional and out of their element in a profession naturalized as male. Engineering programs outside Europe and the United States developed along similar lines of exclusion. The first female engineers in Latin America acquired their degrees well into the twentieth century.[52]

Even among European-descent Americans, the children of Irish and Italian immigrants did not enter engineering until the middle decades of the twentieth century. Any diversity that existed in the US profession came largely from the attendance of foreign students in university programs. Conversely, globally mobile engineering graduates from the United States and Europe carried their gendered and racialized cultural assumptions abroad, where their presence in Latin America, Asia, and Africa often squeezed out opportunities for locally trained technicians. This was the case even as the presence of Western engineers incentivized initiatives to develop local training programs. In regions where imperial or corporate interests had introduced significant numbers of Western engineers, "technical expertise" came to be identified by the educational background, language, and culture associated with engineering programs in US and European universities.[53] Deep-seated norms of social exclusion posed obstacles to the creation of technical training opportunities around the world, especially in colonial and semicolonial settings.[54] Whether imperial, multinational corporate, or local, employers around the world often sought to hire foreign (i.e., Western-trained) engineers, even when locally trained experts were available at lower cost. This preference held even in countries with a long tradition of local engineering talent, including Anglo settler colonies like Canada and South Africa as well as in Mexico, Cuba, or India.[55] More often than not, the local engineers in such places came from elites deeply permeated by Western economic, political, and cultural influence. Culturally embedded exclusions reinforced opportunities for Western engineers and the authority they wielded, but that authority was never absolute. Engineers from other parts of the world proved able to gain access to opportunity, most frequently through attendance at a Western university but also by building on local knowledge. The expansion of the institutions of modern engineering outside the West gradually provided the basis for a nationally endogenous engineering capacity that could both interact with and compete against the early global dominance of Western expertise.

Engineers reflected the cultural norms and aspirations of the emerging, upwardly mobile middle class of the twentieth century.[56] They did not comprise a class themselves but a new and increasingly significant occupational group. Technical expertise of one kind or another became a significant middle-class marker, and diverse occupations wielded claims to specialized knowledge and professional association to establish socioeconomic boundaries above and especially below. Professional engineering institutions and their most visible promoters (e.g., men as ideologically different as Herbert Hoover and Thorstein Veblen) identified engineers as distinct from and superior to skilled workers, mechanics, and other technicians. Engineering associations worked assiduously to emphasize and reinforce this professional identity and status. At the same time, engineers were caught in an ambiguous position among some of the major class and occupational divisions of the era. Rising historically out of the larger mass of skilled laborers, mechanics, and technicians, engineers sought to distinguish their social status by claiming the attributes of a science-based profession: independent, objective appliers of universal laws, apolitical and disinterested possessors of specialized expertise.[57]

But this aspiration had to confront engineers' dependence on organizational interests, whether private or public. Engineers' employment in the hierarchies of competitive, for-profit businesses and the agencies of governments within states and empires tied them closely to those interests. Even those engineers who most vociferously touted their independent expertise—consulting engineers—owed their livelihood to employment by these organizations and thrived as a result of meeting organizational needs. Engineers were nearly always "organizational men" whose careers depended on serving the interests of the organization. Moreover, engineers' technical problem-solving was rarely purely technical but embraced—typically prioritized—issues of authority, control, cost, and profit. Indeed, engineers' work lay as much in evaluating, designing, and managing a socioeconomic technological system as it did in a more narrowly "technical" expertise in, for example, materials science, industrial chemistry, or thermodynamics.

Engineers' role in designing and managing work regimes and labor relations often brought cultural norms to the fore. For engineers charged with maximizing efficiency and minimizing costs within production systems, the management of labor constituted a critical element. The scientific management movement focused on just this issue as it arose from engineering concerns and sought to maximize labor productivity and cut labor costs.[58] The movement grew out of engineers' focus on technical efficiency, propelled by employers' demands for close cost accounting and profit maximization. The scientific management of labor gained enthusiasts in factory settings worldwide but arguably took its most racialized form in contexts where Western engineers managed nonwhite labor, especially in frontier or imperial contexts. Mining operations frequently exhibited the pernicious racialized norms of the profession: with Chinese labor in California, West Indians in the Caribbean and Central America, mestizo or indigenous labor in Venezuela, and Chinese, Indian, or Black labor in South Africa. Indeed, racialized labor regimes were not uncommon in a wide range of engineering projects. Engineers became the architects and managers of technological systems that reinforced and institutionalized race-based distinctions.[59]

Engineering and the work of organizational planning focused on the design of objects and the command of subjects. Popular conceptions of modernity, civilization, and progress became associated in the late nineteenth century with the particular forms of industrial advancement, mechanization, and technological capabilities that emerged in Western Europe and the United States. Engineers came to represent and embody that identity, bounded along lines of gender, race, and class. As thousands of engineers from Western Europe and the United States pursued work opportunities around the world, they carried with them the belief that modern science and engineering were the exclusive products of a masculine, Western tradition. Engineering was what Western men did, they assumed, reflecting the Social Darwinism of the day and explicitly referencing their beliefs about the innate limitations of nonwhite, non-Western, nonmale peoples. Histories of imperial and corporate engineering projects in Latin America, Africa, the Middle East, and Asia during this period amply demonstrate the ways that engineers wielded both technical and cultural authority. Their assumptions of the hierarchy were reproduced, embedded in, and often indistinguishable from implicit assumptions of white male supremacy.

If social norms were deeply embedded in engineering identity and practice, the methods and practice of engineering also reached into social realms outside the workplace. For example, Progressive era reformers adopted the quantitative, technical, systems-oriented, and problem-solving orientation of engineering as a solution to the social questions of the day, with poverty (and what were seen as associated behaviors, including deviancy and criminality) central among them. Social planners and policymakers worked to identify, quantify, and measure social ills and to devise policies that would reorganize (or reengineer) social relations to ameliorate them. Forms of social engineering proliferated globally through the twentieth century as engineers and other technical experts served as government advisers and policymakers.[60]

One manifestation of applying the scientific method and engineering approaches to the organization of work and society came in the "technocracy" movement of the 1920s in the United States. Drawing from the principles of scientific management and the social engineering of the Progressive movement, technocracy's adherents argued that control of industry should be in the hands of technical experts, especially engineers. By extension, so too should the management of the polity. Most famously, Thorstein Veblen suggested that an engineer-led, technocratic society represented the natural progression of social organization. More generally, the prevalence of technical experts (and especially engineers) in governance and policy-making increased globally through the middle decades of the twentieth century worldwide. At the same time, state-led economic planning became increasingly common, most visibly in centrally planned economies but also in developmentalist states. Technocratic economic planning became common after the Great Depression, and engineers played central technical, managerial, and leadership roles.[61]

These trends illustrate the pervasive reach of engineers and engineering in the first half of the twentieth century, only a few decades removed from the origins of the modern profession. In 1870, engineering had been a relatively small, informal profession of highly skilled men who had largely acquired their expertise in the workplace. By the 1920s, tens of thousands of engineers were being trained in university programs in the United States and Europe but also in Japan, India, China, Mexico, and beyond. A few engineers followed career paths that led them into the topmost ranks of organizational management and leadership, serving on corporate boards and as corporate directors, as well as in positions of leadership in public-sector organizations. Engineers reached an early peak of social influence as the profession fully matured in the late 1920s, when some translated the technical problem-solving and systems-planning orientation of their profession directly into positions of public, political, and electoral relevance.

In a moment of striking convergence, men who had built their careers as technical designers, administrators, and planners within organizations came to lead nations and empires as the presidents or prime ministers of the United States, Great Britain, Mexico, India, Portugal, and Colombia, all in the interwar era.[62] Although exceptional in their profile, these men represented the rational application of engineering and technical expertise to politics and social planning. As a social group, engineers had reached a peak of influence within the dominant organizations of the era that they would retain for the remainder of the twentieth century, largely under the radar of most social commentators and later historians. Meanwhile, tens of thousands of their far more anonymous fellow engineers labored at all levels within the organizations of business and states around the world, designing, planning, and managing a new and more highly engineered world.

CASE STUDIES OF GLOBAL ENGINEERING

This book argues that engineering emerged as the largest technical profession of the twentieth century in the decades after about 1870 and was globally engaged from those beginnings. The institutions that defined the profession and its practice—university training, professional associations, the widespread exchange of textual materials, and organizational employment—instantiated a local grounding for the profession but also provided the venues for cross-border and increasingly global interactions. By the middle decades of the twentieth century, the work of engineers was, at every level, indispensable for the work of both private firms and public agencies in the global economy. The eight chapters that follow offer a series of empirical case studies focused on engineers and the work of engineering in places around the world, from India to Mexico, South Africa to London, and linking China to France and the United States. Based on original archival research, each chapter illustrates the locally focused but globally engaged work of engineers, their connections to organizations, and their frequent mobility across organizational boundaries. Each chapter illustrates the central themes of the engineering profession that we have highlighted in this introduction: the essential and ever-present interaction between the local and the global, the dramatic temporal and spatial expansion of engineering during these decades, the overlap and interaction of various temporal and spatial scales, and the ways in which engineering was both socially embedded and socially engaged. Each chapter tells a story from the history of modern engineering that helps us understand its local rootedness, its national character, and its global presence.

Stephen Tufnell examines the construction of a particular space of technical knowledge: the city of London. The chapter shows how a relatively small number of consulting engineers emerged as a critical intermediary link between large-scale capital and the technical organization and expertise required for the operation of globally diverse mass mining operations. Focusing on the case of the Witwatersrand gold district in South Africa, Tufnell demonstrates how the abundance of low-grade pyritic ores at great depths and the capital needs of the technologies necessary for its extraction facilitated the collaboration of an extensive network of globally mobile and university-educated engineers. These men connected their own technical networks of experts with the confidence of investors and designed globally engaged systems of business organization. They occupied a distinct local space—just a few blocks within the city of London—from which they drew on globe-spanning networks of capital, information, technology, and technicians to command mining operations in another specific location, thousands of miles to the south.

Mikael Wolfe's contribution in chapter 3 shows how the expert practice and institutions of meteorology emerged out of the engineering profession in late nineteenth-century Mexico. Meteorologists asserted claims as an applied science with practical applications, especially to agriculture, and drew from transatlantic currents of technical and institutional knowledge. Focusing on newly established national organizations, Wolfe follows a group of meteorological engineers, most of whom had graduated from Mexico's National Engineering School, the oldest technical training school in the Western Hemisphere. In meteorology as in other forms of applied science, Mexico's government drew on cosmopolitan ideas and institutional models in its effort to achieve parity with the industrializing nations of the North Atlantic. Investment in new national institutions in Mexico City led this effort, including the Central Meteorological Observatory. The history of the observatory shows a combination of state-led efforts to modernize the country through the use and diffusion of science and technology, the reinforcement of Western stereotypes regarding the physiological consequences of tropical climates on the population, the constraints of gender roles in the diffusion of technical expertise, and emerging conflicts with locally situated traditional ways of predicting the weather. The national impact of this cosmopolitan institution was in the end constrained by its failure to reach local spaces of agricultural practice by Mexico's farmers, only a few years before the explosion of agrarian discontent in the Mexican Revolution.

Doug Jones's chapter 4 examines the globalized nature of engineering in South African mining but from a very different perspective. Jones shows how labor management practices circulated through engineering networks as individual engineers moved between South America, North America, and Africa at the turn of the twentieth century. Focusing on a group of predominately American engineers working to extract gold-bearing ore in the Witwatersrand basin, Jones describes how their previous employment in El Callao, Venezuela, and the quicksilver mine of New Almaden, California, shaped their technical approach to labor management in South Africa. Engineers' adoption of formal labor accounting techniques, driven by an increasing demand to cut production costs in an era of low-yield mass mining, shaped their views toward labor and led many to identify racially based productivity and cost differences between workers. In the gold fields near Johannesburg, mining engineers argued about the relative costs of white, indentured Chinese, and Black African laborers in ways that became central to shaping British imperial social policies during and after the war of 1899–1902. As they worked to solve local production and cost accounting problems on the Rand, these engineers exhibited the racialized assumptions of their era and drew on a globally linked set of experiences with racialized labor in the mining sector. Jones highlights the intersection of cosmopolitan, transnational, and local scales that help us understand how mining engineers crafted race-based labor management.

Elisabeth Köll's chapter 5 explores global engineering networks in East Asia at a moment of transition from late colonial to national interests, ca. 1900–1950. In China, Japan, and the Philippines, the work of engineers from Europe and the United States reflected the interests of transnational capital as well as of Asian nations and nascent nationalist movements. Köll shows that the formation of a national technical capacity was not a lineal process but a succession of experiments, partial success and setbacks, and the combination of decisions by different local and global actors. Railroads provide an illustrative early case of transnational engineering focused on the development of economic opportunities, where the employment of foreign engineers during the years of railroad construction led gradually to the development of local engineering expertise and capacity, with coinvestment from both foreign capital and local private and public interests. The chapter sketches trans-Asian and global connections and then shows how the formation of technical experts in China arose from multiple scales: local, corporate, national, and imperial. The pages of the *Far Eastern Review* reveal East Asia as a globalized space of constant and complex interaction, negotiation, and exchange between businesspeople, bureaucrats, technical experts, and journalists. It provides a regional perspective on the interaction of local and imperial interests and a window on the emergence of China's technological elite.

Mark Hendrickson's chapter 6 examines the early history of antimony production in China's Hunan province and how local Chinese entrepreneurs acquired the technical and commercial capacity to command global antimony markets for much of the twentieth century. In Hunan's local space, the success of antimony mining depended on the support of provincial leaders allied with the local gentry. These interests dovetailed with early Chinese nation-building interests, which supported the training of Chinese engineers in the United States, France, and England. The adoption of new metal refining techniques initially developed in France proved extraordinarily successful in Hunan. As Chinese antimony production overtook French production, the technical capacities developed in Hunan would alter the global distribution of this vital metal during World War I and through the remainder of the century. Chinese engineers who led the local enterprise went on to become cosmopolitan experts themselves, prominent if not preeminent in their field in engineering circles in the United States and Europe. In contrast to Western consulting engineers who brought cosmopolitan knowledge to local places around the world, the Chinese actors at the center of this story brought Western techniques to China and then back to Western engineering communities.

Marco Bertilorenzi's chapter 7 presents the global profile of French engineers using a unique database of over 2,500 graduates from the École des Mines de Saint-Étienne between 1880 and 1962. The school was established in 1816, just twenty-five years after the founding of the Paris École des Mines, and both were designed to produce technical specialists who would serve national interests in public or private employment. The Saint-Étienne graduates proved as mobile as their colleagues in other countries, as much in the service of a global mining capitalism as in French national interests. French firms and French agencies working in French colonies provided employment for most, but many followed job market prospects across national and imperial boundaries. The near century-long time frame of the data reveals two long cycles in the global mobility of French engineers: rising steadily through the end of the nineteenth century to a peak during World War I, then falling quickly to lower levels by the middle decades of the twentieth century.

Aparajith Ramnath's chapter 8 on the engineer M. Visvesvaraya in late colonial India examines how engineering practiced in an imperial space shaped the early emergence of national economic planning in India. It highlights how national engineering capacities built on British colonial foundations and drew from local expertise as well as from circuits of universal knowledge. Visvesvaraya's career in Bombay, Mysore, and beyond illustrates the interaction between local, regional, and quasi-national spheres and the intertwined nature of technical, bureaucratic, and political considerations in an otherwise "technical" profession. Serving variously as public works engineer, prime minister of a princely state, and public intellectual, Visvesvaraya was one of the earliest proponents of economic planning in interwar India and a strong advocate for advanced technical institutes. His interest grew out of elements of the culture of colonial engineering into which he was socialized, as well as ideas he absorbed on his many study tours to Europe, Japan, and elsewhere. Ramnath's case study illustrates how engineering bureaucracies initially set up by the colonial state for its own purposes gave rise to some of the leaders and technocrats who would push in the early decades of the twentieth century for the construction of an Indian economy based on industrialization, presaging the formal planning initiatives that would be so central to India's national project several decades later.

Aurora Gómez-Galvarriato's chapter 9 uses the history of the Armour Research Foundation (ARF) in the 1950s to show how technical research institutes that were headquartered in the United States could serve as transnational conduits for technical knowledge and local capacity building around the world, albeit within the quasi-imperial context of the Cold War. Examining the case of the ARF (with parallels in the Rockefeller, Guggenheim, and Ford foundations, for example), the chapter highlights interactions between corporate and state spheres, as well as global and national spaces. The ARF began as a development consulting firm located and specializing in the US Midwest. After helping to develop technologies for World War II, it later expanded into international activities consistent with President Truman's postwar Point Four Program. The history of this foundation illustrates the tensions between the international influences of the US economic empire and the nation-building interests of technical experts in the Global South. Although primarily focused on industrial development in the United States, the ARF undertook assessment and capacity-building projects in Latin America, the Caribbean, India, Lebanon, Burma, Saudi Arabia, Afghanistan, and beyond. The chapter describes how the ARF formed specialized teams abroad and collaborated with government agencies in those countries to establish and support locally based technical research institutes. These institutes, born of collaboration with the ARF, later established their own networks of technical specialists, often with lateral, cross-national relationships between them.

In the final, concluding chapter of the volume, we step back to revisit the global dimensions of engineering in its first half-century or so as a modern profession. Drawing from the eight case study chapters, we reiterate our central argument: the modern engineering profession of the twentieth century arose out of a set of institutions that have long been seen to define and bound the profession and the work that engineers do. The evidence suggests, however, that the transnational—indeed, global—mobility of engineering knowledge, institutional forms, and engineers themselves was not simply a consequence of the profession's dramatic expansion but was instead an integral part of its emergence worldwide, an intrinsic part of what the profession is and what engineers do.

The engineered worlds examined in the chapters that follow were shaped by local, national, imperial, and global contexts, as well as by the social structures within which technology and expertise were embedded. The places treated in these case studies are just a small fraction of the corners of the world that suffered dramatic transformations in those years. Germany, Russia, and Japan were central actors in the history of modern engineering, and other territories in the Americas, Asia, Africa, and Eastern Europe also transformed their engineering traditions in the late nineteenth and early twentieth centuries. The case studies in this book examine just a few examples of the growth of the engineering field in those years but illustrate broadly representative global patterns and themes.

As a global phenomenon, the emergence of modern engineering transformed workplaces in the mines and the fields, in laboratories at universities and at chemical plants in firms, across railroad landscapes and on the pages of scientific journals, in colonial as well as postcolonial settings, and through the construction of racialized labor practices and the recording of atmospheric phenomena. Engineering quickly became so omnipresent in the history of industrial societies during the late nineteenth and through the twentieth centuries that its role has been taken as a given. Engineers are often nearly invisible and treated as interchangeable technical inputs. Yet we have inherited, after all, a world fully engineered. The construction and development of the engineering profession—and this outcome—never followed a preestablished path. Engineers' agency and trajectories as major actors in the design and planning of our world remain to be fully explored by historians. This book contributes to that literature and to current debates on the role of technicians, technocrats, and technical firms in political, economic, and social realms.

2 "A MAN OF MANY PARTS": CONSULTING ENGINEERS AND THE ENGINEERING OF MASS MINING CAPITALISM

STEPHEN TUFFNELL

By the turn of the twentieth century, a new and distinctive regime had reshaped global finance, technologies, and ecologies: mass mining. Standing at the center of the technological and financial organization of these vast industrial enterprises was the consulting engineer. But who exactly was the consulting engineer? What did consulting entail? And why did this class of engineer rise to social and financial importance in the regime of mass mining? Contemporary clients and professionals were not sure. "We are accustomed to hear the phrase 'consulting engineer,'" noted one treatise. "In what particular branch of the profession he holds himself out for consultation does not always appear," it continued, "Nor is it necessary that he shall have any particular experience in the art of consulting or advising in consultation. Any man can assume this imposing title."[1] This was only partially true. Consulting was not entirely new, but as this chapter demonstrates, under mass mining, consultants reinvented the role as a new model of professional technical, managerial, and financial intermediary, part engineer-in-chief, part executive officer, part corporate capitalist.

Engineering consulting was an evolving set of advisory practices on the technical, financial, and executive challenges facing mining enterprise. As the engineering profession expanded and was transformed from a craft to a profession in the final third of the nineteenth century, consulting flourished across all its major branches. Scientific consulting emerged on both sides of the Atlantic early in the nineteenth century as a lucrative and, in the United States, the "dominant," expression of scientific entrepreneurship.[2] Historians of science and engineering have used the term to capture the diversity of career paths open to enterprising men of practical science, including government surveying, professional expert witnessing, academic research, public lecturing, or consulting.[3] To obtain consulting positions, an entrepreneurial engineer marketed their two most important assets: their expertise and their reputation. Unlike those engaged in pure science, mining consultants did not always feel the need to abstain from commercial gain.[4] Recognition of expertise and reputation and, with it, social capital were often measured by financial reward for the consulting engineer. The close relationship between capitalism and consulting thus began early. "It was a professional practice pointing toward the union of science and capitalism," Paul Lucier has surmised.[5]

That union was completed under the regime of mass mining. The commodification of engineering expertise took on new forms under its rule, a process in which consulting engineers themselves participated. As one leading consultant wrote in 1908, they had become "the advance agent of capital."[6] By the turn of the twentieth century, mass nonselective mining took hold on mineral fields worldwide as engineers responded to increasingly complex geologies and leaner ores.[7] By removing greater quantities of unmarketable ores from the deposit and then subjecting them to intensive milling and chemical extraction processes, the mining of low-grade ores became more productive and more profitable than selective mining techniques. But increased scales of working required unprecedented scales of capital. For the financiers who invested in mass mining enterprises (or who found that this was the only way to make their investment payable), employing mining consultants was a costly but shrewd investment in human capital and an important nonmonetary asset. The consultant offered their clients organizational knowledge, managerial skill, experience, and, through their address books, access to even larger pools of human capital. Hiring consultants might unlock long-term profit, but in the short term, it also legitimized a culture of financial risk with commodified consulting expertise at its heart.

To examine these processes, this chapter turns to South African mass mining at the turn of the twentieth century. Transforming the Rand was the work of globally mobile engineers, who traveled a well-worn path to colonial enterprise through London. Center of a vast empire of financial services, colonial contracting, and knowledge industries that encompassed the world, London became the global hub of mining consulting. In the imperial capital, financiers and mining consultants collaborated in exploiting and developing

South African gold mines, both financially and materially. At the heart of this chapter are the consulting careers of a cadre of elite, internationally mobile consulting engineers—John Hays Hammond, Herbert Hoover, and Hennen Jennings—and the border-spanning networks they oversaw. These men were self-assured experts, highly experienced, unashamedly self-promotional, fiercely independent, and unapologetically capitalist. All had storied careers in the histories of late nineteenth-century gold mining and reinvented the practice, organization, and professional ethics of consulting. But while they are each well known to historians, few accounts examine their consulting practice—even though their published and private accounts tell us a great deal about their work. And they worked hard. That work meant continual movement between the industrialized ridges of the Transvaal to the offices of investors and multinational corporations in the City of London, connecting networks of technical experts to the confidence of mass mining capitalists. This chapter follows them in that work. The art of consulting lay in their ability to read the landscape, visualize the subterranean layout of the mine, and organize a commercially viable system to exploit it, as told in the first section. Increasingly, these technical arts complemented consultants' evolving role as knowledge brokers in an intangible financial marketplace, the subject of the second section.[8] Consultants engineered the machine of mass mining capitalism in the City of London as they articulated plans to investors, solicited corporate capital to bring mines into production, mobilized networks of human capital to support their work, waged public information campaigns in the technical and financial press, and dealt in insider knowledge.

Consulting blurred the distinctions between advisory and executive functions, as well as the ideal of expertise as a public good and its commodification for the pursuit of private profit.[9] As a result, reflected one observer, consultants were often "caught between the devil of self-interest and the deep sea of loyalty to employers."[10] Histories of engineers have long recognized their "double location as both objects and representatives of corporate power."[11] As we shall see in the final section, however, consultants themselves often argued there was no meaningful demarcation to be made between these boundaries. "The engineer above all things desires permanency and profitableness in mining operations. The capitalist desires the same," surmised one correspondent to the *Mining and Scientific Press*.[12] Although contested, the cultures of risk and consulting legitimized one another. It wasn't simply that the mining consultant united the separate entities of science and capitalism then but that they coproduced the knowledge economy and culture of risk at the heart of mass mining capitalism.

THE MINING CONSULTANT

South African mass mining birthed new economies of scale that prized efficiency but involved huge financial risk. Consulting practice changed as a direct result of this new capitalist organization. Finance and geology combined to effect this transformation. After two prospectors, George Walker and George Harrison, stumbled upon gold-bearing outcrops on Langlaagte farm in 1886, miners rushed to the Transvaal and soon found that the outcrops were just the tip of the iceberg. South African gold stretched 60 miles along an almost unbroken arc from Randfontein in the west to Modderfontein in the East. What the first prospectors did not know, but which the US geologist Ferdinand Becker later confirmed, was that the reef dipped diagonally southward underground, extending the deposit to as deep as 4,000 feet in narrow, fractured seams of "banket," a conglomerate of rocks and minerals in which gold was embedded in low but remarkably consistent amounts.

The central problem facing mining companies and engineers was not where to find the gold or how to remove it from the ground but how to profitably separate the gold from the banket once it was on the surface. Fortuitously, cyanidation, introduced to the Rand in 1890, increased recovery rates from low-grade ores to as high as 90 percent and, in turn, fundamentally changed the economics and organization of deep-level mining. To create the technological and administrative machinery required for mass mining, the preferred corporate strategy among the Randlords was to form holding companies or groups (the two largest of which were Consolidated Goldfields and Eckstein & Co.) that owned, partially or in whole, subsidiaries to spread the risk of deep-level mining. Doing so enabled the group to supply a centralized set of financial, administrative, and technical services at lower costs to subsidiary members—and insulated mining finance houses from the capital shocks of unproductive properties and collapsing share prices. By the close of the nineteenth century, fewer than ten groups controlled 124 companies, and together, they accounted for more than one-quarter of the world's gold output. Group owners were, for the most part, financiers who lacked the specialist knowledge to develop prospects into producing mines and instead bought the services of consultants.

Reimagining mining on this scale required considerable engineering expertise and managerial logic, the qualities most in demand of consultants. "The special knowledge required by these men has to be almost encyclopaedic in its range," wrote British mine expert James H. Curle.[13] Before production began, consultants strategized ore extraction and tailings disposal; oversaw shaft sinking, the stoping of underground levels, the creation of dams and reservoirs from water rights, and the transportation of ore below ground; and, to speed the ore through production at maximum efficiency, erected sorting belts, stamp batteries, tube mills (for fine crushing), water coolers, cyanidation whorls, tailings wheels, engine houses, and winding gear above ground. Once operations began, often between five and seven years later, the consultant organized management hierarchies and the recruitment of skilled and unskilled laborers, as well as consulted with the mine managers, metallurgists, engineers, and machinists in each department on the mine's operations. "Their business," reflected one *Engineering* editorial, "is generally to scheme out plans for carrying out works; to draw up the specifications in which these plans are, or ought to be, described; to superintend the work as it goes on; and, in general, to formulate what they or their employers want to be done, and then to see that it is properly done."[14]

It is worth pausing to consider the scale of these "schemes." Take, for example, Simmer and Jack Proprietary (S&J), part of the Consolidated Goldfields group (see figure 2.1). By far the Rand's largest mining company, S&J embraced 1,206 claims after a series of amalgamations of deep-level companies orchestrated by their newly hired consulting engineer, John Hays Hammond. To operate the S&J's 100-stamp mill in 1895 required a battery manager, senior amalgamator, six amalgamators, two juniors, two men to run the hoisting equipment, two general hands, two fitters, and a carpenter—in addition to Black labor.[15] Under Hammond's guiding hand and the management of Pope Yeatman and his superintendent Victor Clement (all three Californians), S&J was transformed into a huge industrial enterprise whose claims stretched a mile along the Reef and employed 450 white and 3,000 to 4,000 African migrant laborers.[16] By 1899, a battery of 280 stamps (managed by George Osler Smart), each weighing 1,200 pounds and capable of crushing 400,000 tons of ore annually from 10 shafts, three of which extended to 2,500 feet, showed a profit of almost £30,000 monthly.[17] Once milled, 600 tons of tailings were processed daily through the mine's cyanidation works, "the largest of their kind in the world," comprising five vats, each 42 feet

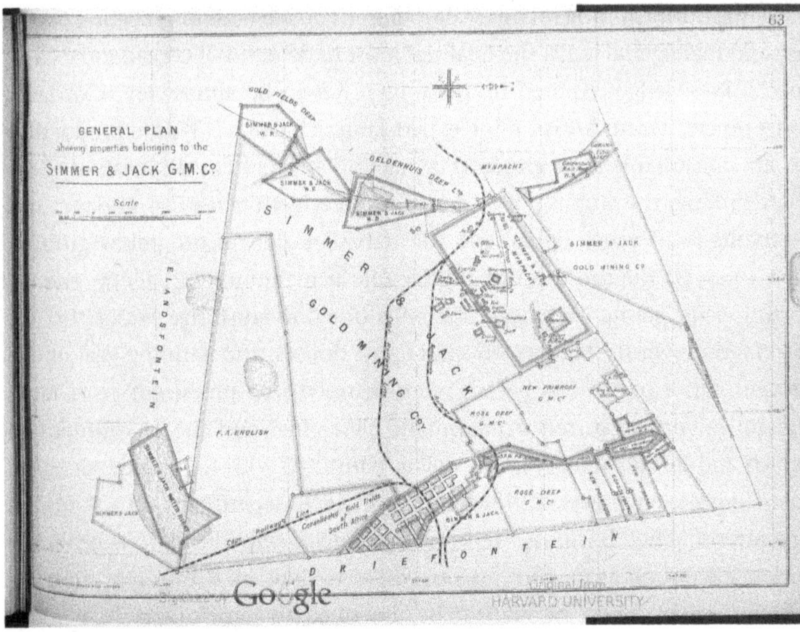

FIGURE 2.1

Simmer & Jack Gold Mining Co., circa 1895. This map shows the S&J before the installation of its 280-stamp mill. The compound for African laborers can be seen near the boilers and cooling ponds, with the accommodation for white foremen and managers some distance away near the outcrops. *Source*: Charles Sydney Goldman and Joseph Kitchin, *South African Mines: Their Position, Results, & Developments: Together with an Account of Diamond, Land, Finance, and Kindred Concerns*, 3 vols. (London: Effingham Wilson & Co., 1895), 63.

wide and 14 feet high, that took between 10 and 11 hours to fill.[18] Established on the site in 1898 was the General Electric Power Company, a subsidiary of Consolidated Goldfields, to supply current to the industrial complexes of S&J, Knight's Deep, Simmer and Jack East, South Geldenhuis, and South Rose Deep mines and to the towns of Germiston and Georgetown.[19] The Central South African Railways weaved through the property, delivering timber, coal, and frozen meat to feed this voracious industrial metabolism.

This was a colossal enterprise, replicated across the Rand. With twenty-nine companies operating 2,456 steam-driven stamps in 1898, it was little wonder that the ridge was veiled in a "greyish-brown haze" peaked with "Alpine summits" of tan-colored "heaped detritus" from cyanide processing and rang to the percussive, never-ceasing "murmur" of stamps that "sounded like the voice of the sea."[20] After a visit to the Langlaate Deep mine, one South African marveled at the winding engines "with their huge, bright steel crankshafts and connecting rods and massive flywheels," remembering that he "afterwards had nightmares involving me and that awe-inspiring machinery."[21]

Within these multiplex organizations, consultants functioned as specializing generalists, men with expert knowledge of mine engineering but with a broad understanding of interrelated specialties. As Hennen Jennings, chief consulting engineer for the Corner House Group (H. Eckstein & Co.), advised, "The mining engineer must have some sound general knowledge of all other branches of engineering inasmuch as in the equipment and running of great mines and metallurgical plants he must make use of the training of engineers in almost all the other branches" and "at least appreciate the foundations of their specialties."[22] For Herbert Hoover, and not without a touch of self-glorification, the mining engineer required "a wide knowledge of certain branches of civil, mechanical, electrical, and chemical engineering, geology, economics, the humanities, and what not; and in addition to all this, engineering sense, executive ability, business experience, and financial insight."[23] This was a formidable skill set. As mine corporations grew through diversification, purchasing ancillary enterprises like dynamite factories, electrical companies, coal mines, brickworks, meat processing, equipment manufacturing, and railroads as subsidiaries, the executive work of consultants only increased, and they became more than mere mining engineers but engineers of technical systems.[24]

Practicing consultants agreed that under mass mining, their role had taken on more executive functions. "To be successful he has to possess a skill and a versatility required in few other professions," wrote Hugh Frederick Marriott in his 1925 work *Money and Mines*.[25] Marriott's reflections were shaped by more than two decades working on the Transvaal, where he arrived in 1891 as assistant consulting engineer to Hennen Jennings at Eckstein & Co., the Rand's largest gold producer before becoming his successor in 1905. In a chapter devoted to the topic, Marriott ruminated at length on the consultant's role and its relationship to the new forms of industrial and capitalist organization of the Rand. "To define the exact duties of a consulting engineer is difficult," he reflected, but "in the type of organization, favored by the modern group system, controlling a number of separate mines belonging to different sets of owners, the consulting engineer exercises executive,

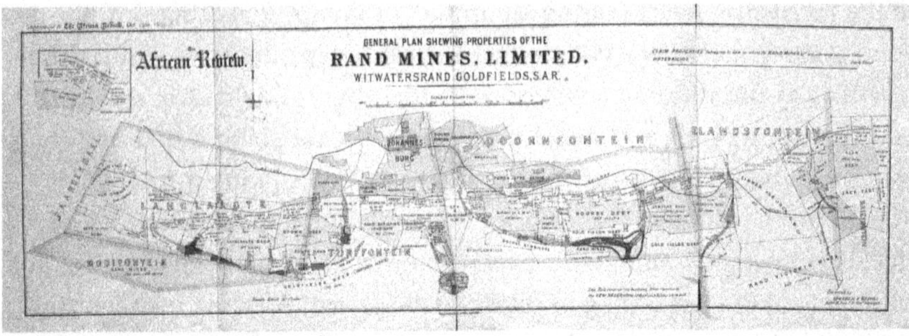

FIGURE 2.2
General plan of Rand Mines Ltd. Chromolithograph after Edward H. V. Melvill, 1895, illustrating the extent of mining company properties and the critical role of mining consultants who provided centralized technical services to the firm's subsidiaries that ran both outcrop and deep-level mines, the sheer extent of which can be seen above. At the time the map was made, Rand Mines had just one mine in production (the Geldenhuis Deep), but by 1898, this had risen to nine. *Source*: Wellcome Collection, https://wellcomecollection.org/works/ngr7jjwk/items.

as well as consultative, functions." He was the "chief technical authority" and formed "the link between the administration and the operating staffs at the mines."[26] As Marriott's mentor, Hennen Jennings, put it, "In these days of impersonal corporation ownership, the closest and most intelligent link between capital and labour is that of the engineer."[27] The consultant was the lynchpin of the mass mining groups. "Under the group system," wrote Marriott, "the consulting engineer has a special value, in that he has the opportunity of knowing what is being done at the different mines, and is therefore able to put the managers in touch with each other on common problems to point out how difficulties affecting one mine have been solved in another."[28]

Professional identity was never based on technical proficiency alone, however. For consultants, credentials were established by spending time in different roles within the multinational mining enterprise—from the mineshaft to the management office—and across different mineral fields. As a matter of course, the consultant was male and white.[29] The professional identity of the mining consultant amalgamated masculine notions of technical mastery, expert autonomy, and imperial notions of civilization and progress. "Essentially the mining engineer is a pioneer," mused Hammond, "Alone he makes his way through the difficult conditions imposed by the promoter or capitalist who employs him, and success or failure hangs on his sole judgement."[30]

Engineering narratives were replete with the claim that expertise was embodied in autonomous, mobile, white manhood.[31] Mine engineers were not exceptional in this regard, but their peripatetic careers gave them a prominent place in the cultural romance of the pioneering engineer.[32] "The consulting engineer must be prepared at a moment's notice to go to any part of the world," wrote one typical account, "whether it be the tropics or the Arctic Circle it is all the same" to the mining engineer, the account continued, making him "truly a man of many parts."[33] Global mobility was thus both real and idealized. In the iconography of the male technical expert, mobility was a surrogate for white male authority, but the genuine mobility of consultant engineers also equipped them with two of their greatest assets—hard-won knowledge of mineral fields across the globe and their access to a great store of human capital in the form of social and professional networks they created en route.

Global mass mining depended on this hypermobility, as did the culture of engineering expertise and professional practice that was itself redrawn around the new technical system of mass mining. Mechanisms of organization and diffusion contributed to the creation of strong professional networks among leading mining engineers. Technical schools were one component of the international spread of expertise and a form of cultural capital accumulated by elite engineers to publicly demonstrate their professional and social standing. As chapter 1 in this volume illustrates, the number of technical schools and students grew rapidly from 1870 in the United States and Europe and enjoyed international student enrollments, with many new graduates then joining professional associations to augment their technical credentials. The membership rolls of professional organizations therefore also grew across the world (see table 2.1). Concerning mining, the American Institute of Mining Engineers (AIME), founded in 1871, and London's Institute of Mining and Metallurgy (IMM), founded in 1892, were the most prestigious of these national societies. By 1905, AIME's membership had reached more than 3,600, representing ferrous and nonferrous mine engineers in addition to chemists and geologists.[34] In contrast, the IMM's membership stood at 1,799 in 1909, rising to 2,000 four years later—in both cases, a small proportion of the number of practicing engineers but evidence of the growing culture of professionalism.[35]

Professional societies—local and international—were modeled closely after one another and sustained the global flow of experts and expertise through shared memberships, reciprocal honorary memberships, and swapping of

Table 2.1
Institutes of mining, 1855–1898.

1855	Australian Miners' Scientific Association, Sandhurst, Victoria
1859	AMSA becomes the Bendigo Miners' Association
1860	Ballarat Mining Institute
1871	American Institute of Mining Engineers (AIME)
1887	Mining Society of Nova Scotia
	Engineering Institute of Canada
	Witwatersrand Chamber of Mines
1889	Institution of Mining Engineers, London
1890	Engineering Institute of Canada, Toronto Branch
1891	Ontario Bureau of Mines
1892	Institution of Mining and Metallurgy, London (IMM)
	Association of Mine Managers, South Africa
	South African Association of Engineers and Architects
1893	Australasian Institute of Mining and Metallurgy
1894	Chemical and Metallurgical Society of South Africa
1897	American Mining Congress
1898	Canadian Institute of Mining and Metallurgy

Source: Compiled by the author.

publications, a process that was a strategy of professionalization and exclusion.[36] Local professional networks were integral to the coproduction of mining expertise, ensuring that knowledge created on the frontier of extraction was ensconced in metropolitan professional settings. In South Africa, the Witwatersrand Chamber of Mines (1889), the South African Association of Engineers (1891), the Association of Mine Managers (1892), the Chemical and Metallurgical Society of South Africa (1894), and the South African School of Mines and Technology, founded in Kimberley in 1896, were all important venues in the generation and diffusion of expert knowledge of the geology, extraction, and refining of gold.[37]

Engineers further participated in the global culture of expertise by attending international conferences, contributing to the technical press, and writing autobiographies. All these activities burnished professional reputation, cultivating the markers of manhood and prestige necessary for engineers to distinguish themselves in the competitive entrepreneurial world of consulting and often as public intellectuals.[38] A great deal of the ideas shared in these venues also pertained to the financing of gold mines. Indeed, engineers insisted that they possessed "character" and were "trustworthy" and "honorable" individuals—key idioms of capitalist cultures of risk in the late nineteenth century.[39] Technical acumen and good business sense meant little if not supported by moral and masculine vigor. Mining consultants therefore spent a great deal of intellectual energy establishing the boundaries of professional practice and the ethics of their relationship to mining capitalists. In some instances, as we shall see, the culture of expertise curated by consultants was one that normalized close connections between specialist knowledge and the high-risk financing of mass mining.

GLOBAL NETWORKS OF MASS MINING

As the South African landscape was dismantled, a new landscape of mass mining capitalism was erected in the City of London. Engineering consultancy thrived here in its "golden age" between 1890 and 1914.[40] In that period, some 8,408 mining and exploration companies were registered in the UK with a combined capitalization of roughly £1,045 million (of which £250 million was actually issued).[41] From an annual average of £8 million in the 1880s, investment in mining surged to around the £20 million mark for much of the 1890s, before dropping back to around £10–15 million per annum between 1902 and 1914.[42] Less stringently regulated than its competitors in Paris and Berlin, the London Stock Exchange (LSE) offered ready marketability for leading mines, competitive lending markets, and a rapid system of international arbitrage (especially Paris).[43] London was also the world's leading bullion market, equipped with high-quality refining facilities, and through the Bank of England maintained the remarkable stability of the gold-backed pound. Even though much of this investment was never put into gold production, as early as 1898, British-owned companies produced 60 percent of the world's annual output of gold and created a mining investment market that produced new forms of corporate organization to finance and manage the global mineral industry.[44] Mining consultants were central to the global knowledge economy sustaining this extraordinary boom.

While mass mining operations may have been truly global in scope, it depended upon a tight-knit, at times remarkably local, world. Among the foggy, gas-lit streets of the Inner City around the Bank of England and LSE, the cultures of capital and consulting were interwoven. Here mining investors, consultant engineers, and the specialist mining press were headquartered cheek-by-jowl with the offices of mining finance houses, company promoters, and stockbrokers—scrupulous and not. Just as imperial civil engineers established themselves as the "Great George Street Clique" close to Westminster and the government offices that could secure them lucrative contracts, global mining consultants gravitated toward the City.[45] Here they took advantage of a supporting web of financial services, including merchant bankers, insurers, credit reporters, and accountants, and accessed and influenced the flows of market information that converged there. Finch Lane was home to the editorial offices of the *Mining Journal*, founded in 1835 by the London stockbroker Henry English, whose specialist reports complemented the general investing advice of the financial press. For those dealing in copper, life revolved around the London Metals Exchange, established in 1882, just a stone's throw away. A walk northward along Old Broad Street from the LSE was London Wall, home to some of the city's largest investment banks, where they jostled with the offices of mining syndicates and consultants, the editorial offices of T. A. Rickard's *Mining Magazine*, and, from 1892, the headquarters of the Institute for Mining and Metallurgy. Walking south from the LSE, one reached St. Swithin's Lane, the epicenter of gold rush finance: the Rothschilds. For the international mines they invested in, Rothschilds acted as financial advisers, metal brokers, gold refiners, insurance agents, shipping brokers, and bankers. Here was surely the most lucrative gold field in the world.

Mining capitalism required a steady flow of reliable information into the city. Consultants produced that information in the form of technical reports and assays. Some consultants acted as independent generalists, such as the firms Hill & Stewart, James Brothers, Williams & Co., and Pellew-Harvey & Co., all based on London Wall or nearby on Old Broad Street, as did "celebrity" engineers like Hoover (after leaving Bewick Moreing in 1909) and Hammond, who established offices on London Wall and Old Jewry, respectively (seen in figure 2.3). Most consultants developed specialist mineral expertise, and their practices could be quite small, consisting of an office and perhaps one or two senior engineers and a few apprentices. Others were regional specialists and were sizable operations. This was especially so of well-known consulting firms, which took on mine management in addition to surveying such as John Taylor and Sons for southern Spain and India; Bewick Moreing & Company in Western Australia; Tarbutt, Janson and Son in West Africa; Sir Robert Williams & Co. in the Central African Copperbelt; and Wernher Beit & Company in Southern Africa. The sheer scale of the work involved for these firms was often staggering. By 1897, for example, Bewick Moreing was

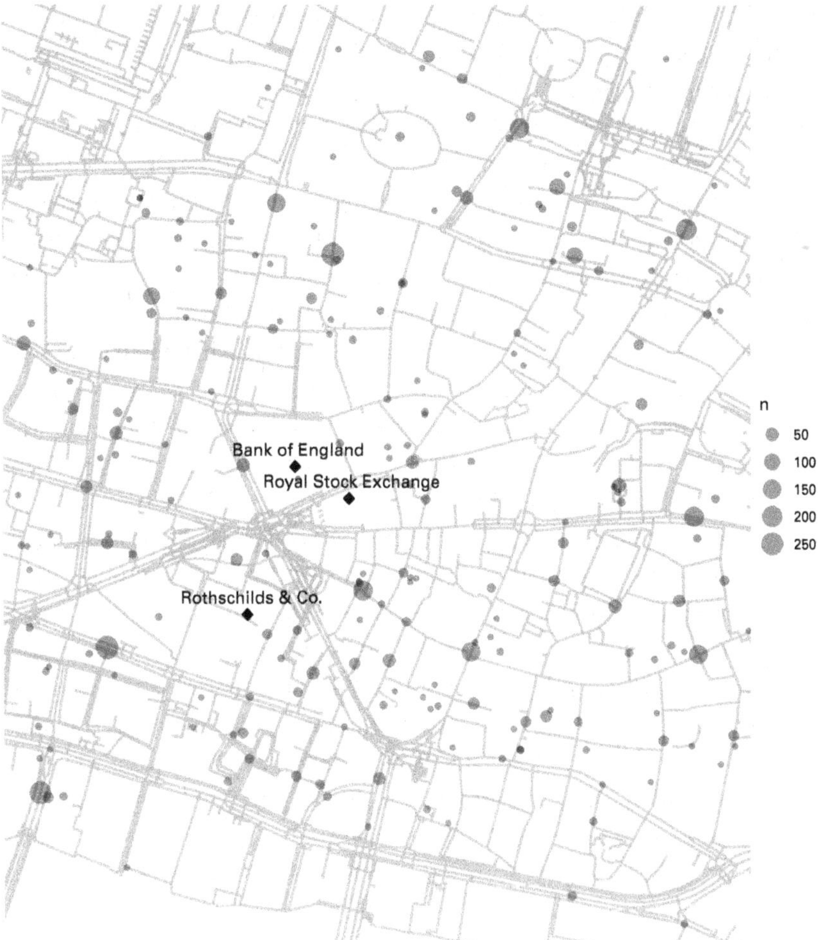

FIGURE 2.3
Location of mining company offices in the City of London. *Source*: Israel G. Solares and Edward Beatty, Blueprint for Modernity Engineering Database, accessed through https://github.com/solaresig/Blueprint-for-Modernity or engineeringhistoryproject.org.

consulting engineer to sixty-four Western Australian gold mining companies registered in London and, by 1904, produced 37 percent of the state's gold and employed almost 20 percent of its goldminers.[46] Mining consultants were thus one part of a great flow of expertise into London that made the city "the centre of the mining world."[47]

Metropolitan spaces were integral to mining's global reach. In New York, mining consultants were part of the city's expanding economic elite, drawn to its dense networks of business contacts, technical experts, and financial elites clustered around Wall Street and the New York Stock Exchange (NYSE) and Consolidated Stock Exchange of New York.[48] New York was the financial epicenter of the "American World of Copper," connecting the city's capitalists to the frontiers of wealth in Montana, Utah, and the Southwest.[49] Phelps Dodge, the American Smelting and Refining Company, and the "Father of the Porphyries" Daniel Jackling's Utah Copper Company and Ray Consolidated Copper Company all had headquarters in Manhattan. As in London, an undergrowth of financial services thrived in the city to support the activities of these firms. Specialist dealers in mining stocks such as Graham & Horton and Waller & Hirshfeld, assayers, credit rating agencies, wire services, and the technical press established offices on Broadway, Pine Street, and others close to the Consolidated building on the corner of Exchange Place and Broadway.[50] The Ingersoll Rock Drill Company could be found at Park Place, neighboring the offices of the *Engineering and Mining Journal*, while the Chicago-based specialist in mine machinery, Fraser & Chalmers, took offices on Fulton Street. Consultants marketed their expertise here too, many acting on behalf of subsidiaries owned by the international groups located in London or for US mining corporations. On his return to New York in December 1899, for example, John Hays Hammond took offices in the Mills Building, on the corner of Broad Street and Exchange Place, from where he orchestrated the mining empires of the Mills and Guggenheim families in Mexico and the interests of London's Venture Corporation in the United States.[51] Capital could not be any closer since the L-shaped Mills building wrapped around the offices of J. P. Morgan, who would later team up with the Guggenheims to develop copper mines in Kennecott, Alaska.

A network of mining cities stretched across the globe. Some acted as gateways for financial capital. Dubbed "the hub of the mining world" by one of the city's leading capitalists, San Francisco was home to bespoke mining markets with savvy investors, a leading technical press, and the headquarters of extractive empires that stretched across the American West and Pacific World.[52] British banks established branch houses and agencies in the city to invest in western mining, and between 1860 and 1914, an estimated 584 joint-stock companies were registered in London to exploit western mines.[53] "London is a lake of money," reflected the *Mining and Scientific Press* in September 1895, "bank full, and running over."[54] In turn, San Francisco's hinterland was subordinated to an urban extractive empire overseen by local capitalists.[55] Regional mining stock exchanges across the American West, such as in Denver, and in British settler wests, such as Johannesburg and Melbourne, Montreal and Vancouver, did a brisk business.[56] These regional exchanges were far from insignificant in the global context. Although nowhere could match Imperial London for the sheer concentration of human and financial capital, these urban bridgeheads were part of an international network of engineering consultancy services, each with their own wells of human capital, that connected bourses in Philadelphia, Boston, Paris, Berlin, and Amsterdam to major ports and inland mineral cities like Butte (MT), Deadwood (ND), Virginia City (NV), and Bisbee (AZ) in the United States, Cananea (Mexico), Nova Lima (Brazil), and Broken Hill and Kalgoorlie in Australia, to name just a handful.

London, though, was the frontier of extraction for mining finance. By 1895, the South African and Westralian markets dominated the daily rhythms of the LSE, not only on the bourse but also on "the jungle" or informal Kerb Market of Throgmorton Street. Share prices rocketed as these markets boomed, their "Circean revels" proving "irresistible" to speculators.[57] By 1899, the par value invested in the Rand mining industry amounted to £25 million, and by 1914, it had reached £125 million as the Rand produced 40 percent of the world's gold.[58] This was a boom in company promotion. To take advantage of the investing public's appetite for mining shares, the number of joint-stock mining finance houses, often called exploration companies, rocketed. In 1895, at the height of the South African and Westralian booms, 40 percent of the 961 newly created mining companies were created for the purposes of exploration and company promotion.[59] With the phenomenal growth of company promotion came new investment practices that were wide open to fraud and disinformation. Many of these firms were little more than speculative endeavors overseen by charismatic financiers. But some exploration companies systematized the search for and development of new mines, providing a steady supply of new prospects for promoters. This was carried out to great effect by the Rothschilds-backed Exploration Company, a joint-stock venture founded in 1885 by the American consulting engineers Hamilton Smith and Edmund de Crano, that directed the Rothschilds toward valuable mining properties worldwide. Smith and de Crano identified the properties through careful research that enabled them to "pick the eyes out" of mining markets and enrich the inside few with "promotion plunder."[60] The company's adroit combination of social connection, financial invention, and engineering expertise can be seen in one attempt by Hammond to procure finance for a Mexican mine when he told its prospective operator, Charles D. Lane, that after making a social call to Smith and de Crano, they were

> very favourably disposed to entertain your . . . proposition especially when I told them what I know of your standing etc. but they have stipulated that you obtain a larger option on the property, so as to enable them to send their expert, upon whose report they will float the property. While having perfect confidence in you, they say that in going to the public, they must have a complete engineers report, and that engineer must be one of their selection. I told them that I would arrange to have [Frederick W.] Bradley make a report which would greatly enhance the value of the property when presented to the public.[61]

Consultants were therefore prominent among the "new breed of intermediaries" bringing mines to the market.[62] Their chief marketable commodity was expertise. As one mining investment manual advised, "Professional advice can now easily be obtained. There are many experienced mining engineers now resident in London, whose services can be obtained as consulting engineers to boards of directors . . . and who can supply the technical knowledge unnecessary in the ordinary director."[63] An engineer's report was essential to any new company prospectus, legitimizing high levels of financial risk with technical expertise. But their advice came at a cost. At the Exploration Company, Smith and DeCrano each took home salaries of £5,000 a year and 1.5 percent of all the business conducted by the company, as did Joseph Harry Lukach, who joined as a consultant in 1896.[64] Hammond was rumored to be taking home a salary of $75,000 at Consolidated Gold Fields. Jennings, a protégé of Smith's, was equally well compensated. Mining financiers also made regular use of so-called founder's shares and vendor's interests, which granted exclusive rights to a specified part of a company's profits before ordinary shareholders received dividends. The remainder of the shares, usually a small proportion, were then offered to the public at prices well above par. Once the capital had been raised, company mongers then filleted the assets for themselves, leaving the investing public with the "shrivelled carcass."[65]

But what exactly were financiers investing in? In short: human capital. Hammond, Jennings, Hoover, Smith, and others called on deep reserves of professional contacts thanks to their global careers. In their executive role, consultants recruited skilled experts to oversee the development of deep-level mines. Shortly after his appointment to Consolidated Gold Fields, Hammond enriched his trusted inner circle with plumb positions. "Advise you strongly to come here prospects are undoubtedly good," Hammond wrote Thomas Leggett, an engineer of great repute who had experience managing mines in the western United States, Mexico, Honduras, and Colombia, in a typical missive, adding, "Will pay you £3,500 . . . expect to do better later."[66] Hammond also secured the services of Pope Yeatman, George Washington Starr, Victor Clement, Ernest Abram Wiltsee, Harry Howard Webb (Hammond's eventual successor), Thomas and William Wallace Mein, Robert Mayo Catlin, and Richard A. Parker, who collectively had experience of working mines in California, Nevada, Mexico, Alaska, and the famed El Callao mine in Venezuela.[67] "I get the expert engineers who are really worth their salt," Hammond bragged.[68] Jennings brought with him trusted colleagues from El Callao to Eckstein & Co. Likewise, the Exploration Company was "instrumental" to the spread of American engineering talent overseas.[69] Consultants thus brokered the great movement of expertise between mining regions in the Global South. "In no other part of the world are collected so many engineers and chemists of the first rank, constantly engaged upon the problem of decreasing the working costs and increasing the percentage of extraction," reflected the British consultant-turned-financier Lionel Phillips while senior partner of Eckstein & Co.[70]

Engineers had considerable power to create markets and capitalized on the city's unrivaled opportunities for social networking. Leading consultants dined at spots favored by investors like "Maeby's" chop house and the City Athenaeum (commonly known as "Thieves Kitchen") on Throgmorton Street, keeping an ear to the ground for financial intelligence. At investors' meetings, they performed their expertise to stockholders and directors, where they cultivated financial and social capital. After returning from the Rand, Hammond spent the autumn of 1894 in London, acting as a "high pressure salesman" to "convince mining investors of the attractiveness of deep-level shares" with diagrams of the outcrop and deep-level mines and charts of his calculations of yield per claim, expenses, and profit, establishing his trustworthiness to potential investors.[71] Maintaining the public image of mining houses was a crucial part of the engineer's expert role. To do so, some consultants were regular contributors to the financial press. Herbert Hoover, for example, was a prolific pseudonymous contributor to the *Financial Times*, *EMJ*, and *Mining Magazine*, assuring readers that the new breed of engineers had introduced efficiencies into mining operations and reformed speculative excess.[72]

A consultant's report could be the difference between a bull or bear market. Hammond recalled of his time surveying the longed-for "Second Rand" in southern Rhodesia that "it was of supreme importance to Rhodes that my report be favourable." It was not. The financial stakes were self-evident. "On my findings would hinge the investment or the non-investment of enormous sums of British capital."[73] But the information consultants provided only grew in asymmetry alongside the scale of mass mining capitalism. As a result, engineers participated in creating a knowledge economy that enriched those on the inside as it impoverished those on the outside, and it often eroded the independence of the consultant. To ensure the loyalty of engineers of repute—and so steady access to valuable stores of human capital—firms made strategic use of shares. Hamilton Smith, for example, was offered four or five shares at $150 below the market value when he was appointed chief engineer of one Corner House mine in 1893. "Smith is a very good friend of ours, and we could afford to be generous and take credit for letting him have shares," the firm's executives concluded.[74] Likewise, when Rand Mines Limited was launched in February 1893, four mining engineers were together allotted 15,000 shares in total. Invited in on the "ground floor," that is, before the initial public offering, these engineers stood to make substantial profits once the company began trading publicly. And so they did. The value of their stock rose 500 percent within a month.[75] As the firm's director Herman Eckstein explained, "In the long run it will pay better to let other people earn something as well. It will bring grist to the mill in the shape of timely information."[76] The directors of these firms invested not only in technical expertise but also in privileged access to information that placed them in "positions of absolute power" to manipulate futures markets.[77] "Everyone has heard of the system of buying information from people in more or less responsible positions in the mines, and acting on it in the market," reported the *Pall Mall Gazette* in 1904.[78] Indeed, it was entirely routinized in the relations between consultants and capitalists.

RISKY SPECULATION AND PROFESSIONAL OBLIGATION: CULTURES OF CONSULTING AND CAPITAL

What are we to make of the close relationship between engineers and the cultures of risk in mining finance? Some historians have argued that following the spectacular collapse of share markets in 1896, a process of rationalization set in as the major gold mining firms turned their attention from company-mongering to production. Favoring a steady flow of working capital to overcome the technical challenges of mass mining, major mining finance houses attempted to distance themselves from the manias of the "circus," with the result that technical expertise supplanted speculation.[79] Consultant engineers were thus central to the management of risk. "The organisation of expert advice and information," write Turrell and van Helten, "went a long way towards reducing the risks associated with mining."[80] Indeed, tangible results accrued as the value of gold production rose from £5,180,090 in 1893 to an excess of £15 million in 1899.[81]

Managing risk was certainly the aspiration of the new institutions of mining, the technical press, and a vocal group of mining experts. To protect the public against the worst excesses of mining's culture of financial capitalism, the international technical press policed the boundaries of professional behavior and posed as the protector of the investing public. From their offices in New York, Rossiter Raymond and Richard Rothwell, editors of the *Engineering and Mining Journal* (*EMJ*), proclaimed independence from all "entangling alliances; with mining speculations and stockbrokers," declared the *EMJ* the "enemy of all fraudulent mining ventures," and vowed to expose the swindlers and confidence tricksters plaguing the mining industry who reduced mining investment to a "form of gambling in which the 'outsider' stood but little chance of winning."[82] Such regulatory impulses, argues Edward Balleisen, were part of a growing culture of fraud detection and surveillance in the United States that attacked the routinization of crony capitalism and the spread of disinformation in financial markets.[83]

Given mining's global reach, there was a transnational dimension to this surveillance. Thomas Arthur Rickard, himself one of the world's leading mining consultants, took up Raymond and Rothwell's cudgels as editor of the *EMJ* (1903–1906 and 1922–1929), as owner-editor of San Francisco's *Mining and Scientific Press* (1906–1909 and 1915–1922), and founder-editor of London's *Mining Magazine* (1909–1915). Rickard targeted both the professional dilemma of collaboration with capitalists and the financial presses' opaque relationship with speculators and promoters. "If the profession of mining-engineering has reached no higher philosophy that that of Wall Street and Throgmorton Avenue," Rickard editorialized for the *EMJ*, "it has become obtuse indeed."[84] From his desk on London Wall, Rickard waged war against the "incurably amateurish, superficial, ignorant, and pontifical" financial press and the "guileless scribblers" and "hired purveyors of misinformation . . . as competent to write on aviation or dairy farming as on mining."[85] Rickard recommended a variety of solutions, ranging from the proper scrutiny of engineers' CVs by the investing public and a vigilant technical press to the complete abstention from promoting mines, speculating in shares, or receiving contingent fees. At threat was fragile public confidence in the hard-won expertise, independence, and professional standing of the mining engineer. "Risky speculation undermines moral and professional obligations," Rickard asserted.[86]

Such preventative strategies should be viewed as integral features of the ongoing public scrutiny of the line between illegitimate gambling and fair enterprise on both sides of the Atlantic.[87] Opportunities for deceit, information asymmetries, and fraud exist in any financial market. But they were perhaps particularly salient in late nineteenth-century mass mining markets where technological breakthroughs, innovative financial organization, eye-watering levels of capitalization, and the "fevers" and "manias" embedded in the cultural perception of mineral rushes lent themselves to high-risk speculation. By policing the boundaries of reputable professional and legitimate economic activity, these commentators attempted to exclude both reckless swindlers and charlatan engineers from the world of mining finance, and by doing so, they in turn reinforced the respectability of some forms of investment and the reputability of professional expertise.[88] Rickard was plain: the problem was not the mining profession, which was "the natural protector of those who place their money in mining ventures," but the corrupting effects of founders' shares and investments in properties one was contracted to advise on.[89] Other engineers agreed. "When a man prays to be delivered from temptation he should not invite it," reflected Henry Cleveland Perkins, a veteran of DeBeers diamond mining and the Exploration Company, on the subject of consultants owning shares in the company that employed them.[90] "There is no moral wrong in an engineer holding an interest in a property which he has examined; there is not even any moral obligation which could require an honest man to decline contingent fees, granting that he believed his judgement would not be warped by the self-interest at stake," Rickard reflected in the *MSP*. But was that possible? Rickard pleaded "deference to the prevailing Adamic weakness."[91]

Yet, using the same moral vocabulary as their critics, many corporate engineers reached the opposite conclusion: financial participation was the guarantor of professional integrity. Consultants were often explicit about their capitalist ambitions. In his address on "Professional Ethics" as president of

AIME for 1908, Hammond was plain: "a mining engineer engaged in such consulting practice should aim to become himself a capitalist, and a purchaser and developer of mining properties."[92] Fresh from a five-year contract with Guggenex working as the highest salaried employee in the United States (1903–1907), recently a nominee for the Republican vice presidency, and now a professor of mechanical engineering at Yale (1902–1909), Hammond took an expansive view of the "Mining Man" as an engineer-financier.[93] "Should he acquire business experience, he might become avowedly a promoter, in a high and worthy sense of that term," he continued, "and thus assist in driving out the unscrupulous promoters who are really parasites."[94] A year later, Hammond expanded on these views in a commencement address at the Colorado School of Mines. "There is now interposed between the mining engineer and the investors a class of men known as promoters," Hammond told the Class of '09. "Not only are the majority of men in this class incompetent and ignorant about mining matters," he continued, "but they are very often unscrupulous as well."[95] Should the "Mining Man" take the place of the promoter, "mining investments would be placed upon a more conservative basis, thereby greatly minimizing the risks of such investments."[96] This was, to borrow Richard White's term, the "morality of consequences."[97] The engineer's social function—protecting investors from the sharp practices of middlemen—justified his reward. In this way, holding and speculating in shares on properties examined was merely the side effect of sound professional practice rather than its antithesis. "In helping themselves," as White put it about the self-perception of early Gilded Age financiers, "they helped investors."[98] In fact, then, as Ian Phimister and Jeremy Mouat have forcefully demonstrated, mining engineers often merely reconfigured, rather than reduced, risk.[99]

Debate raged in the technical press. Reformers saw only hypocrisy in this combination of personal enrichment and ethical virtue. "He thinks that he speaks as a mining engineer," scoffed Rickard, "but he talks as a promoter. . . . A regrettable retrogression."[100] But Hammond was far from alone in his expansive view of the engineer as financier. Hoover leaped to Hammond's defense, reiterating his central claim that the "evils in the mining business can be more quickly remedied if the whole, or a larger portion of the personnel of the industry, forming the chain between the mine and the capitalist . . . are men of technical and practical training."[101] Seeking to cement his reputation as a responsible engineer while building a new one as a public intellectual,

Hoover posed in the *Mining Magazine* throughout 1909 as the "Professional Speculator," attempting to police the boundaries of meritorious speculation and reckless gambling, noting that there was "no common ground" between the engineer-financier and the company promoter whose "business is based on the yield from human credulity; mine upon realizing minerals into metal. He [the promoter] would succeed equally well as a 'bookie' or running a yellow journal."[102] Such a sleight of hand deflected attention from Hoover and his fellow consultant's ruthless exploitation of the information gap between themselves and the investing public.[103]

In the late nineteenth century, mass mining capitalism and modern mining consulting were coproduced. Together, consultants and capitalists achieved impressive feats, mobilizing vast resources of human and financial capital for the exploitation of the Earth's resources on an unprecedented scale. Consultants were far from being instrumental tools of mass mining capitalism's expansive reach, however; they generated, articulated, and sometimes hoarded the engineering knowledge upon which it rested. Adaptable and imaginative, precociously talented, and unashamedly capitalist, mining consultants transformed the community and work of modern engineering. The modern, engineered world system depends upon minerals in staggering quantities.[104] And so, engineers and consultants continue to be key agents in reordering world economic production around new scales of mass mining capitalism. Engineering consulting firms have boomed in that context. They are now much larger, offer a greater range of consulting services, and are more multidisciplinary, more multinational, and more collaborative.[105] They have also been key contributors to the myth that the world economy can continue to overcome the Earth's natural limits.

3 "ENGINEERING SCIENTISTS" IN MEXICO, 1876–1911: THE CASE OF METEOROLOGY

MIKAEL WOLFE

> Scientists and engineers have different goals. Scientists seek to describe and understand the natural world. Engineers consider various criteria and constraints in order to design solutions to problems, needs and wants that better the lives of humans, animals and/or the environment.
>
> —Amielle Major, "4 Simple Ways to Explain the Difference between Science and Engineering," https://blog.eie.org/4-simple-ways-to-explain-the-difference-between-science-and-engineering

The Boston Museum of Science's explanation in 2018 of how scientists and engineers differ dates to the nineteenth century, when the natural sciences emerged as distinct disciplines. Unlike scientists, however, engineers had existed as an identifiable class of constructors and designers, especially of military technology, in Europe since the mid-fourteenth century. Indeed, the word *engineer*, etymologically derived from the Latin for engine, long preceded the word *scientist*, which was likely coined in 1834 to replace "cultivator of science" and "natural philosopher."[1] Therein lay the real and perceived difference between the two by the nineteenth century: the engineer designs and makes things, while the scientist produces knowledge about the natural world and its laws and therefore *how and why* the things that engineers make and design work or not. In practice, however, there was much more overlap between the two. After all, to engineer anything, engineers had to acquire some knowledge of the natural world, while to generate more sophisticated knowledge of the natural world, scientists had to engineer instruments, laboratories, and other technologies—or at least adapt the ones they procured from engineers to their methodologies. Consequently, engineers and scientists often needed the other—or needed to *become* the other—to successfully do their work.

In this chapter, I explore this historical dynamic in late nineteenth- and early twentieth-century Mexico by examining the emergence of meteorology during the Porfiriato (the rule of Porfirio Díaz, 1876–1911)—a period of authoritarian but relatively stable rule in Mexico compared with the preceding half-century of postindependence internecine conflict, war, and foreign invasion. Díaz's "revolution" in 1876 determined to modernize Mexico along North Atlantic lines, adopting positivistic precepts and programs under the slogan of "order and progress." As such, Díaz invested considerable state resources in new institutions of science and technology, largely emulating current models in the United States and Europe, and sought to attract foreign investment to industry and infrastructure that required advanced technoscientific capabilities that Mexico did not yet possess.[2]

My methodology in this chapter is primarily qualitative, drawing on newspapers, journals, published books and pamphlets, conference proceedings covering meteorological and related topics, and secondary literature.[3] My principal argument is that Mexican engineers viewed themselves as "engineering scientists" responsible for the advancement, promotion, and professionalization of a variety of emerging science and technology fields.[4] Historian of science Edwin T. Layton Jr. explained that in the United States, which Mexico sought to emulate,

> Engineering science often differs from basic science in important particulars. Engineering sciences often drop the fundamental ontology of natural philosophy, though on practical rather than metaphysical grounds. Thus, in solid mechanics, engineers deal with stresses in continuous media rather than a microcosm of atoms and forces. Engineering theory and experiment came to differ from those of physics because it was concerned with man-made devices rather than directly with nature. Thus, engineering theory often deals with idealizations of machines, beams, heat engines, or similar devices. And the results of engineering science are often statements about such devices rather than statements about nature.[5]

Although engineering science in Mexico, like in the United States and other countries according to Layton's description, describes well the relationship between engineering, on the one hand, and sciences like physics and chemistry, on the other, the boundary between engineering and meteorology was arguably even more blurred than with these others because the "laboratory" for meteorology was the global atmosphere. Before the development of computer modeling and satellites after World War II, meteorology relied on an array of instruments, such as thermometers and barometers, and a burgeoning infrastructure of communications and transportation (mail, telegraph, printing, mapmaking) to provide weather forecasts. Engineers therefore played a vital role in the advancement of meteorology because they could draw on their training, skills, and experience in devising new technologies to solve practical problems in communications, agriculture, navigation, and infrastructure-building exacerbated by inclement weather conditions characteristic of specific climates. In late nineteenth-century Mexico, engineers were instrumental in establishing the science of meteorology, working explicitly in a transnational and transatlantic context.

Although all engineering scientists were men in Mexico because of the existence of an overt, systemic sexism common throughout the world, in this chapter, I also discuss the important role that a few prominent women science teachers and educators played in publicly diffusing technological advances in meteorology.[6] I begin the chapter with President Benito Juárez's 1867 Education Law, which incorporated liberal and positivist precepts from elementary to university education curriculums for Mexico City. Freshly victorious over French occupiers who had supported Mexico's conservatives in a long-running civil war, the Liberal Juárez saw secular education as key to modernizing Mexico along the lines of the North Atlantic industrializing economies. Mexican intellectuals, most with engineering or law degrees, traveled extensively in the United States and Europe and returned with ideas for reforming Mexican institutions. Nearly a decade later, the Liberal Mexican state consolidated under Díaz would expand on Juárez's incipient modernization effort and establish many more government and professional science and engineering-focused organizations and agencies, including the Mexico City–based Central Meteorological Observatory in 1877. As with many other such agencies, the federal government selected an engineer to direct this observatory.

As engineering scientists, these agency directors found that the steady but slow progress of accumulating techno-scientific capacity and knowledge through national and international data collection and networking, lobbying, and dissemination was difficult and frustrating. In the case of the Central Meteorological Observatory, directed by the engineer Mariano Bárcena, the steady but slow progress culminated in the founding of a National

Meteorological Congress in 1900. At the first Congress meeting that same year, Mexican meteorologists, many of whom were educated as engineers, shared their accomplishments as well as commiserated about the enduring challenges of their underdeveloped science, especially compared with their European and North American counterparts. Yet the highly dedicated but still mostly amateur meteorologists who attended these congresses, among whom engineering scientists played leading roles, made further if halting progress professionalizing and advancing their science through the outbreak of the Mexican Revolution of 1910. The most visible and tangible result of that progress was the impressive increase in meteorological stations and observatories throughout Mexico, as we see in figures 3.1 and 3.2. Engineering scientists played leading roles in this achievement, roles they would maintain for decades into the postrevolutionary period.[7]

HOW ENGINEERING SCIENTISTS RAISED METEOROLOGY'S PROFILE

On December 2, 1867, Benito Juárez decreed the Organic Law for Public Instruction in the Federal District (Mexico City), which would serve as a model for the entire nation. The preamble justified the law in terms of "spreading the enlightenment of the people" as the "surest and most effective means of educating [the people] and of establishing freedom in a concrete way as well as of respect for the Constitution and the law."[8]

Influenced by the positivism of the French philosopher of science Auguste Comte, the drafters of the law sought to overhaul Mexico's public educational system by secularizing and focusing it more on "practical" subjects, especially science, engineering, math, and technology (the rough equivalent of modern STEM *avant la lettre*).[9] To that end, the law declared the founding of schools devoted to specific fields, which included schools of medicine, surgery, and pharmacology; agriculture and veterinary medicine; engineering; and natural history (what would later become ecology and biology). Common to their curriculums was the study of various branches of mathematics, applied physics, general chemistry, and elements of natural history. Meteorology was initially included in the schools of medicine and agriculture but not in engineering. However, in the early 1880s, the government expanded the law and incorporated meteorological studies into the curriculums of telegraphy, geographic engineering, topographic and hydraulic engineering, industrial engineering, and metallurgic engineering. At first glance, requiring that the industrial and metallurgic engineering curriculums include meteorology appeared unnecessary. But besides the fact that variable weather and climate affect industry and metallurgy in terms of material procurement and durability, there was another reason why meteorology became standard across several engineering curricula from 1867 to 1883. In the 1860s, mining and metallurgical engineering was the most advanced and well-staffed technical field, and these engineers could—and would—play a central role in advancing—indeed engineering—the burgeoning science of meteorology.

President Porfirio Díaz established a central meteorological observatory in March 1877 a few months after taking power in November 1876. In part to symbolize its importance to the nation, the observatory was placed in an old guardhouse at the top of the National Palace under the charge of the powerful Ministry of Development, Colonization, Industry and Commerce (hereafter "Fomento"). The minister of Fomento, Vicente Riva Palacio, appointed the engineer Mariano de la Bárcena as director of the observatory, which Bárcena directed until his untimely death in 1899. Bárcena's life and work exemplify the polymath nature of engineering in late nineteenth-century Mexico and how and why he and others became "engineering scientists."

As a child growing up in Guadalajara, Bárcena was always interested in collecting rocks, plants, and insects. He pursued those interests at the National Preparatory School in Mexico City, founded under the 1867 education law. He had to take a two-year hiatus in Guadalajara from his studies due to the political instability stemming from the Liberals' war against the Conservative-backed French occupation from 1863 to 1867 before returning to Mexico City to complete his degree. In 1871, he enrolled in the Special National School of Engineers, the successor to the Royal School of Mines that was founded in 1792 under Spanish colonial rule. This institution had long been the major STEM school through Mexican independence and postindependence civil conflict and war (and was one of the first technical schools in the Western Hemisphere). In 1883, it would be rechristened again as the National School of Engineering before being incorporated into the National University (today's UNAM) in 1910. At the School of Engineers, Bárcena excelled at his studies, including mineralogy and geology for survey engineering in materials, especially metals testing. Upon graduation, he joined the Ministry of Fomento and quickly rose in its ranks, representing Mexico at the 1876 Centennial Exposition of US Independence in Philadelphia. This experience positioned him to become the founding director of the new meteorological observatory in 1877.[10]

It is unlikely that Bárcena formally studied meteorology as an engineering student because meteorology was not yet incorporated into the curriculum of engineering fields in the late 1860s and 1870s. It is possible, however, that he became familiar with meteorology while studying "geographic engineering," which included "physics of the globe" as part of its astronomy curriculum, a subject that meteorology would later be paired with. Regardless of his formal coursework in engineering, Bárcena sustained his childhood interests in collecting rocks, plants, and insects by cofounding the Mexican Society for Natural History in 1868 and publishing articles, essays, and entire books on subjects in geology, physics, botany, zoology, and paleontology, among others, throughout his career. He also became a member of several scientific organizations, including the Geographical and Statistical Society, which granted him an honorary membership in 1872.[11]

Five years hence, in one of the Central Meteorological Observatory's first publications addressed to the Fomento Minister Riva Palacio, Bárcena initiated the institution's first efforts to gather national meteorological data. In his capacity as the new director, Bárcena requested that Fomento supply the necessary equipment, such as barometers and thermometers, for direct data collection. He also asked that Fomento "circulate invitations addressed to engineers, pharmacists, doctors, agriculturalists, and other enlightened people for them to send this Observatory temperature, rain, and other data that they can gather in their localities."[12]

Notably, Bárcena's request did not include sending invitations to meteorologists since few professionally trained ones existed in Mexico. Indeed, the cofounder of the "Antonio Alzate" Scientific Society and instructor of science courses at the nation's Military School, Rafael Aguilar Santillán, compiled a bibliography of meteorological studies, published in 1890 by the Society's journal. He found that there were few published meteorological studies since 1769—the year that the Spanish priest and scientist José Antonio de Alzate (the Society's namesake) had begun recording daily temperatures—through 1877. Besides Alzate, only seventeen other individuals were known to have collected meteorological data before 1877. Among them, five were engineers and the rest were physicists, doctors, military geographers, naturalists, or agriculturalists.[13]

After 1877, according to Aguilar Santillán, there was an exponential increase in published studies on meteorology and related subjects through 1889, when he conducted this survey. He counted eighty published in the major journals, newspapers, and other publications that he researched in various institutes and libraries, of which approximately thirty-three were published by civil, mining, topographic, and geographic engineers. Moreover, many of these engineers had been appointed as directors of state, regional, or municipal meteorological observatories throughout the country (see figure 3.1).[14] At the same time, the national organization engaged in a direct relationship with several foreign and international meteorological institutions. As Bárcena noted in Fomento's annual report to Congress in November 1877, the Central Meteorological Observatory had recently joined the international meteorological service based in Vienna at the invitation of the US Weather Bureau in Washington, DC. As a result, the Central Observatory served as a national node for the subnational "auxiliary" observatories, and Bárcena noted that because of the need for massive data collection domestically and internationally, "it should not be forgotten that Meteorology demands painful toils that constitute a kind of slavery."[15]

It is thus no surprise that meteorology became more prominent in engineering school and program curriculums by the early 1880s. Bárcena played a leading role in this process as an engineering scientist, not only for meteorology but also for several other sciences. Two of Bárcena's meteorological works that stand out in this regard are the coauthored *Comparative Meteorological Studies*, published by Fomento in 1885, and the single-authored *Statistical Essay on the State of Jalisco* of 1888 on the geography and climate of his native state, for which he served as secretary of state and then governor for several months in that same year.[16] In the 1885 study, Bárcena and his coauthor Miguel Pérez, his subdirector at the central observatory, acknowledged that meteorology was a young science in Mexico that required global awareness and cooperation: "Atmospheric movements observed in any one place are not isolated occurrences, nor do they originate in that same place, but are rather the result of causes sometimes residing in very distant places and in turn of an origin from phenomena that can spread very far." Noting that "if science begins when method begins . . . now is when . . . the data of meteorologists [must] begin to be ordered methodically." They were optimistic about the science's future: "It is also certain that, although in diapers,

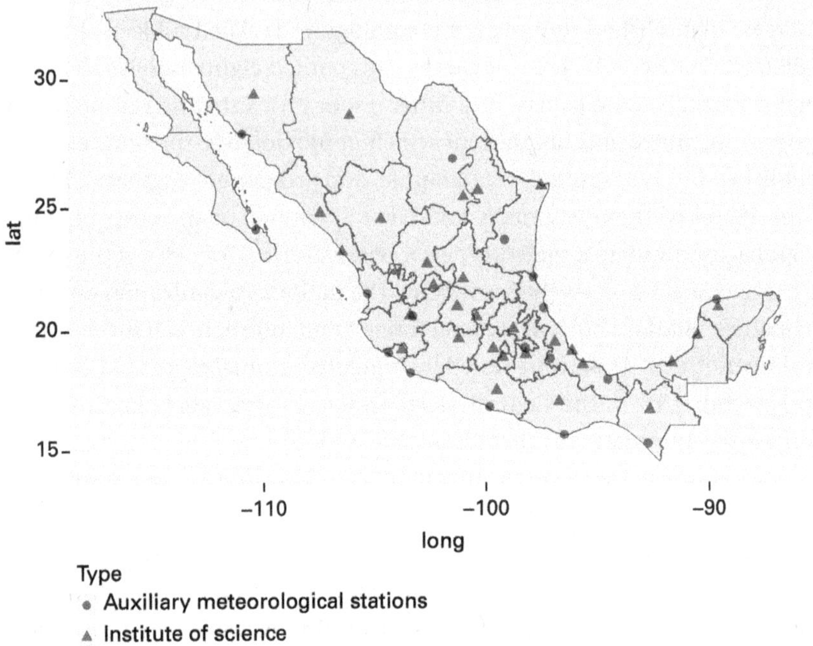

FIGURE 3.1
Meteorological stations and science institutes, circa 1877. Elaborated by Israel G. Solares based on *Instrucciones que comunica el Observatorio Meteorológico Central a los observatorios foráneos* (México: Impr. de F. Díaz de León, 1877).

Meteorology is marching at a speed much faster than other branches of science in their infancy."[17]

Despite their optimism, Bárcena and Pérez attracted criticism for collecting data for years before 1880. After all, meteorology was a science for forecasting weather, not one intended to delve into its past. But as they explained, "The knowledge of a regime of any order of natural, social, and political phenomena is not of instantaneous creation and would be false in all respects if it didn't have the study of the past as a solid foundation. For it to deserve such a name, History does not start from the moment it is written, but rather fixes its attention in a very specific way, on the previous eras that it studies and analyzes, and our work is nothing more than a special kind of history that deals in its way with a certain order of events."[18]

Their defense of meteorology's historicity was an indirect reference to climatology as meteorology's complementary sister science of collecting, cataloging, and analyzing long-term meteorological records for distinct patterns. Little did their critics seem to know that publishing four years of data was insufficient for forecasting purposes. It is in this sense that Bárcena and Pérez then compared the young science of meteorology with what they saw as the more mature science of astronomy: "When [meteorological] science has reached such a high degree of perfection as astronomy, it is not only convenient but necessary to anticipate the weather, predicting with extreme precision the time and the march of the phenomena. But when science is being formed, like meteorology, it is not only convenient, but forced, to look at the past and not abandon it, and to group its elements in order to become a relatively exact science."[19]

Although the distinction between meteorology and astronomy was clear in Bárcena and Pérez's minds, some Mexicans still conflated the two sciences. A few years later, in December 1892, for instance, the director of the Astronomical Observatory, the civil engineer and architect Ángel Anguiano, wrote a letter to the paper *El Tiempo*, republished in *El Siglo Diez y Nueve*. Anguiano complained that the official federal gazette, *El Diario Official*, ran a piece criticizing the observatories directed by him and Bárcena as costly and wasteful. Anguiano pushed back by pointing out that the astronomical and meteorological observatories were distinct, and each in a different way provided vital services to the nation at relatively low cost.[20] As with Bárcena, Fomento had tapped Anguiano for his engineering and architectural skills to establish a modern center for astronomy, deemed essential to Mexico's progress vis-à-vis the industrializing North Atlantic countries.[21] Given the country's paucity of actual scientists, including meteorologists and astronomers, both Bárcena and Anguiano became leading engineering scientists charged with developing the necessary intellectual, institutional, and physical infrastructure for the professionalization of the sciences more generally in Mexico.

Anguiano, Bárcena, and other government observatory directors had their admirers and supporters. The *Gaceta Médica de México*, for example, in May 1894 praised Bárcena's study of Mexico City's climate, determining high, low, and average temperatures as well as rainfall data, it argued could help with containing water-borne intestinal diseases and diarrhea.[22] Similarly, in August 1898, the Mexican English language newspaper *The Two Republics* praised Mexico's National Health Board, founded in 1891, for obtaining a "medical climatology" of the Valley of Mexico.[23] The favorable coverage in both papers reflected the global popularity of medical climatology during the nineteenth century, which emerged out of medical geography, a field that searched for correlations between disease etiology and climates. Medical climatology revived Hippocrates's hypotheses about the effect of "seasons and soils, winds and water, aspects, shelters and exposures, altitudes, temperatures, vegetation, the habits, food and drink of the inhabitants" on health while also perpetuating the racist and colonialist climatic determinism of many European Enlightenment figures such as Montesquieu.[24] Most Mexican mestizo and white elites rejected the implications of European climatic determinism for themselves but not necessarily in their views of indigenous Mexicans.

Prominent among such mestizos was the engineer Agustín Aragón, who, like Bárcena, studied at the National School of Engineering but specialized in topography and hydrography. Aragón worked in several government posts, taught meteorology along with other subjects at the Schools of Mechanics and Engineering, and served as a congressional deputy for ten years. He was an impassioned positivist and prolific author, writing numerous books on scientific, political, and social topics. He expressed his views on climatic determinism in the government-commissioned series of books *México: su evolución social*, published between 1900 and 1902 and edited by the eminent positivist educator, government minister, and founder of the National University, Justo Sierra.

In his contribution to Sierra's monumental series, Aragón emphasized the importance of studying natural history, including climate, to understand Mexico's social evolution, but he cautioned that "the influence of climate in the evolution of social organisms has been exaggerated by certain writers to an extreme that seems unreal since they have omitted the fact that this influence can be largely countered by human efforts." Rather, he argued that "the same people, on the same soil, can be savage to civilized, and what the climate engenders is the variety of efforts for the achievement of one same end. The land is worth what the man who cultivates it is worth. The high plateau of Anáhuac, unproductive in its appearance, and supplying us with wheat, is the granary of many coastal regions." Yet Aragón disparaged indigenous peoples by claiming they were "inept at invention, but very apt for imitation." This alleged indigenous intellectual inferiority to mestizo and white Mexicans did not mean they were physiologically inferior, however, for, as he qualified, "on account of the lifestyle they lead, they can withstand the elements in an amazing way." In other words, he claimed that indigenous peoples were inherently better acclimated to their environments, revealing his selective and racialized climatic determinism.[25]

Just a few years before Aragón wrote these words, in 1896, Bárcena had written a long essay in Mexico's premier agricultural journal, *El Agricultor Mexicano*, that did not explicitly single indigenous people out or essentialize them the way that Aragón would. Instead, Bárcena's goal was to convince all Mexican farmers to embrace meteorological science in their agricultural planning. He began by praising "traditional" methods of weather forecasting, such as reading the skies and stars, determining wind directions, tracking the lunar cycles, and observing animal behavior in their environments. He then gently pointed out the geographical limitations of these methods, which could not be applied beyond a farmer's local area when meteorological phenomena originated elsewhere. By applying scientific methods with a network of observatories able to rapidly communicate weather patterns across vast distances through the telegraph, farmers could significantly improve their forecasting abilities, he argued. Since the invention of barometers and thermometers, "the relative appreciation of heat, pressure, air currents, atmospheric moisture, and other physical phenomena, has no longer relied on individual sensation, because with the help of these instruments one can study said phenomena, thereby forming the scientific study of meteorology."[26] As ethno-meteorologists and climatologists have pointed out in recent decades, however, indigenous knowledge of weather and climate did not, and still does not, rely on "individual sensation" but rather on complex communal epistemologies that have long incorporated and syncretized European scientific knowledge.[27]

METEOROLOGICAL GROWING PAINS

For all the progress that Mexican meteorology based on Western engineering science had made, however, there was also a fin-de-siècle frustration at the pace of its progress and the limited scale of its adoption. For example, a pamphlet titled *El clima de la República Mexicana en el año de 1896* [The Climate of the Mexican Republic in the Year 1896], published by Fomento in 1900 and written by the overseers of the magnetic-meteorological department at the National Astronomical Observatory, began with praise for meteorology and climatology's accomplishments in emulating European countries. But they then lamented, "We still have a long way to go before we reach the level of the countries that fight with courage to tear off the veil with which the fickle meteorology conceals even its mysteries." They warned that weather knowledge and forecasts were still insufficient across Mexico's vast territory for peasants to be told to stop planting their fields because a possible cold spell might ruin their crops or for sailors to be told not to make their transatlantic journeys because of the possibility of stormy seas. As such, they called for more and better scientific instruction in schools, including in meteorology.[28]

Heeding such calls, the first National Meteorological Congress, held in Mexico City in 1900, was the most significant development in the professionalization of meteorology since the Central Observatory directed by Bárcena was founded in 1877. In 1899, Bárcena had passed away prematurely and was replaced by the engineer Manuel Pastrana, who acknowledged the limitations of both the Central Observatory and its regional auxiliaries. Specifically, Pastrana noted that making observations at the fixed hours of 7 A.M., 2 P.M., and 9 P.M. local times, as Bárcena had instituted, was "useful only from the point of view of the climatology of each one of the places where the observations are made but of little or no use for the study of dynamic Meteorology, which should serve as the foundation for the principal objective of meteorological observations: weather forecasting."[29]

Pastrana and other engineering scientists played prominent roles in founding the National Meteorological Congress, accounting for thirteen of its thirty-one organizers. Among the thirteen, one was the honorary president and two were honorary vice presidents (out of three vice presidents). The presenters at the Congress came from across Mexico and held various titles, from engineer to professor (*profesor,* meaning teacher and rarely university professor at this time) and even priest. Some presented their technical work of designing and improving instruments for weather observation and experimentation, while others spoke more generally about the past, present, and future of the meteorological profession. It is the latter that I will highlight, focusing first on the contribution of the engineer José María Romero. Romero was a delegate from the National Agricultural School, where he taught physics and meteorology and also served as a deputy in the Mexican Congress.

At the meeting, Romero gave his colleagues a longue-durée overview of the development of meteorology focused entirely on Europe. His attitude toward traditional meteorological knowledge was quite different in tone from Bárcena's 1896 article in *El Agricultor Mexicano.* Although praising the ancient Greeks for their discovery of basic meteorological elements, Romero characterized post–Greco-Roman meteorology as one of mostly ignorance and religious superstition because it considered the atmosphere as mysterious and "governed by the capricious will of the gods or superior beings" to whom one had to make offerings or pray so that inclement weather would not cause harm to people and property. Doing so conflated meteorology with astronomy, he lamented, leaving weather forecasting "subject to the gravest errors and the most absurd superstition," in contrast to Bárcena, who in 1885 saw astronomy as an advanced science to which meteorology should aspire.[30]

Romero stated that not until the "discoveries of modern physics and chemistry and the invention of the thermometer, barometer, hygrometer, and other instruments of observation" was a new path to meteorological studies paved with "the knowledge of the physicochemical constitution of the atmosphere and with the means to verify methodical experiences through which the laws that govern weather variations were induced." In other words, the work of meteorology necessarily involved engineering scientists making and designing new technologies *and* advancing knowledge of the natural world.

Romero credited engineering scientists' discovery of meteorological laws for providing an important service to Mexican maritime navigation as well as to agriculture. They had advanced "the science of weather forecasting," as Romero called meteorology: once its principles and laws were confirmed through experimentation, "the new discoveries and apparatuses and the perfection of already acquired instruments facilitated the observations and extensions of the results in a major way."[31]

Romero saw 1867, when the liberals led by Juárez triumphed over the French occupation and Mexican conservatives, as the year that meteorology as a science began to bear fruit in Mexico. He noted that the federal government had prioritized meteorology, "expanding the field of observations and informing the directors of roads and the engineers employed in public works that, upon executing topographic, geodesic, astronomic and other kinds of operations under their charge, they carry out observations of barometric pressure, temperature and rain, and they collect data related to the climatology of the regions they cover." He then listed the many accomplishments of meteorology from 1867 through 1900 for cities, agriculture, industry, and commerce in Mexico and for government agencies, such as the Geographic-Exploratory, Border, Hydrographic, Geodesic, Geologic, and others, that helped to realize them across the nation "under diverse scientific aspects and by a large number of engineers and men of science that have made meteorological studies in various points of the country during the construction of public and private works under their charge." Romero then reiterated the complementarity, if not the interchangeability, of scientists and engineers—indeed, engineering scientists—in the rise of Mexican meteorology: "We should not deny in any way the utility for the man of science of the sum of meteorological observations collected until now; engineers make frequent use of them, and in foreign observatories the scientific data serves as a base for the research of the laws that rule the general movements of the atmosphere."[32]

For all this progress, however, Romero acknowledged the greatest challenge facing the public adoption of meteorology, reiterating the critique regarding public engagement: "The mass of the agricultural and industrial population of the country cannot use the meteorological records to the benefit of their enterprises, because they lack the necessary scientific instruction to understand and compare them, and because they may not even know of their existence, by virtue of the fact that official publications that contain them, although they circulate within government offices and agencies, still don't penetrate among the multitudes that populate the fields; as a result, the intellectual movement which animates the cities of the Republic doesn't reach [the rural masses], except in small waves." Despite Romero's condescending characterization of Mexico's rural population, he nevertheless conceded that not all local popular knowledge was useless:

> It is true that the diligent peasant, from the experience he acquires from constant observation of the proper signs of his locality, is accustomed to predicting atmospheric variations. He follows certain rules deduced from the direction of winds, the condition of the sky, cloud formation, the indications which plants and animals provide, and the application of rural adages and proverbs preserved by tradition, but he lacks the powerful support of scientific observations, without which it is impossible to obtain, with certain probability, weather prognostication . . . the ultimate scientific result of the meteorology that the farmer and industrialist are greatly interested in opportunely knowing.[33]

This "regrettable deficiency," as he called it, "in the practice and application of meteorology," caused the rural population and some landowners, like the ancient Greeks and Romans, to naively pray to Providence for the well-being of their fields. The engineering scientists of meteorology thus had their work cut out for them.

Encouraging the government and private entities to systematize meteorological data, including the creation of weather maps, Romero praised the work of engineer Agustín M. Chávez, who was the director of Federal Telegraphy. In the telegraphy central office, Chávez founded a meteorological observatory (presumably in coordination with the Central Meteorological Observatory) and supplied it with "numerous modern instruments of notable perfection." He also founded the telegraphic bulletin, a publication in which daily weather maps appeared for the first time in Mexico. The maps contained technical weather data based on daily observations from telegraphy stations and "all the points of the country by telegraph or telephone." The problem, however, was that the telegraph and telephone did not yet reach most of the country, and even if they did, outside of the cities, most Mexicans were illiterate.[34]

To address the problems identified at the first National Meteorological Congress and the critiques of Romero and others, in 1901, the Central Meteorological Observatory became the headquarters for a new National Meteorological Service and organized a Second National Meteorological Congress. Sixteen of the second Congress's forty-two organizers were engineers, and they adopted nine resolutions on the "vulgarization" (simplification) of meteorological knowledge for mass dissemination. The first resolution requested the Federal Education Ministry, state governments, and principals of private schools to adopt elementary meteorology in primary school curriculums while the seventh resolution requested that the directors of state observatories and physics teachers immediately provide explanations to local newspapers whenever "notable meteorological phenomena" occurred to debunk popular misconceptions and "absurd theories" about the weather.[35] The composition of the committee tasked with forming the Association for National Meteorology by order of a different resolution was the same one in charge of the committee for the Simplification of Meteorological Knowledge.

The president of both committees was "Señorita Profesora" ("Ms. Teacher") Raquel Sánchez Suárez, a delegate from the Mexican Society for Science Education and one of the few women organizers of the Congress. Sánchez's committee appointments reflected the limited career advancement that women had gained by the late nineteenth century, before which they had been barred entirely. The historian of geographic education in the Normal School for Teachers in late nineteenth-century Mexico, Rodrigo Vega y Ortega, called this the "feminization of children's education," or allowing women with means to pursue "less prestigious" and remunerative careers, such as primary teacher, pharmacologist, obstetrician, and nurse, than male-dominated ones such as law, medicine, and engineering.[36] Reflecting the starkly gendered boundaries of professions that limited women's career options, the government tapped women as the preferred communicators but rarely active members or creators of STEM-related fields and initiatives.

Fulfilling this role at the National Meteorological Congress, Sánchez Suárez presented on "the barometer and weather forecasting," providing a historical overview of the invention and development of various weather instruments from Renaissance Italy to Mexico in 1900. She then self-referenced an article she wrote and published in the February 1896 issue of the *Monthly Bulletin* while she was a student in a meteorological class at the Normal School for Teachers, titled "Forecasts Provided by Barometers." The article argued that barometric readings were erroneous in a "great number of cases" and proceeded to show why, using mathematical calculations and statistics. Like her male colleagues, she concluded by advocating for the scaling up of meteorological observations throughout Mexico's vast territory.

Sánchez Suárez was part of a group of women science teachers who wrote articles on various scientific and engineering topics for lay audiences during the Porfiriato, especially from the 1890s, when a new generation of women was being trained as teachers in far greater numbers. Their writing style was often far more readable than that of many male engineering scientists. For instance, in a June 1895 article titled "Atmospheric Studies," a science teacher waxed almost lyrical about the virtues of meteorology, a passage from which is worth quoting at length:

> Meteorology, one of the sciences that is most useful to man, and that frequently attracts his attention, is easily accessible to anyone who wants to discover its secrets and investigate its causes; it can truly advance of its own accord, since it is within the reach of people at any educational level. Its domain is immense, for it does not cease to investigate the innumerable phenomena that appear to us at every step both within the bosom of the atmosphere and in the depths of the oceans; its effects are clear and every day it makes us notice with greater interest the powerful influence it exerts on all our actions, while protecting us from the endless dangers that it has been able to prevent through observation and study. The navy, agriculture, hygiene, and consequently our health, are subject to its laws; at the same time, it supplies us with powerful instructions to prevent epidemics, and finally, to top off its work, it solves one of the problems that for the longest time has vexed the wisdom of the sages and the very first physicists: "weather forecasting."[37]

Despite their talents and skills as science communicators and promoters, not a single woman was enrolled in a dedicated science or engineering school until just before the outbreak of revolution in 1910. As a result, it was unthinkable in Mexico, and for decades thereafter, that a woman would ever run a government agency like that of the Central Meteorological Observatory.

A TALE OF TWO METEOROLOGICAL PAMPHLETS

The National Meteorological Congress's two back-to-back annual meetings in 1900 and 1901 received much press attention and praise from Mexico's politicians, including President Porfirio Díaz. Three years later, in preparation for Mexico's participation at the Universal Exposition (World's Fair) held in St. Louis in 1904, the National Meteorological Service published a pamphlet in English written by Pastrana. His tone was upbeat and positive, seeking to make a good impression of Mexico's meteorological progress to the world. Pastrana naturally began by flattering his ultimate boss, the "energetic" Díaz, who had been the "beneficent founder of the peace" in Mexico since 1876. Thereafter, Pastrana focused the pamphlet on the progress of the meteorological service's organization, national coverage, personnel, infrastructure, and instrumentation.

Although Pastrana tried to spin his report as positively as he could, he could not avoid acknowledging some of the ongoing challenges that the service faced. One was the federal nature of the republic. Even with an authoritarian leader who had been regularly "re-elected" (as the only permissible presidential candidate) for nearly twenty-five years, Mexico's thirty-one states and districts could not be forced to subordinate the management of their meteorological services to the National Meteorological Service run by the Central Observatory. Pastrana explained that

> the division of the meteorologic service into sections, per the political division of the country, is an absolute necessity since it is the only way in which the governments of the states can be induced to establish them on their account. The division into sections in climatologic or flowing regions [watersheds] would be more scientific and easier for the climatologic studies of the country and for predictions of the weather. This division, however, could only be adopted if every station were federal—that is to say, established and supported by the Federal Government, and not by the governments of their corresponding states.

Despite having to organize the National Meteorological Service according to state-level political divisions, Pastrana expressed confidence that the same results would be obtained as if the service were organized according to climatological criteria because the central office in Mexico City would eventually be able to remap the data, although "with more difficulty." Most of the rest of his pamphlet consisted of descriptions of state and local observatories along with brief biographies of the people who ran them, many of whom were engineers, as evidence that the observatory system was up to the task.

Two years after the St. Louis Universal Exposition, for which he had produced an upbeat English-language pamphlet, Pastrana updated it in Spanish for his superiors at Fomento. Writing in his native language for a domestic audience, he expressed far less optimism by emphasizing the deficiencies of the existing service and its need for more resources and investment. Among the deficiencies was the lack of telegraphic coordination between the US Weather Bureau's office in Galveston, Texas, near the northern Gulf Coast of Mexico and the meteorological service in Mexico City. Although it fell on the latter to adjust to the different intervals of the technologically superior US observations, the central observatory had been unable to do so for years due to the improperly installed telegraphy stations in the north and the "lack of good practice" of the Mexican telegraphists who often committed "notorious errors" in data collection. Quite apart from this glaring deficiency, Pastrana was frustrated that this kind of observation appeared to be "the only mission" of the meteorological service. He pointedly took Fomento to task on this point in a long missive:

> Given the sheer number of losses that they cause farmers and industrialists, shouldn't snowstorms, frosts, and hailstorms be the preferred objects of [the service's] studies, to be able to predict and forecast them opportunely [so] all interested parties can avoid their disastrous effects? And the rains, fertilizing our fields when they are proportionate to the needs of the crops, increase our wealth, [but cause] rivers to rise unexpectedly when they [the rains] are excessive, [and] overflow when they are opposed, sowing death and destruction in their path—don't they deserve a special study so they can be forecast with sufficiently advance notice, so the farmer can plant his fields with a good chance of success, and so that, through conveniently adapted and opportunely used engineering works, the devastating effects of large river flows can be avoided?[38]

Pastrana then provided far more detail than he did in his English pamphlet about the efforts that he made to convince the governors of Mexico's thirty-one states to set up and fund their state-level meteorological observatories. His efforts included traveling across the country and meeting personally with anyone in a governor's administration who would give him an audience. As of 1906, nineteen states had accepted to set up an observatory; nine of them had done so, one was in the process of setting one up, and nine others planned to do so forthwith. That still left thirteen states and districts without observatories, which he warned made the National Meteorological Service's ability to predict weather on a national scale deficient. He thus called on the state legislatures to pass laws compelling the adoption of the National Service's guidelines and instructions to standardize them nationally, which he argued was more effective than federal decrees.[39]

Not much more progress was made in this regard, however, through June 1910, when Porfirio Díaz rigged his seventh reelection after he jailed the principal opposition candidate, Francisco I. Madero, on spurious charges. Díaz then presided over a grand celebration of the long-planned centenary of independence in the capital that took a substantial portion of the government's annual budget. *El Centenario,* as it was called, included a national scientific and artistic competition sponsored by the Mexican Judicial and Legislative Academy, to which numerous writers, artists, scientists, and others submitted their work.

Much like the first generation of women science teachers writing in the 1890s, "Ms. [Señorita] Professor" Isabel Gonzalez García wrote an essay for the competition on the history of Mexican meteorology from 1810 to 1910, on behalf of the Astronomical Society of Mexico. Her narrative of a century of Mexican meteorology was one of continual progress, especially since 1876, praising Díaz and the various male engineering scientists that his government had appointed to institutionalize and professionalize the science. The institutional manifestation of this story is apparent in the greater number of establishments visible in the map in figure 3.2, compared to the 1877 map in figure 3.1.

One woman, moreover, played an important role in González García's narrative: Rafaela Suárez, the director of the Normal School for Women Teachers. After Luis G. León, professor of physics and chemistry at the school, proposed in 1894 to Suárez the establishment of a meteorological observatory, the latter took the initiative to found one without requesting financial or other assistance from the Secretariat of Justice and Public Instruction. She feared this male-dominated federal institution would obstruct the project and so instead relied on school funds and donations of equipment from the physics classroom to get it off the ground.[40] Joaquín Baranda, Mexico's secretary of justice and public instruction, attended the new observatory's inauguration, about which he was pleasantly surprised. He remarked, "This is the way I like things done, by one's initiative. If the School Director had proposed

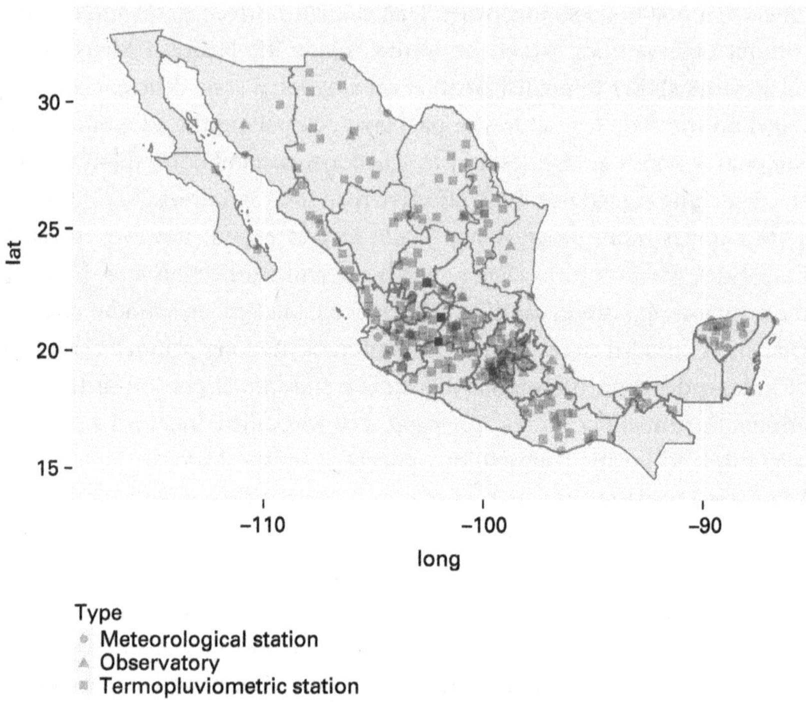

FIGURE 3.2
Meteorological stations and observatories in 1910. Elaborated by Israel G. Solares based on Isabel González García, *Los progresos de la meteorología en México de 1810 a 1910; estudio presentado en nombre de la Sociedad Astronómica de México* (México: Tip de la Vda. de F. Díaz de León, 1911).

ex officio the establishment of the observatory, I probably wouldn't have given the matter much importance." As González García recounted the story, Suaréz's predictions of gendered impediments from the federal government proved right while her in-house efforts to overcome them bore fruit. Yet overt sexism persisted. In 1908, at the urging of the National Meteorological Service, Fomento sent two young Mexican men, not women, to the United States for a year to study the operations of the Weather Bureaus in Washington, DC, Chicago, and Boston, followed by a stint in Europe, during which the men won a prize from the Belgium Astronomical Society. By 1910, they had returned to Mexico and were imparting their newfound knowledge to Mexico's central meteorological observatory. With an engineering scientific eye but without a corresponding title, Isabel González García acknowledged a major problem with the national observatory: it was in "very bad condition" on top of the National Palace and needed to be moved to a larger area at ground level, where its various departments could be housed, surrounded by gardens, as was the case, she noted, with US and French observatories, where field observations could be carried out more easily.[41]

CONCLUSION

Professor González García's essay was published in January 1911, just two months after the outbreak of the Mexican Revolution. Within five months, Díaz would resign and hand power over to an interim president until free elections were held in October of that same year. Madero, the opposition candidate who Díaz had jailed and who called for the revolution that eventually toppled the president, easily won the first free election in decades (albeit with the franchise restricted to men with means). Madero would govern for a tumultuous year and a half before being overthrown and assassinated in a bloody coup with the crucial assistance of the US ambassador. Thereafter, civil and revolutionary war became even bloodier, reaching cataclysmic proportions in 1915 before gradually winding down by 1920. During this decade of violent instability, Mexico's engineering scientists did their best to continue their work. In June 1913, the National Meteorological Service was moved from the National Palace to Tacubaya, where the Astronomical Observatory had been relocated since 1883.

During the brief occupation of Mexico City by Pancho Villa and Emiliano Zapata's revolutionary factions in December 1914, the observatory was closed and did not reopen until September of 1915 under the order of revolutionary factional victor Venustiano Carranza's subsecretary of Fomento, the engineer Pastor Rouaix. Rouaix replaced Pastrana with the engineer Joaquín Gallo, continuing the tradition of engineers directing the meteorological observatory as well as other government scientific agencies. Remarkably, despite major advances in increasingly specialized meteorological knowledge and technology, including radar and then satellites by the 1960s, engineers still directed the Central Meteorological Observatory in its various institutional incarnations until 1971—the year that a professionally trained meteorologist with no engineering background finally took charge. The reign of the engineering scientists, at least in meteorology, had ended.

4 CORNERSTONE OF THAT NEW IMPERIALISM: US MINING ENGINEERS AND LABOR MANAGEMENT IN SOUTHERN AFRICA, 1890–1910

DOUGLAS R. JONES

At a 1907 meeting of the South African Association of Engineers, held in the offices of the Transvaal Chamber of Mines in Johannesburg, the Association's vice president, a mechanical engineer from Ontario, Edgar Laschinger, reflected on the first two decades of gold mining on South Africa's Witwatersrand Basin.[1] Given the sheer scale of the work involved in extracting and processing the Rand's ore, Laschinger would have surprised no one in the room when he observed, "One of the greatest questions in South Africa has been and is, labour." Far more surprising, at least to historians, is his argument that "to an overwhelming extent, this is an engineering problem." The Canadian engineer's remark suggests how essential highly mobile professional engineers were to both the "machinery" and "human machinery" of the world's foremost gold mining region, and his career encapsulates many of the themes of this book.[2] As Beatty and Solares show in chapter 1 in this volume, engineers like Laschinger were newly professionalized, mobile, oriented toward specific outcomes like cost-efficient ore extraction, and, as this chapter demonstrates, central actors in a dramatic episode of imperialism and economic globalization. Perhaps most significantly, Laschinger and the other engineers active on the Rand are a key example of how engineers' technical and cultural authority—in this case, over the Earth and over certain humans—was inseparable from white supremacy. This chapter will show how the way engineers approached the so-called labor problem in the mines was decisive in cementing racial capitalism in South Africa. In doing so, the chapter offers a clear but cautionary example of the ways in which turn-of-the-century professional engineers were called upon to solve far more than technical problems of industry or the ways they understood racialized labor as one more technical problem to solve.

While largely forgotten by history, in his time, Laschinger was known for solving the issue of water flooding the deep mine shafts and was generally well regarded by his peers in the South African Association of Engineers, nearly all of whom worked in one way or another for mining interests. Led primarily by engineers from the United States, the Association's discussions are noteworthy for the way engineers applied the same calculating, objective formulas to the labor capacity of the human beings who might comprise the mines' workforce as they did to technical mining objects like winding cables for deep shafts. Many of these same engineers began their careers by managing racialized migrant labor forces, first in the US Southwest and then in Venezuela in the 1870s and 1880s. This chapter focuses on the efforts of mining engineers to lower working costs in the gold mines in and around Johannesburg by reducing reliance on higher-paid white labor drawn from British settler colonies. US mining engineers came to shape British imperial policy toward South Africa in the lead-up to, and aftermath of, the South African War of 1899–1902. They recommended the importation of indentured Chinese labor—as had been done in the US South and West—and recommended a workforce of majority African miners on the basis that formerly enslaved people had proven what Black workers could accomplish in the wake of emancipation in the United States. The chapter concludes by considering how engineering rationality, epistemology, and values, combined with engineers' enduring position as "boundary objects" between capital and labor, allowed a small group of technical experts to engineer such dramatic social transformations.[3]

US MINING ENGINEERS ON THE RAND, 1889–1899

Hennen Jennings arrived in Johannesburg in December 1889 at the beginning of an economic slump in gold mining. Jennings and other American engineers, such as Hamilton Smith, Thomas Mein, George Webber, and Henry Perkins, and British engineers, like F. H. P. Creswell, had previously worked outside of the United States for British mining capital at the El Callao mines in Venezuela, and Gardiner F. Williams had already paved the way for US mining engineers in his role as manager of the De Beers diamond mine in Kimberley. Jennings was part of a cohort of about seventy-five US mining engineers employed on the Rand, themselves just one component of about 1,500 US citizens living in the Transvaal by the eve of the South African War.[4] By the middle of the 1890s, US mining engineers had come to manage over half of the gold mines of the Rand. Unlike the case of diamonds, vast deposits of low-grade ore meant that the route to profitable gold mining on the Rand was to control costs ruthlessly and relentlessly.[5] While mining engineers made numerous advancements to extraction and refining technologies through the 1890s, labor was the most variable cost factor and thus the central question that British mine owners tasked US engineers with solving. As mine magnate Cecil Rhodes understood in the middle of the South African War (1899–1902), "The future really turns on labour and supplies and those depend on the war."[6] This was especially true in the case of deep-level mining, as the English engineer E. J. Way pointed out: "This labour question is the question where low grade mines are concerned."[7] US mining engineers were crucial to resolving the labor issue.

Such was their success that one US engineer, Harry H. Webb, on the occasion of Jennings's farewell dinner in 1903, noted that the total output of gold from the Rand in the three years before Jennings's arrival was 642,000 ounces. On the eve of the South African War, in the single month of August 1899 alone, output was 460,000 ounces—roughly twenty-five times what it had been a decade before the Yankees began pouring in.[8] Indeed, barely more than a month into Jennings's time on the Witwatersrand, mine magnate Lionel Phillips was already writing to his partners Julius Wernher and Alfred Beit in London to make moves based on Jennings's advice. Phillips wrote that the firm needed to move into deep-level mining—which was in fact the long-term future of mining on the Rand—and cited Jennings as the authority on the matter.[9] A few weeks later, Jennings was already putting his experience in the United States to work, complaining to Phillips about the inefficiency of surface sorting and crushing operations, which had been done far more cost-effectively in the Black Hills of South Dakota.[10] Two years later, when Jennings contemplated switching firms, Phillips wrote in a panic to Eckstein that this must not be allowed to happen. If neither Jennings nor another American, Henry Perkins, were available, he wrote, "I *must have* a high class man and *soon*."[11] Later in life, Phillips recalled how fortunate he was when, having been on the Rand only three months after building his fortune in diamonds in Kimberley and still inexperienced, that his staff was strengthened by the arrival of Jennings, "an American engineer of eminence" who had "high moral principles and used meticulous care in securing and marshalling facts."[12]

While engineers were successful in the 1890s at maximizing the technical efficiency of mining operations on the Rand, several sticking points remained for the industry, chief among them the supply and management of African labor. In 1897, engineers entered an ongoing conflict between British imperialism and largely foreign mining capitalists, on the one hand, and the Zuid-Afrikaansche Republiek (ZAR), an Afrikaner settler state led by Paul Kruger, on the other. US mining engineer John Hays Hammond recognized this tension earlier that year, writing that while he considered South Africa's prospects promising, the "labor problem" would not be resolved until all of South Africa's provinces were unified and the British–Afrikaner divide bridged.[13] In an effort intended to preserve his state's independence from Britain and provide the appearance that the ZAR was considering the needs of the mining industry, Kruger convened an Industrial Commission in 1897, with six members of the legislature serving as commissioners. Mining experts who testified were deeply critical of the Kruger government's mining policies. Responding to a situation where many mines had stopped operating for lack of profit in the previous year, the commissioners' report aimed at lowering costs. The report focused primarily on reducing labor expenses since labor constituted 50 to 60 percent of typical operating costs. Above all, the report called on the state to increase the supply of African labor by expanding recruitment to Portuguese territory and raising compensation given to chiefs.[14]

Historians have generally agreed that the testimonies and recommendations offered to the commission sketched a blueprint for British imperial rule of the Transvaal following the South African War.[15] What has been less appreciated is the role of US mining engineers in setting that agenda.[16] The significance of engineers, and Jennings in particular, was a fact clearly impressed on contemporaries: the commissioners themselves noted that other witnesses seemed to be following Jennings's lead and relied on his cost and efficiency calculations regarding labor, while the correspondent for the London *Financial Times* enthusiastically parroted Jennings's conclusions, approving of the "enormous mass of statistical information" from the "eminent mining engineer."[17] The praise came from fellow engineers as well: Hammond wrote to

Jennings to express his "high appreciation" for the "capable representation you have made to the government."[18] Jennings's testimony was reprinted in the London financial journal *South Africa*, as well as included as an appendix to *Present Day Metallurgical Engineering on the Rand*, published by the London *Mining Journal* with the rather accurate title, "Economics of the Transvaal Gold Mining Industry."[19] To those with capital at stake in Rand mines, what Jennings had to say was important.

The presence of engineers at the commission hearings would have been hard to miss: of thirty-three witnesses who gave testimony between late April and early June, 1897, fully a third were engineers, most specializing in mining, while some worked in allied fields like mechanical engineering. Of those, six were from the United States, including Jennings, his younger brother Sidney, Louis Seymour, Thomas Leggett, William Hammond Hall, and Robert Mayo Catlin. Another, George Denny, was Australian but had extensive professional experience in the United States, Canada, and elsewhere. The remainder were British, but one, Edward J. Way, had been assistant to Hammond previously.

Of the thirty-three witnesses, Jennings's testimony was the longest, taking place over three days and running to forty-six pages of printed text. His younger brother Sidney, who had been in South Africa as general manager of the Crown Reef Gold Mining Company for less than a year, had initially laid out some of the basic issues. Sidney Jennings testified to the excessively high cost of coal and dynamite thanks to the ZAR's concession policy and called for higher reliance on and lower wages for Black workers, which he believed could be implemented if the current laws restricting the sale of alcohol to Black mineworkers and pass laws to prevent absenteeism were properly enforced. If African workers were prevented from drinking and absconding, he testified, they could be taught to do the "lower forms of work that are now done by white men," cutting costs while encouraging more mine development as well as white immigration. This "would be a permanent benefit to the country in many directions."[20] Catlin similarly had not been long in South Africa, and his primary suggestion to the commission was a defense of the bonus system, whereby African miners could be paid well above their usual wages if shaft sinking proceeded at a correspondingly high rate. Catlin demonstrated that even taking the largest bonus as an example, the cost of shaft sinking with the bonus system in place, expressed in pounds sterling per foot, was in fact, "by long odds, the cheapest system in vogue."[21] Catlin's attitude toward how labor was managed was typical of the kind of calculated rationality practiced by US engineers on the Rand: "My object is to get the shafts down as rapidly, and at the cheapest possible cost; and I do not care whether I have to pay an ignorant man or an educated man. I believe I work for the interest of my shareholders, and on this system would gladly pay a man £200."[22]

More than any other witness, Jennings had come prepared with a mass of evidence, but his recommendations could be easily summarized: expand recruitment and train African workers to do the "intricate work demanded of" them.[23] As first president of the South African Association of Engineers and Architects and as a long-time resident by Johannesburg settler standards, he was a well-known figure. Jennings began by reaching back to the engineers and geologists who, in his mind, had laid the foundation for future prosperity: his old mentor, fellow Kentuckian Hamilton Smith, who first pointed out the need for the best possible engineering and management to make the mines pay, given the low concentration of gold in the reefs.[24] Smith had estimated in 1893 that the Witwatersrand goldfields contained £325,000,000 worth of gold. Three years later, George Becker of the US Geological Survey more than doubled that estimate to £700,000,000, which he estimated would last for approximately ninety years.[25] Jennings pointed out that this estimate was already out of date and was no doubt even higher in consideration of newly explored areas. Jennings made some recommendations based on his experiences in Venezuela and the United States. In Venezuela, working costs had been reduced considerably by first improving machinery, which he said was already the best in the world on the Rand, and, second, by "the encouragement of negro labour obtained by the West Indian Islands." At first, he said, these workers were of little use, but with training and experience, they were able to replace over 90 percent of white workers.[26] Jennings suggested similar approaches to the labor question in South Africa. He believed that the "efficiency" of Black labor could be increased twofold through training, a claim he held was "further strengthened by the well-known evolution of the negro in America since the abolition of slavery."[27] Further, Jennings recommended that the "peculiarities of the different African tribes should be taken into consideration" for the purposes of recruitment. "East Coast" miners, he claimed, "take better to mining than the higher class native—Zulus and Basutos." The latter, in his view, were "not willing to work underground to the same extent."[28] In closing, Jennings impressed upon the commission the necessity of encouraging Black "labour to the utmost extent, realising its vast importance to the prosperity of these Fields and to this Republic."[29]

Initial reaction to the commission by mining observers was overwhelmingly positive. In London, prices for South African mining companies rose swiftly. Jennings's employer, Corner House, was enthusiastic: Phillips remarked that "we all agree with, and are most favourably impressed with, the evidence" given by Jennings and J. P. FitzPatrick.[30] Governor of the Cape Colony Alfred Milner was also enthused and optimistic. In Pretoria, Schalk Burger wrote on behalf of the commission to President Kruger, recommending with few exceptions all of the major points put forward by the US engineers. Burger recommended measures to lower the cost of living for white workers, encourage the growth of industries besides mining that could employ whites, and develop various measures to increase the supply of Black labor, including an expansion of the area migrant laborers were drawn from, noting that "the chief supply must come from the East Coast (Portuguese territory)," as Jennings had advocated.[31] The commission only stopped short of recommending African "locations"—segregated urban areas—on the Rand.[32] Burger had evidently been convinced by the mass of facts and statistics offered by the technical experts at the commission, noting that the ZAR was "still in its infancy" but had an opportunity to learn from other countries "where these problems have for decades exercised the minds of their leading citizens."[33]

Although the ZAR did make some improvements toward modernization and state support for industrial mining, the mining industry was not satisfied that the commission's recommendations had been sufficiently implemented. As Jennings recalled later, although he saw the commission as "broad-minded and favorable to the mining industry, it was not thoroughly acted upon by the Government."[34] The mining industry showed no awareness that it was, in the estimation of one historian, simply "beyond the capacity of the existing ZAR government to implement the recommendations."[35] Mining stocks slumped, with the London correspondent for the *Engineering and Mining Journal* reporting that "there is no doubt that the hitch in the adoption of the Transvaal mining commission's recommendations is the cause of this lull."[36] Mine owners began to take matters into their own hands, reducing Black wages by 30 percent, marking the most significant cost reduction of 1897.[37] In his 1898 survey of Rand metallurgical practice, English mining engineer John Yates wrote that "the time has arrived when our able Engineers and Managers can improve the workings of the mines but little further," and moreover, "unless the Government affords the much needed relief from present burdens, the brilliant future which as so often been predicted for these Fields will never be attained." This impression that it was governmental hindrance that caused the "prevailing depression on these Fields" led Yates to take the unusual step of including Jennings's 1897 testimony as an appendix to the book, endorsing it as "a clear, reliable, and comprehensive review of the economics of Rand Mining."[38] Jennings's views, then, continued to guide mining industry demands.

POSTWAR RECONSTRUCTION: WAR, EMPIRE, FINANCE, AND CHINESE LABOR, 1900–1903

It was after the South African War that US mining engineers made their most lasting impact on the racialized contours of South African society. Between 1902 and 1907, the Transvaal was under the imprint of a "conquest" state with the ability to intervene in civil society in ways that would not otherwise have been possible.[39] While some historians have noted in general terms that the demands of US mining engineers in the 1890s shaped or foreshadowed subsequent developments, far less appreciated is how a number of them returned in the years after the war to advise the British imperial state how to mine gold and reorganize labor efficiently—and hence society—along lines most conducive to that end. Central to this plan was once again Jennings's vision. He had sketched out in the late 1890s a vision of a mine labor force reliant on a small number of white supervisors managing a predominantly Black, migrant workforce. Now that the Rand was firmly under British control, there was greater potential to make it look more like the racially segmented labor force of the New Almaden mercury mine in California or the El Callao in Venezuela. As in the postemancipation United States, Jennings viewed Chinese labor as a key element in this transition.

Writing to his employer, German-born British mine magnate Hermann Eckstein, Jennings sketched out a strategy for the postwar mining industry that was consistent with the demands engineers had made of the ZAR at the 1897 Industrial Commission. "When we get working," Jennings wrote in October 1900, "I anticipate that important reductions can be made in working expenses. The chief of which will be due 1st to Native Labor; 2nd to Coal; 3rd to dynamite; 4th to General Supplies." Jennings proposed a 40 percent reduction in wages to African workers compared to prewar rates, yielding a "reduction equivalent to 2/6 per ton [of ore] milled." African workers themselves, he noted, were "willing to work for the Boers for small wages" but expected a return to prewar wages upon British victory. This was of no concern to Jennings: "This is a difficulty which will easily be surmounted under the firm rule of the military authority." This sort of technical and cultural knowledge—Jennings at least appeared to have his finger on the pulse of the mine workforce and was able to enumerate precisely how his plan would reduce the all-important figure of working costs per ton—was invaluable to the imperialists and capitalists about to reshape South African society.[40]

Once the war had ended, mine owners complained that the exigencies of war had inflated African wages, and not enough were showing up looking for work. As one engineer, Henry Webber, correctly assessed, "The fate of the gold mining industry on which the prosperity of South Africa had been built was in the balance."[41] In addition to the postwar labor shortage, mine owners were also keen to become "independent of Portuguese goodwill" for a major source of labor, an option pursued vigorously after 1897.[42] The British Colonial Office also made efforts to support the mines, looking far and wide for labor sources. One British official, having been refused permission to recruit in West Africa, fruitlessly wrote to officials in Barbados in search of laborers.[43] Looking for alternative solutions, Randlords once again turned to Jennings for advice.

Jennings's subsequent recommendation to recruit indentured Chinese laborers—a significant departure from the status quo—was the first such proposal in South Africa.[44] Four years after he had left for London in June 1898, Jennings returned to the Rand in the service of the mining firm Wernher, Beit & Co.[45] It did not take him long on the ground to write back to his employer in London to suggest that the Rand be worked by indentured Chinese laborers, leaving Black South Africans to work the land. In doing so, Jennings drew upon a widespread postemancipation practice in the United States of replacing newly emancipated and empowered Black labor with indentured Chinese labor.[46] He may also have been abreast of mining engineer Herbert Hoover's glowing reports on the industry of Chinese workers from his experience managing mines there. In 1899, Hoover wrote to his brother Theodore Hoover, another mining engineer, that he would "break records in [working] costs here when we are started—for labor is only

10 cents per diem and superior men to Transvaal n******."[47] Jennings and others drew on a range of sources to guide the decision to turn to indentured Chinese labor on the Rand.

British mining magnate George Farrar, head of East Rand Proprietary Mines, became the leading public voice advocating for Chinese labor importation and eventually introduced a bill for passage in the legislature.[48] On August 20, the Chamber of Mines appointed a private and confidential committee with the primary goal of evaluating different races' suitability for gold mining. The chamber wrote to thirteen engineers and mine managers asking for their experience managing Black and Chinese labor in various contexts around the world and got nine replies, including from five US engineers.[49] Generally, managers called for better food and living conditions for Black mine workers and for a proscription on alcohol, which they considered a serious issue under the previous government. Leggett, a Columbia-educated mining engineer from New Jersey, allowed that "ka**** beer"—likely referring to sorghum beer or *umqombothi* in Zulu—could be permitted if properly controlled. Leggett drew on what he described as his "own experience with a similar class of labour in Central and South America," which, he said, "has shown me that this can be done and that it is advantageous to do it."[50] Leggett, like most of the other engineers, also called for the introduction of piece work and for longer contracts, though he cautioned that contract length should be flexible, given that "Transvaal Msutu, Zulu, or Southern Basutu" (referring to Africans from British colonial possessions) were only willing to "work for a few months, then return to their crops and come back again to work," whereas "East Coast Native[s]" (from Portuguese colonies) could come for one to five years. It is not clear that Leggett or any of the others were aware of the intense degree of coercion Portuguese colonialism subjected African men to, but they were well aware of their apparent pliability.[51] Leggett's views appear to be more informed by his experience in the United States when he warned that "any attempt to work the mines by white labour only is sure to be unsuccessful." As he saw it, "Wherever white and black labour come into contact, the white laborer will always refuse to do certain classes of work performed by the black." Instead, Leggett's solution to the present "paralysis" on the Rand was to increase the efficiency of Black labor already in southern Africa.[52]

Leggett's assistant engineer, Fred Hellmann, who first arrived on the Rand from the United States in 1895 and had published on Rand mine valuation, allowed that it was possible to build a large force of skilled white labor but held that it would take an unreasonably long time to do so.[53] Even then, based on Hellmann's view of racial dynamics, the only way it would be possible would be if Black labor were "entirely eliminated from the working classes of the mines." Hellmann similarly called for longer engagements, lamenting that "the curse of the Industry lies so far in the fact that the mines are constantly breaking in new boys," a proposal consistent with other calls to build the skill and capacity of African workers, as Jennings had initially recommended in 1897. Hellmann endorsed the abilities of Chinese workers based on his experience in California, calling them "the most efficient labourers in the world." If it proved impossible to recruit sufficient African labor, Hellman believed, importing Chinese labor could work, provided that steps were taken to "safeguard the interests of the [white] community, which [Chinese laborers] would otherwise menace."[54]

While Hellman allowed that "the introduction of an alien race into any country is not a thing to be undertaken lightly," it was the Australian on the committee, George Denny, who most accurately apprehended the intensity of the white backlash that would happen if Chinese labor were introduced on the mines of the Rand.[55] Denny wrote that the "violent opposition which is aroused here by the mention of Chinese labour is due entirely in my mind to prejudice and perhaps ignorance." In Denny's view, white anxieties in Australia and the United States about Chinese immigrants becoming "undesirable Colonists or citizens" were overblown. He argued that the Chinese should simply be denied citizenship and compelled to return to China at the end of their contracts, just as they were in the "Dutch West Indies," where, he noted, the mining industry is "operated entirely" by Chinese laborers. Still, Denny's prediction proved to be remarkably accurate: he wrote that while Chinese laborers would lower working costs on the Rand, "it could not be effected without the most serious disturbances amongst the white workmen, leading to such embarrassment to the Government that the whole effort would probably prove abortive."[56]

It was another member of the committee, in fact, who ensured that the "whole effort" of importing Chinese laborers "prove[d] abortive." Creswell was a British mining engineer, educated at the Royal School of Mines, and, like many of the engineers from the United States, had also worked in Venezuela.[57] Creswell's extensive report to the private committee stands among the earliest expressions of white labourism coming from within the Witwatersrand gold mining industry and demonstrates how engineering authority and reasoning could be applied to numerous ends.[58] Creswell had worked alongside US engineers for years and shared many of their habits of thought and concerns, yet he came to a completely opposite conclusion. Creswell, too, worried about the "demoralising" impact of "cheap" Black labor on white workers and that the migratory nature of African workers made it difficult to cultivate much in the way of skill, in his view. Creswell proposed a radical shift involving "freeing ourselves from dependence on the native by on the one hand establishing a white working population and on the other hand by using more labour saving appliances." White immigrants from Britain and Scandinavia, he believed, were more intelligent than Africans, for whom "the idea of labour is utterly alien." Creswell allowed that Chinese laborers represented the "only other alternative with the K*****" but cautioned that "under the circumstances [Chinese labor] seems hardly worth dealing with," presumably referring to white labor's opposition to them in gold mining countries around the world. Contrary to some US engineers, who were well known for their opposition to organized labor, Creswell denied that a white workforce would lead to any increase in work action or strikes. Notably, Creswell based his recommendations on an experiment utilizing white labor he had been running at the Village Main Reef mine, where he was employed as general manager, and he marshaled statistics to defend his claim.

Controversy over the validity of this experiment and its implications would roil the community of expatriate engineers on the Rand and have lasting implications for the future direction of what became South Africa.[59] More immediately, however, the Chamber of Mines ignored Creswell's evidence and his defense of white labor and instead ordered "that steps be immediately taken to send a competent person to China to ascertain what number of Chinese Coolies would be available for service at the mines, the wages they would require, term of engagement and the cost of getting them here?"[60]

Concurrent to the private committee, the British Chamber of Mines appointed Hennen Jennings chair of a committee of fifteen mining engineers to formally advise Colonial Secretary Joseph Chamberlain on the pressing labor problem ahead of his visit to the Rand. At least nine were from the United States: Hellmann, William Honnold, Harry Webb, George Webber, George Hoffmann, Pope Yeatman, Frank Carpenter, and the two Jennings brothers, Hennen and Sidney. Unlike in 1897, at least one South African–born Afrikaner, the German-educated mining engineer Robert Kotzé, was included on the committee.[61] The final report, presented to Chamberlain in January 1903 as "A Descriptive and Statistical Statement of the Gold Mining Industry of the Witwatersrand," read a great deal like Hennen Jennings's testimony in 1897. Once again, working costs figured as "the cardinal factor for consideration," and accordingly, labor costs were of "dominating importance." Solving this "problem," the engineers reasoned, was "not only the main question for the mines, but also seems vital for the advancement and prosperity of the whole of South Africa." While claiming to have the imprimatur of scientific experiment, the recommendations were based entirely on the engineers' assessment of the colonial and racial situation. Noting that while they believed "the white race is the superior race mentally, and that the black should recognise it," they argued that, "as a mere muscular machine," an African who had spent sufficient time at the mine to gain skills "is the equal of the white man." With what was surely the memory of emancipation and Reconstruction in the United States in mind, the engineers simultaneously held that if Blacks worked alongside whites, whites would "become declassed and ashamed of themselves." Experiments in replacing Black labor with unskilled whites were deemed unsatisfactory by the mine managers. In any event, and in comparison to the United States, "The native races have not, as in some countries, been swept away by contact with civilization." Instead, the engineers envisioned a future South Africa that was not wholly white, as others demanded, yet one where "the greatest good and harmony" would "prevail only if a distinct line is drawn, and it is recognised that the white is master and the black is man." The report continued, "For expansive prosperity, whether from the aspect of the shareholders in the mines, or the white employees thereof, we believe that an abundant supply of cheap labour drawn from the coloured races is of supreme importance, and without this aid there do not appear to be any great potentialities for the shareholder, the white mine employee, or the country at large." The engineers had three specific recommendations: increase measures at compulsion, expand the area of labor recruiting into more British territories in Africa, and, "as a last resource," import indentured Chinese laborers.[62] In an interview thirteen years later, Jennings mused, "It is interesting, in looking over this most exhaustive report with its tabular statements, to see how closely the future has established some of the predictions made therein."[63]

Chamberlain, who was staying at Hennen Jennings's prewar home in suburban Parktown during his visit to Johannesburg, took the message to the intercolonial conference in Bloemfontein held in March, where it was resolved that recruitment of laborers would expand north of the Zambezi River and beyond South Africa. This step enlarged the recruitment area and potential supply of African labor and thus fulfilled a goal the mining industry had held for years but was unable to solicit state support for under the ZAR. On the question of Chinese labor, while the conference absolutely rejected the "injurious" permanent settlement of "Asiatics," it did allow for some to be introduced if "industrial developments positively require" it.[64]

Two months after the Chamber of Mines had written to canvass the engineers' private views on the labor question, Jennings began to worry about Creswell's experiment at the Village Main. Calling the matter a "crisis" for the industry, Jennings criticized Creswell's "Village experiment of white labour" by attacking its veracity as applied science, politely calling it "unsatisfactory." He tasked his younger brother Sidney with investigating. The younger Jennings found that with unskilled white labor, working costs per ton amounted to 63 shillings per ton, which yielded 43 shillings worth of gold per ton, equaling a loss of 15 shillings, amounting to £3,000 per month to finance Creswell's experiment. The elder Jennings hoped that he could prove Creswell to be in error and force him to "publicly acknowledge it, so as not to supply him with a martyr's crown. He is, however, very set and fanatic in his views."[65] In January, as word was spreading, Jennings wrote to both Milner and Chamberlain to remind them he had been an engineer at Village Main, followed by his brother, and assure them that Creswell's figures were wrong.[66]

The Transvaal Labour Commission repeated many of the points made by the engineers to Chamberlain, although the commission's ambit extended beyond the mining industry to consider the needs of settler agriculturalists and other industries. Proceedings took place over thirty-two days, from July 21 to October 6, 1903. The majority report of the commission emphasized the severity of the labor shortage, which, they noted, arose "in its present form" in 1835, since up to that point, "slave labour was extensively employed in South Africa."[67] The postwar mining industry, however, was most severely affected by the labor shortage: the current number of 129,000 workers was already insufficient and was forecast to worsen when, in five years' time, the mining industry would need an additional 196,000 laborers. The majority report did not mention Chinese laborers by name, but they hardly needed to. By then, Farrar had made his speech in favor, and more recently, Scottish engineer H. Ross Skinner had already returned from his trip to China, California, and British Columbia to investigate the use of Chinese labor, announcing it to be suitable.[68] So when the majority report expressed their final finding in blunt terms—"There is no adequate supply of labour in Central and Southern Africa"—everyone knew what they meant. Conversely, the minority report had the exact opposite finding—"there is sufficient labour in Central and Southern Africa for present requirements"—which could be "supplemented and superseded by white labour." In closing, the minority report sharply criticized Hellmann in particular: the mineral wealth of the Transvaal, they believed, should be developed "in the interests of the [white] people of the Transvaal," not for capitalists in London or Paris. This belief formed the whole basis for their conclusions, which Hellman had called "a pernicious one."[69] The split report signaled that, on the eve of the arrival of the SS *Tweeddale*, carrying the first group of Chinese laborers who arrived on the Rand on June 19, 1904, white settler society on the Rand had split into two camps, each of which could trace their origin to a conflict over the efficiency of white labor at the Village Main.

THE RAND AND CHINESE LABOR, 1903–1907

Between 1904 and 1910, over 60,000 Chinese miners came to work mostly underground in South Africa's mines. In addition to resistance from the Chinese workers themselves, the effort caused an uproar of protest from white settlers as well as their sympathizers in Britain, propelling the British Liberal Party to power in the 1906 election over their opponents who had authorized the measure. Anticipating the furor over the "Chinese Question" of allowing or excluding Chinese from British settler colonies, mining engineers quietly sought to use the time bought by the project to engineer a more lasting solution to the issue of labor on the mines.[70]

Mining engineers attempted to enumerate two distinct new routes to a profitable workforce. The first was embodied by Creswell's calculus that concluded white labor was more profitable. The second was already underway before the last indentured Chinese were repatriated. In many ways picking up from where Jennings was headed in 1897 at the commission, this second route to profitability aimed at demonstrating that "native" labor was the best option. The Rand's largest mining concern, the Corner House group, once again turned to a US consulting engineer, this time Ross E. Browne.

Browne came from the United States to South Africa in January 1904—six months before the first Chinese workers arrived—and stayed until September of the next year. This was an extensive consulting trip on behalf of the Corner House group. Browne was held in high estimation by other mining engineers. Hammond called him "one of the best of our American mining engineers."[71] Hennen Jennings got his first steps into the mining engineering profession by working as Browne's assistant at the New Almaden mercury mine in California, and it was through Hennen Jennings that Corner House hired Browne.[72] An internal letter read of Browne, "He will probably stay a month or two here with Mr. [Sidney] Jennings in order to go through the mass of information which is available and which will form the basis of the investigations which Mr. Browne himself will undertake on the Rand. Advice in regard to the great problems which have to be solved in the future is the main object for which we desire to have the services of this eminent engineer."[73]

Browne's visit to the Rand with his longtime assistant Hoffmann focused on working costs and lasted twenty months at an annual salary of £5,000. After his trip, Browne wrote to Corner House, "I believe we are in full agreement about the desirability of educating the general public of the Rand up to an understanding and appreciation of the vital importance of a radical reduction of working costs per ton of ore."[74] Browne and Corner House agreed that the report should be discussed at the meeting of the South African Association of Engineers such that it receive wide reception and make its way to every engineer and manager on the Rand.

Browne's 1907 report ran to over fifty pages of text, formulas, charts, and graphs. It is one of the clearest expressions of the technical and cultural authority claimed by engineers. Browne's vision for the Rand is worth quoting at length:

> South Africa with its preponderating numbers of blacks is not the country for cheap white labour. The black man is there to stay, and must be dealt with. He must be controlled, and efficiency of control demands utilisation of his work. The assignment of similar work to whites and blacks would result in lowering the standard for each, and breaking down the barriers which are essential to the supremacy of the white race. A distinct line of separation in the duties of skilled white and unskilled coloured labour is of paramount importance. Fortunately for the mining industry this division of labour, with efficiency on both sides of the line, will lead to the best possible results.[75]

The report was essentially an extended comparison of various costs and practices between two mines in California with those on the Rand. One of the most significant variables was the ethnic and racial makeup of the mine workforce. The Rand mines employed white, Chinese, and African workers, "while the labour in the California vein-mines is exclusively white." Browne was critical of the Rand workforce in general, commenting that "most of the labour, especially underground, is unambitious. No general spirit of competition prevails. The standard measure of a day's work is low." His sharpest criticism, however, was leveled at the "merely supervisory" nature of white labor, which took up most of the time of white workers. Worse, because of high living costs, those whites were far overpaid: "Compared with the miner of California, the average white miner of the Rand is less skilful, does less effective work, and receives nearly double the pay. The wages of machine-drill men are on the Rand 25s., in California 12.5s."[76] Browne's analysis of the comparative costs (and implicitly, the comparative productivity) of white, African, and Chinese labor in different mining and refining work exemplifies the way in which he and other mining engineers brought engineering logic to the race and labor question, as illustrated in figure 4.1.

Like his protégé Jennings had argued in 1897, African men could be useful workers, if only the state would allow them to be: "Many of the coloured men are capable of undertaking work requiring a certain amount of skill and intelligence, but this is not permitted under the regulations. As it is, he is a fairly good muscular machine, and, under proper training and control, his work constitutes much cheaper labour (per unit of work) than may be had in California for vein-mining purposes." Given that white labor was too costly and drove up the working cost per ton of ore and that African workers could be employed for so little, Browne recommended maintaining "a distinct separation in the work of the white and coloured labour": "The coloured man will do all the work requiring no special skill, and selected coloured men will under white direction do the work of actual handling and running of machine drills underground."[77] Maintaining a color bar in this way reflected the changed political and labor climate compared to Jennings's 1897 recommendations.

Initially, Browne's report was kept confidential and inside Corner House, although Phillips admitted he had not yet "managed to get through all the appendices yet."[78] Instead, Phillips wrote in June 1906, "We are pressing the working cost and efficiency question with our own managers, without inviting any public discussion upon the Ross Browne lines yet."[79] While there is no direct evidence giving a reason for the June 1907 discussion of the report,

TABLE "H."

Labour employed per ton of ore—the number of shifts, or the fraction of a shift of one man's labour. The reciprocal of the figure given will indicate the number of tons per man per shift. The figures include a due proportion of the numbers indirectly employed in the workshops, in power development, etc., and are greater than the numbers directly engaged in developing, mining, milling, etc.

1. White—		
In Developing, per ton developed	0·036	shifts.
In Mining (excluding developing), per ton mined	0·238	,,
In Milling, per ton milled	0·054	,,
In Cyaniding, per ton milled	0·038	,,
In D and E Accounts, per ton milled	0·017	,,
Total White Labour, per ton milled	0·417	,,
2. Native—		
In Developing, per ton developed	0·108	,,
In Mining (excluding developing), per ton mined	1·021	,,
In Milling, per ton milled	0·143	,,
In Cyaniding, per ton milled	0·138	,,
In D and E Accounts, per ton milled	0·049	,,
Total Native Labour, per ton milled	1·600	,,
3. Cape and Coolie—		
In Developing, per ton developed	0·001	,,
In Mining (excluding developing), per ton mined	0·063	,,
In Milling, per ton milled	0·014	,,
In Cyaniding, per ton milled	0·008	,,
In D and E Accounts, per ton milled	0·003	,,
Total Cape Labour, per ton milled	0·096	,,
Total Labour—		
In Developing, per ton developed	0·145	,,
In Mining (excluding developing), per ton mined	1·322	,,
In Milling, per ton milled	0·211	,,
In Cyaniding, per ton milled	0·184	,,
In D and E Accounts, per ton milled	0·069	,,
Total Labour, per ton milled	2·113	,,
Average Wages per Shift—		
White	18·86	shillings.
Native (Kaffir)	2·66	,,
Cape and Coolie	5·26	,,
General Average	5·98	,,

FIGURE 4.1

Comparative analysis of mining and milling labor, 1907: Browne presented comparative costs (Table G) and required volume (Table H, reproduced here) of labor in the Witwatersrand mines and mills, by racialized labor groups. *Source*: Ross Egerton Browne, "Working Costs of the Mines of the Witwatersrand," *Journal of the South African Institution of Engineers* XII (1907), reprinted in Mining Industry Commission, *Minutes of Evidence, with Appendices, Part IV. Presented to both Houses of Parliament by Command of His Excellency the Governor* (Pretoria: Government Printing and Stationery Office, 1908), 1601.

it must nevertheless be understood in the context of the economic recession begun in 1906, growing protests from unemployed Afrikaners, and the ongoing white miners' strike that had begun in May of that year.[80] The onset of the recession also meant that South Africa was increasingly less lucrative as a destination for mobile US mining engineers, as Honnold lamented in 1906. Further, the "vital" question for mine owners and engineers was when the new deep levels of extraction would start becoming profitable.[81]

Browne's paper was discussed at the Association of Engineers, with commentary from Robert Kotzé. Kotzé had been on the Rand since the 1890s and would soon be appointed as government mining engineer of the Transvaal, a post he held from 1908 to 1910 on the recommendation of his close friend and future South African prime minister Jan Smuts and of the Union of South Africa until 1926.[82] Kotzé's largely technical remarks on Browne's report sought only to refine and improve upon Browne's research rather than overturn it. Kotzé agreed that engineers "should utilize the native where the work is of a more or less mechanical nature," but white workers should be reserved for "positions where superior intelligence is the first requirement, and where capacity for muscular work is of secondary importance." Ultimately, Kotzé concluded, "If the scheme advocated be adopted for experiment and carried to a successful issue, it is to be anticipated that our native labour difficulty will be rendered much less acute."[83] The assembled corps of engineers expressed no doubt that applied science could solve the most urgent and intractable problem facing settler society.

Meanwhile, Creswell attempted to use Browne's report to further his own case for white labourism. In response, the Chamber of Mines felt compelled to publish a letter from Browne making it clear that this was not, in fact, the case. Writing from California, Browne maintained that "it is already established that a more efficient use of the K**** labor will result in lowering the costs materially. There appears to me no other practical way of effecting an adequate reduction. The white community will get the lion's share of the product in any event, and the smaller the costs, the greater the product." Phillips replied to Browne with a report from the commission, lamenting that "you will see that very little progress has yet been made in reconciling white men to unskilled work." Phillips concurred with this vision, a plan for gold to subsidize the livelihoods of white South Africans to offset their lack of income from mine labor: "By far the larger proportion of the gold produced provides sustenance for the inhabitants of the country. It follows, therefore, that the gradual reduction of working costs, which must result in an increased scale of operations, and in a prolonged life for the mining industry, is of great consequence, not only to the mining industry, but to South Africa."[84]

At the same time, the 1907 Mining Industry Commission—tasked with determining the best way to increase the number of white mineworkers—featured testimony from US engineers Hellmann, Sidney Jennings, Webber, and Honnold. Creswell seized the opportunity to cross-examine them. Creswell asked Webber, "If you were able to get a white man as cheaply as a native, have you any tendency to take the native?" Webber replied, "That would depend. I should say if we had a white man that would work as cheaply as a native there would be no difference in our working costs, but unless you can make a clean sweep and substitute all coloured for white, or all white for coloured, it would not be policy to mix up the different classes."[85] Webber's opposition to white labor ran counter to the public position of the newly elected Het Volk government, but privately the engineers' prescription for the industry continued to shape policy. Prime Minister Louis Botha assured Phillips that the report's recommendations would not be put into practice.[86]

Inside Corner House, internal figures updated Browne's numbers in 1909, showing that working costs had indeed decreased as a result of adopting Browne's suggestions.[87] Browne's report was still top of mind for Corner House in 1913. Considering what to submit to the upcoming Economic Commission, Phillips wrote to his employer, "Ross Browne's comparison, for instance, between the labour used in California and the labour employed here, is to some extent apposite and it is of course not being lost sight of; but there may be other instances in the world that will be useful."[88] Browne himself noted that year that the Rand had surpassed California in terms of output.[89]

In his outgoing presidential address in 1909, Laschinger surveyed the recent history of South Africa's gold industry, noting that major developments had been "conceived and carried out mainly by engineers." Referencing the US efficiency engineer Harrington Emerson, Laschinger congratulated the industry on its reduction of working costs and cast the Chinese labor "experiment" in terms of efficiency. Laschinger maintained that without exception, mine managers agreed that the most important outcome of the presence of Chinese "was that of increasing the efficiency of the native by raising the standard." He also noted that the growth in gold production showed how "the whole economic aspect of a country's resources may be changed, and its possibilities enlarged."[90] The combination of a putatively "efficient" and largely Black workforce, supported by a state committed to the industry, was envisioned by Jennings and others in 1897, and now it had come to be put in place. After the era of direct imperial control came to an end with independence for the Transvaal in 1907 and the Union of South Africa in 1910, state support of the mines increased. One mine owner in 1909 commented that "since General [Louis] Botha's government came into power in this colony it has done its utmost to assist the mines in their recruitment of labor."[91] This arrangement was left largely unchanged until the end of apartheid and arguably persists to this day.

CONCLUSION

Conflict between workers, engineers, the capitalists they served, and the state characterized mining from the start but erupted most dramatically with the Rand Revolt of 1922. For that insurrection of white mine workers, "the defeat was utter."[92] In the long run of the twentieth century, the distinguishing feature of the South African economy was the continual low cost of segregated Black labor in mining and agriculture—the very object of the mining engineers' debates dating back to at least Jennings's 1897 testimony.[93] Engineers' practical, objective-driven way of thinking, which calculated and analyzed humans and nonhumans interchangeably, formed the blueprint for a state that became nearly synonymous with racial capitalism in the twentieth century.[94]

In line with a key theme of this volume, this chapter has demonstrated that US mining engineers on the Rand were equally capable of solving the technical problems of ore extraction as they were of devising deeply unequal and racialized social structures like twentieth-century South Africa's. This chapter also echoes the volume's argument about the relative historiographic invisibility of engineers, showing how even in the voluminous literature on the origins of South African racial capitalism, mining engineers have often been noticed by historians but rarely considered central to the story. Moreover, this chapter has shown that only by focusing on the global mobilities of knowledge and people that are an essential characteristic of engineering and engineers in this period—in this case, from California to Venezuela to South Africa—can historians begin to apprehend the significance of an exceptionally itinerant profession.

5 ENGINEERING MOBILITIES ACROSS EAST ASIA: CHINA AT THE INTERSECTION OF LOCAL, REGIONAL, AND GLOBAL INTERACTIONS

ELISABETH KÖLL

From the mid-nineteenth century onward, the training and employment of engineers in China became a global affair in terms of frequent and sustained transnational interactions. In the late Qing (1644–1911) period, the imperial government hired foreign engineers to work and train the first generation of Chinese engineers at the newly built shipyards in Shanghai and Fuzhou. Around the same time, the first cohort of Chinese students was sent abroad to study engineering at universities and technical schools in the United States and bring their knowledge back to China. When foreign financial syndicates joined forces with the imperial government to start mining and railroad projects at the turn of the century, foreign engineers joined these operations in various professional capacities. With great demand for engineers due to increasing industrialization and modernization projects, Chinese students sought education abroad, especially in Japan, then the powerhouse of industrial and scientific education in Asia. However, by the mid-1920s and throughout the 1930s, the Republic of China (1911–1949) made significant strides in establishing technical schools and new universities that were now able to train the country's engineering talent. These academic and professional institutions became the hallmark of education initiatives during the Nanjing decade (1927–1937) under the firm control of Chiang Kai-shek's Nationalist government. Chinese engineers began to lead infrastructure and industrial projects in progress all over China while also creating academic and professional network structures similar to those connecting engineers in Europe, Japan, and the United States.[1]

This chapter addresses the theme of mobility by outlining distinct waves of engineering professionalization and knowledge emerging in China from the late nineteenth century, when the Qing empire cautiously engaged with industrial and military development as a strategy to counter Western imperialism, all the way to the Japanese invasion in 1937. During those decades, China evolved as a space of vibrant engineering mobility, with Chinese and foreigners moving back and forth for study, training, and employment against the background of increasing economic globalization and imperialism framing regional developments in East Asia. The case study of China echoes findings in other chapters of this volume by demonstrating the rapid expansion of professional engineering through institution-building driven by the modernization agenda of an evolving nation-state.[2]

At the same time, China's case also highlights the significant impact of semi-colonialism in shaping the trajectory of engineering education and professionalization, creating competitive tensions as well as productive exchanges. In many ways, this chapter complements Mark Hendrickson's case study of China's success in antimony mining in Hunan province by discussing engineers' trajectories and mobilities in the broader transnational context of East Asia before World War II. China's need for infrastructure, industrial, and commercial development shaped the demand for and supply of engineers, directing their professional mobility across local, regional, and transnational spaces. Chinese engineering mobility also manifested itself via the integration of global engineering practices, networks, and associations adapting Western-oriented standards and norms. However, certain distinct Chinese practices found their way back into Western engineering knowledge due to increased professional and knowledge exchange among the international engineering communities at work in Asia.[3]

Outlining the professional evolution of engineers in China, the chapter introduces the imperial government's first but insubstantial attempts to employ foreign engineers and send abroad small cohorts of Chinese students for training. However, engineering education gained momentum only with the founding of the Republic and the transition to a modern education system. As the analysis of engineering magazines, newspapers, and archival correspondence shows, professional engineering education in China took off during the late 1910s and early 1920s when every industrial and commercial sector competed in the search for domestic engineering talent. Ironically, once increasing numbers of graduates were seeking employment in the mid-1920s, they could not immediately be accommodated by the private sector and thus were channeled into government employment instead.

Shifting the geographical focus more broadly to Asia, the chapter goes on to highlight employment patterns of engineers as documented in the *Far Eastern Review*, the most widely circulating commercial and engineering publication in Asia at the time. As the analysis of the journal's reports and correspondence of commercial partners demonstrates, political and economic agendas driven by distinct national interests and imperialist agendas also impacted the mobility and flow direction of engineering talent. With the integration of Asia into the globalization of the engineering profession in the early twentieth century, the final section of the chapter focuses on institution-building through the establishment of professional networks and associations in China. Here publications like the *Journal of the Association of Chinese & American Engineers* present a useful source to explore issues of global convergence and national divergence in knowledge transfer, professional status, and identities.

This chapter argues that China's modern history of engineering should not be viewed as a linear trajectory from imported foreign knowledge to building a modern education system, resulting in the development of an independent, inward-looking engineering sector. Instead, the examples of professional training, mobility, and knowledge transfer among Chinese engineers highlight a continuous, substantial level of global interactions and collaborations framed by external ruptures such as political regime transitions, semi-colonial conditions, military conflict, and economic crises. Validating a recurrent theme in this volume, Chinese engineers were locally focused through their work in industrial and commercial projects across many provinces but also maintained global connections through their associations, secondment abroad, and professional knowledge exchange. As the trajectory of engineers and their mobility in China was closely aligned with national efforts for economic modernization, this chapter primarily relies on examples of railroad engineers for its analysis because they represented (together with mining engineers) the largest group of civil engineers in the country. Whether trained abroad or in China, these engineers' educational and professional trajectories exemplify their integration into China's modernization projects and, even more important, into the evolving global engineering community.

UNMET DEMAND: ENGINEERS IN LATE IMPERIAL CHINA

The role of engineers and their contributions to industrial innovation and economic growth of nation-states is widely recognized in the discussion of emerging modern technology, business, and industries in Europe and North America during the late eighteenth and nineteenth centuries.[4] In the case of China, where political and institutional constraints meant that industrialization took a much slower path, the engineering profession emerged and made its impact considerably later. The first small cohort of Chinese sent abroad to study in the United States and Europe was associated with educational missions sponsored by the Qing government in the 1870s and 1880s. This cohort of about one hundred students was part of China's so-called Self-Strengthening Movement and pursued a rather narrow range of educational goals oriented toward Western scientific knowledge such as military technology and other related subjects that would benefit China in competition with the West. The mission found an early end due to concerns of conservative Chinese officials about the students' exposure to American lifestyles in addition to an increasing anti-Chinese public climate in the United States.[5] Although the first student cohort included the famous Yale-trained railroad engineer Zhan Tianyou, "the size of the early group was too small and the academic training of its members was generally inadequate to make it a significant new force in Chinese society."[6]

In the context of China's Self-Strengthening Movement, the Chinese government also began to sponsor enterprises that combined military with industrial and commercial applications.[7] Government-sponsored, provincial shipyards like the Jiangnan arsenal in Shanghai became the cradle of engineering expertise, in terms of both academic teaching and applied training. The Fuzhou shipyard, established in 1866, was managed by a combination of foreign technical personnel and Chinese officials under the leadership of a director-general appointed by the court.[8] Introducing engineering training under foreign supervision reflected the imperial government's agenda to increase the efficiency and productivity of the shipyard to produce modern naval vessels. Addressing the acute lack of skilled labor upon which engineers' work depended, the shipyard established a separate School of Apprentices where students were taught French, arithmetic, science, and engineering, with the goal of training them as future foremen in the arsenal's workshops.[9] However, by the 1890s, only twenty-nine engineering students had graduated from the shipyard's comprehensive training program, with five of them finding employment in the slowly developing sector of railroad construction.[10]

The so-called second wave of Chinese students going abroad was quite different and took place during the early years of the twentieth century when attitudes toward Western learning changed, especially after the abolishment of the imperial examination system in 1905.[11] State and society were now in favor of acquiring new knowledge beyond military technology and science. Many Chinese students chose to obtain Western knowledge in Japan, which had consistently followed the path of modernization by selectively adapting and integrating Western learning into its education projects since the 1868 Meiji restoration. In 1906, over 8,000 Chinese students studied in Tokyo, most of them pursuing degrees in subjects like law, medicine, education, and military science. By comparison, at the time, only about 300 Chinese students studied in the United States.[12]

However, the Boxer Indemnity Scholarship Program began to shift the trend of Chinese studying abroad toward the United States. The program was an indirect result of the international military alliance's victory over Chinese troops in the Boxer Uprising of 1900 and the resulting outsized demands for reparations in the imposed Boxer protocol. By 1908, several Chinese diplomats with educational experiences in the United States succeeded in negotiating to channel the remission of excess funds from the Boxer Indemnity into scholarships that would fund the education of Chinese students in America.[13] Between 1909 and 1929, the program funded about 1,300 scholarships for studies in higher education degrees at American universities.[14] The motivation for establishing these scholarships went beyond pure educational philanthropy. As historian Madeleine Hsu argued, the scholarships allowed Americans "to smooth over hurt feelings while using none of their own resources by bringing Chinese to study in the United States, which would teach Chinese to appreciate American civilization and friendship and thereby enhance American influence in China over that of Japan."[15]

Despite these efforts, the demand for engineering talent and knowledge by new mining, shipbuilding, and railroad construction initiatives taking off around 1900 remained urgent but unmet. For example, in 1898, the Pekin Syndicate, which had obtained mining concessions in Henan and Shanxi provinces, recruited an engineering expedition consisting of six Royal Engineer officers "on loan by the government of India" under the leadership of J.G.H. Glass, chief engineer and secretary to the government in the public works department of Bengal.[16] The goals of the expedition were to assess the value of the coal and iron ore deposits and to determine whether the mines could be connected via railroads to waterways for the transport of mining products. For the actual surveys, Glass hired ten Indian surveyors who had trained in the survey department of the Government of India.[17] The issue concerning the lack of Chinese engineering professionals was so acute that it even found its way into China's first promulgated Commercial Laws in 1904. As the section on revised mining regulations stated, "At present China has very few Mining experts, and holders of permits must therefore be allowed to employ foreign engineers."[18]

Most of the first railroad companies in China expanding after 1900 were established as Sino-Foreign syndicates.[19] Addressing the lack of domestic engineers and skilled labor, some railroad syndicates tried to lay their own tracks for educating engineers by initiating the building of a localized knowledge infrastructure through company-based railroad schools and teaching materials.[20] These company-based institutions employed foreign engineers to train Chinese engineers and skilled labor in the specific context of competition between foreign syndicate partners and the imperial government for control over financial and political "spheres of interest." Not surprisingly, that control became tied to control over "spheres of knowledge" through the preferential use of country-specific technical standards, language instruction, and railroad equipment.[21] The early period of engineering training in China thus elucidates not only the challenges of cross-national and cross-cultural knowledge transfer but also the political and economic competition among foreign powers and with the Chinese government as well as between private and government-sponsored business interests more generally.

Compared to Japan, foreign engineers had a much deeper and longer engagement with China. Consulting engineers or inspectors from the United States and Britain who brought their knowledge as former engineers with country-specific expertise to the table were particularly valuable to Japanese trading companies supporting the expansion of the rail sector in Japan. For example, in 1878, Americans like J. U. Crawford were invited as civil engineering consultants to help with railroad construction in Hokkaido, assisting the founding of local railroad companies and the railroad industry. However, most significant is the fact that Crawford never spearheaded the construction efforts per se. According to historian Naofumi Nakamura, apart from their technical expertise, these foreign consulting engineers only provided "essential routes of knowledge transfer in cross-regional commercial management," which benefited the speedy and successful expansion of Japan's rail network around 1900.[22]

Japan managed to gain independence from direct Western expertise in most areas of industrial and infrastructure development by quickly building robust training and academic infrastructures that allowed Japanese engineers to take over the positions from Western advisers when "the weaning process began in earnest after the turn of the century."[23] In terms of equipment, however, Japan continued to buy locomotives and related materials from the West. Not only expert manufacturing according to requested specifications, but also a strong direct sales representation enabled especially American firms to sell locomotives to Japanese railroad companies with great success.[24]

Within Asia, Japan emerged as the early adopter and leader in industrial development by successfully acquiring and mastering advanced technology in the railroad industry as well as other industrial sectors. To establish academic engineering training, the Japanese government assumed a much more proactive role compared to the Chinese imperial government and founded the Imperial College of Engineering in Tokyo in 1893. The government-sponsored training was intended to provide the Ministry of Public Works with engineering graduates to be employed in its industrial projects.[25] According to Naofumi Nakamura, the college not only became the leading institution for engineering education in Japan but also gained fame for its "sandwich method" of instruction, which alternated the study of theory with practical engineering training in a highly successful manner.[26]

Contrasting the trajectory of Japan's modern engineering history with that of China is significant as it points to the impact of different political circumstances on the use of foreign expertise, domestic engineering training, and the establishment of national technological standardization. Working for a sovereign nation-state after the Meiji restoration, Japanese officials were able to exercise "selective control over the emerging features of the country's technological development by their strategic and contractual management of hired foreign aides."[27] Belonging to the category of foreign aides, it is not surprising that American and British engineers experienced the collaboration and knowledge exchange with their Japanese hosts in different ways according to their individual project involvement. However, most important is the fact that the Meiji government almost exclusively sent away the hired advisers and engineers after project completion by not renewing their contracts.[28] By contrast, most construction and mining operations in China involved Sino-Foreign syndicates operating under semi-colonial conditions, which limited such actions on behalf of the Chinese government. Syndicate partners financing the railroad or mining projects cemented the presence of their foreign engineers by negotiating contracts that ensured the continued employment of foreign experts until all loans were fully repaid to the foreign partners—a process that often lasted for decades and, in some cases, well into the 1940s.[29]

SECURING LOCAL SUPPLY: TRAINING CHINESE ENGINEERS

The Chinese students of the second cohort studying in the United States after 1916 had a much better experience in terms of belonging to a larger student group, improved training opportunities, and easier reintegration into the Chinese economy after their return. During the Republican period, the largest number of study-abroad students came from prestigious educational institutions such as Qinghua College in Beijing, St. John's University in Shanghai, and Beiyang University in Tianjin. One considerable advantage of these institutions was that they had foreign teachers on the faculty who helped prepare the students for their study abroad.[30] Continuing with educational reforms and efforts to promote and integrate modern science and engineering into the curriculum, the Chinese government established Jiaotong University, also known as the University for Communications and Transportation, in 1921. This new university was made possible by merging three colleges in Tangshan, Tianjin, and Shanghai that were known as pioneering institutions offering industrial and technical education at the turn of the century. By the late 1920s, Shanghai Jiaotong University was the most prestigious campus among the three, with strong departments in electrical and mechanical engineering as well as railroad administration. Reflecting the increasingly close link between the Nationalist government's economic modernization agenda and the university's academic mission, the minister of communications served as president of the university.[31]

Considering these new academic institutions and the increasing prominence of state-sponsored technical education, Chinese students graduating in the 1920s were now better qualified and able to pursue employment opportunities according to their specific engineering qualifications. In fact, many were now prepared to penetrate the upper stratum of chief engineers and managers in Chinese railroad and industrial companies that previously had been dominated by foreign engineers. However, the growing number of engineering graduates also generated a problem because the railroad network and thus positions of employment did not expand proportionately at the same time. In fact, substantial new line construction and rail sector expansion that could have absorbed these graduates did not occur until the early 1930s. This led to a stunted phase in professional mobility that Harold Stringer, an adviser to the Chinese railroad administration, noted in the *North-China Herald* in April 1929 with the complaint that "all departments have been compelled to return an increasing number of returned students, for whom there was no adequate employment. For 16 years there has been no railway expansion. Railway men rust in routine."[32]

The new asymmetry between increasing numbers of Chinese engineering students and the availability of appropriate positions created substantial difficulties for graduates with diplomas from Chinese and even overseas engineering schools to find employment. However, the asymmetry caused by the time lag between the graduation of newly minted engineers and available positions in the railroad sector was a phenomenon not just limited to China. This asymmetry also could be found in the case of Japan, where in the 1880s, graduates from the Imperial Engineering College in Tokyo did not find employment right away due to the lack of suitable positions in their engineering specialty. According to historian Wada Masanori, it was difficult even for those graduates to find appropriate work until about 1895, despite their prestigious degrees and specialized training.[33]

As a result of this unexpected labor market development, employment in various parts of the Nationalist government's expanding bureaucracy, headquartered in the new capital of Nanjing from 1927 onward, became an ever more appealing career choice for Chinese engineering graduates. Career information included in the Ministry of Railways' employment records from 1929 also reveals other important trends in the emergence of the Chinese engineering profession: in 1911, the first generation of employees with overseas degrees working for its predecessor, the Ministry of Communications, had mostly degrees in law, commercial science, or economics from Japanese universities. By contrast, in 1929, only one of the fifty-two employees in the Ministry of Railways' operations department had a degree in railway science from a Japanese institution.[34]

Apart from the increasingly tense political relationship between China and Japan and a steadily growing flow of Chinese students seeking all or part of their education in North America and Europe rather than in Japan, other factors related to the issue of work experience and professional opportunities contributed to this development. In the early twentieth century, the railroad sector played a leading role in the US economy and industry, creating many options for eager and well-trained engineering graduates to work in and manage rail companies and related industries. The economic and business focus of the railroad sector helped to create an excellent environment for advanced academic study and research as well as ample apprenticeship opportunities for aspiring Chinese engineers who were able to gain relevant work experience before their return home.[35]

It is not surprising that the demand for Chinese engineers would lead to not only new schools and universities but also increased public promotion of the possibility for Chinese students to receive engineering training "at home," so to speak. For example, on October 12, 1911, the English-language newspaper *The China Press* carried an advertisement with the bold headline, "Would you like to become an Engineer?" which promised: "You can get the same modern technical training, here in China, that you would get if you were attending one of the good technical colleges in Europe or the United States. You do not have to leave your home or give up your present position; for you study in your spare time."[36] The advertisement by the China Agency I.C.S. went on to inform the reader that no entrance exams were required, promising that "if you can write and read simple English, you can make a success of any I. C. S. course" and that "there are great opportunities in China at the present time for ambitious young student engineers."[37] Based on the nature of the advertisement, we can assume that the offered courses were not geared toward training professional engineers in competition with prestigious university programs. However, the search for students so publicly indicated urgency and the need for preparing technical talent to join the labor market in engineering more generally.

The challenges and progress of China's engineering education also became a frequent topic in publications ranging from general newspapers to professional magazines. Foreign engineers with training and supervising experience in China frequently gave public addresses that commended Chinese engineering graduates for their dependability, accurate calculations, and excellent performance of instructions in the office or the field. Foreign engineers serving as teachers and supervisors also came to realize that certain negative comments by their colleagues about Chinese engineering graduates preferring to hold white-collar jobs rather than doing manual work in the field were due to "faults of the system under which they [the graduates] have been trained than to any inherent lack of ability to become good engineers."[38] In response to this argument, engineering professors and advisers recommended improving the training in Chinese engineering schools through increased exposure of each student to industrial work during their studies. They also were in favor of extending coursework with an option to select between a five- and six-year-long course, which would allow Chinese students to engage more deeply with their subjects, especially since they had to perform all their studies in English as a foreign language.[39]

ENGINEERING MOBILITIES ACROSS ASIA

Japan's annexation of Taiwan in 1895 and Korea in 1910 created specific employment opportunities for Japanese engineers serving the colonial ambitions of their government.[40] With the Japanese takeover of Manchuria in 1931 and the creation of Manchukuo as a puppet regime, the push for industrial development and the expansion of coal extraction became cornerstones in the preparation for Japan's future war agenda and approach to managing the domestic economy.[41] Throughout the 1930s, the combination of railroads, intensive mining, and ironworks turned occupied Manchuria into a center for heavy industry and engineering.[42] Large mining operations like the Fushun colliery, taken over by the South Manchurian Railway in 1907 as a subsidiary, became the anchor of the mining efforts and attracted foreign as well as Japanese engineers. As historian Victor Seow has shown, in the early 1920s, operations like Fushun's gigantic open-pit mine became "an icon of Japanese industrial modernity" where Japanese and American engineers engaged in heated debates about the most appropriate mining technology and extraction methods.[43]

To identify a few general trends of engineering mobility within the expansive and diverse geographical region of East Asia, reports in *The Far-Eastern*

Review: Engineering, Commerce, Finance (hereafter *FER*) provide useful information for our analysis.[44] The *FER* was published in English from 1904 to 1941, first in Manila and later also in Shanghai and Yokohama. The voluminous monthly issues of this journal covered primarily engineering and commercial news, contained advertising of related equipment and services, and reported on new technologies, labor practices, employment opportunities, engineering firms, and professional organizations across Asia. In addition, the journal published summary reports on specific political and commercial market conditions in various East and Southeast Asian countries, including Japan, China, Taiwan, Korea, the Philippines, Malaya, and Siam.[45]

There is no doubt that the *FER* coverage provides an uneven depth of information with its preferential focus on certain Asian economies based on their strong investment ties to Europe and the United States via engineering and commercial projects. George Bronson Rea (1869–1936), the publisher and editor-in-chief who also had a track record as a journalist, engineer, spy, and lobbyist, solely financed the publication of the *FER*. We can assume that pragmatic and financial interests must have driven the journal's selection process of topics that would interest a broad professional readership as well as the journal's advertising partners. Advertisements placed by foreign firms constituted a substantial part of each monthly issue and, presumably, the revenue. With a print run of about 3,000 copies per issue, about a third of the copies went overseas, mostly to the United States, with commercial, educational, and government institutions as subscribers.[46]

Long features like the 1908 report on China's state of railroad development contained aspects of economic, political, and commercial intelligence. The articles provided detailed information about every line construction finished, in progress, or planned, including information on the financial syndicates and foreign engineers involved in the projects. Detailed maps, including topographical maps, photographs of equipment and construction sites, and translations of contracts between syndicate partners, provided comprehensive introductions to important engineering projects and their host countries. We can assume that this information would have been of interest to engineers already working in Asia, engineering and consulting firms, equipment manufacturers, financial institutions, trade commissions, industry lobbies, and investors exploring potential opportunities abroad.[47]

As editor, Rea exerted a strong influence on the journal, which for the first twenty-five years was focused on detailed intelligence and data related to engineering, business, financial, and economic projects and activities.[48] However, as Rea developed an increasingly pro-Japanese vision of Asia's future trajectory, *FER* reports on political developments in territories controlled by Japan and its expansionist aspirations in Asia need to be read carefully with Rea's ideological attitude in mind.[49] According to Leslie Eaton Clark's biography, Rea considered his opinionated, long editorials a cornerstone of the journal as "only engineers or others directly interested in a business venture in the Far East would trudge through tedious pages on subjects like soybean cultivation."[50]

The content of the *FER* mirrored different stages of industrial development and the pursuit of political and economic interests by American, British, and European companies, thereby de facto framing their engineering and commercial interests as part of the West's modernizing mission in Asia. For example, in the decade after 1904, reports on American-led engineering projects such as "Approved Water System for Manila" and "Municipal Improvements, Manila" point to the interests of the United States in the Philippines, which it had seized for $20 million following the Spanish–American war. It is not a coincidence that the *FER* started out being first published in the capital and port of Manila, which Senator Henry Cabot Lodge patronizingly characterized as "the prize and the pearl of the East. . . . Rich in itself . . . it will keep open to us the markets of China and enable American enterprise and intelligence to take a master share in all the trade of the Orient."[51]

Tracing the mobility patterns of engineers by following their professional journeys in Asia, a survey of *FER* articles indicates that by the early 1900s, American engineers began to seriously compete with European engineers for employment even outside the Philippines. American engineers could be found at work, especially in mining and construction projects in Siam, on the Malay peninsula, and in China. Not surprisingly, as noted above, Japan did not feature in reports as a "receiver" country of Western engineering expertise but as a competitive manufacturer of industrial equipment and services. China's 1911 revolution and the greater accessibility of its market were marked by hundreds of articles on planned, projected, and ongoing rail construction projects, which involved engineers from Europe, Britain, and the United States. However, by 1913, articles on "The American Engineer" and "The Young Chinese Engineer" were published in the same issue, indicating the rise of a new cohort of engineering talent now emerging from within China.[52] Due to the impact of World War I, German engineers and commercial interests lost their considerable influence, especially in China, whereas British interests and engineering projects received renewed attention. As a result, the *FER* published correspondence related to the British Engineering Association and gave prominence to the role of British engineers who contributed their expertise to the engineering school of the University of Hong Kong.[53]

By the 1920s, *FER* issues began to include reports reflecting the mobility of Japanese engineers who found employment in Manchuria, China, and Siam and introduced advanced technology and equipment to local mining, steel production, and sugar refining. These reports underlined Japan's increasing role as an economic powerhouse and producer of advanced technology, but they also endorsed the broader strategic interests of the Japanese government to expand its market reach and secure new supply chains of natural resources within Asia.[54] George Rea, who made no bones about his strong pro-Japan stance and hatred of surging communist activities in China, defended his editorial views by stating that "Japan is not only our best customer in Asia, but next to our own nationals, the best salesman of American goods in China."[55] Against the background of increasing economic globalization and political expansionism driving national strategic considerations, *FER* reports on the Soviet oil industry, aviation developments, and hydroelectric plant construction indicated regional political power shifts while also articulating economic visions related to national competition for strategic leadership in Asia.[56]

Overseas Chinese entrepreneurs in East and Southeast Asia had employed Western engineers in their businesses since the nineteenth century and continued to do so without particular preferences for the engineers' nationality. For example, entrepreneur Lim Peng Siang had received a British education in Singapore and afterward entered the family business that necessitated the employment of foreign engineers and boilermakers. Rice processing and coconut oil manufacturing were the backbone of Lim's business conglomerate Ho Hong, which was set up and organized by a Scottish engineer while an English engineer took charge of establishing the Ho Hong oil mill. Both engineers stayed on as supervising engineers and managers of these business entities once the construction had finished. Lim's collaboration with Western engineers continued once his conglomerate expanded into shipping operations and cement manufacturing, which was designed and set up by an American engineer.[57]

A glance through the issues of the *FER* reveals that it took some time for Chinese engineers to be regularly featured in the journal. In the late 1910s and early 1920s, hardly any articles commented on Chinese engineers outside of Sino-foreign ventures but focused on the emergence of Chinese engineering works and machine shops, which started as small repair shops and then expanded into the manufacturing of machinery and machine tools. These machine shops, concentrated especially in the Shanghai area, employed Chinese engineers under sole Chinese management.[58]

By the late 1920s, *FER* articles increasingly began to report on railroad projects progressing now under full Chinese engineering control. For example, in 1932, a feature report on the Hangzhou–Jiangshan railroad listed Chinese in charge of the project in the roles of managing director and engineer-in-chief, associate director, bridge engineer, mechanical engineer, and division engineers in charge of construction, proudly characterizing the operation as "100% Chinese," with most of the engineers having graduated from American colleges.[59] The *FER* also covered the activities of prominent Chinese engineers like Ling Hongxuan (H. H. Ling), not only because he supervised the construction of the Canton–Hankow railroad as managing director, but also because of his high profile as contributor to international engineering publications, including the *Journal of the Association of Chinese & American Engineers*.[60]

Engineering mobility between China's distant regions and provinces became a frequent phenomenon in the late 1920s and 1930s as school and college networks were often decisive factors in the appointment of lead engineers directing large regional infrastructure projects. For example, the famous bridge construction engineer Mao Yisheng reported that he was teaching at Beiyang University in Tianjin in 1933, when Du Zhenyuan, his former classmate from the Tangshan Institute and head of the Zhejiang railroad bureau at the time, invited him to lead the construction of the Qiantang river bridge project at Hangzhou. Accepting the invitation that Du Zhenyuan also issued in the spirit of national pride and trust in China's own engineering talent, Mao Yisheng moved his career from the north to the south of China.[61] This form of "domestic" engineering mobility deserves proper recognition as we have to consider China's vast geographical space with distinct regional diversity in language, climate, and culture. Engineers moving for project employment from a province in the north to the south or southwest of China can be compared to engineers making a professional move from Germany to Italy or Spain in the context of Europe's geography.

However, Chinese engineers not only moved around the country for work assignments within China but were also sent abroad on professional secondment by the Nationalist government. For example, the Ministry of Railways began to send experienced Chinese mechanical engineers to Britain to ensure that machinery on order, especially locomotives, was built to the correct technical specifications for Chinese railroad lines.[62] In 1934, a list compiled by the British Locomotive Manufacturers' Association contained the names, length of tenure, educational background, and sponsors of the Chinese engineers and students placed at British manufacturing companies such as the Vulcan foundry, the Hunslet engines company, or the North British Locomotive Company. These arrangements, some in the form of secondment for knowledge exchange and others in the form of apprenticeships, ranged from three months to several years of employment.[63] As the personnel files also reveal, some of the seconded Chinese engineers were employees of the Ministry of Railways, several had engineering degrees from Jiaotong University, and many of them had prior work experience from their time in the locomotive departments of the Shanghai–Nanjing railroad, the Qingdao railroad, and Hangzhou railroad. Several other Chinese participants in the program had engineering degrees from Glasgow University, Cambridge University, University College, and Imperial College of Science & Technology in London.[64]

The secondment sponsored by the British Locomotive Manufacturers' Association offered additional training and professional exchange for Chinese engineers and, by extension, benefited the government-controlled railroad sector. However, the program was certainly driven by commercial considerations on the British side. The association was keen on countering "their [the Chinese] American bias" with regard to locomotives and rail equipment and hoped to instill a preference for British equipment among the program participants.[65] In the association's opinion, "Chinese engineers trained abroad invariably had a preference for machinery made in the country in which they received their training."[66] In an even blunter statement, the association expressed its hope that "if capable young Chinese could be brought to the United Kingdom and trained here they would return to China as propagandists for United Kingdom practice and materials."[67]

Finally, the celebration of successful engineering schools and programs across East Asia and their aspirations in terms of educational mission became a returning feature of *FER* articles in the 1930s. Not surprisingly, the *FER* repeatedly featured Jiaotong University in Shanghai as the premier training ground for China's engineers. Employing a teaching staff of 210 faculty in 1933, Jiaotong University trained students for the needs of the Ministry of Railways and other government institutions, and it was no secret that it had become relatively easy for Jiaotong graduates to secure employment in industry, government, and commerce.[68]

The University of Hong Kong was another premier academic institution receiving praise for graduating students in engineering and applied science. In fact, in 1933, the engineering school's graduates included even a female Chinese student.[69] C. A. Middleton Smith, then the university's dean of the Faculty of Engineering, perceived the mission of engineers trained in the British colony to contribute to civilization and, more specifically with regard to China, help create a "more healthy and more prosperous nation."[70] This sentiment echoed earlier comments by the governor of Hong Kong, Sir Cecil Clementi, who in 1929 had called for Hong Kong University engineering graduates to consider employment in China. He envisioned the colony becoming "intimately associated with future engineering developments all over China through the medium of young Chinese engineers, who will have been trained in Hong Kong and then go into all the eighteen provinces of China, taking with them an affection for their alma mater and a true comprehension of the thoroughness and the high ideals of British engineering."[71] Supporting Chiang Kai-shek's political program of economic reconstruction and the embrace of technology and science by Republican China's elites more generally, Middleton Smith became a fervent proponent of publicity promoting engineers and engineering to highlight their important role in transforming Chinese society.[72]

EMBRACING GLOBAL PRACTICES AND NETWORKS

Once civil engineering had become an established academic discipline with a variety of subfields taught in universities and technical institutions, engineering degrees became a new currency expressing professional success and social status in China. According to a "Who's Who" register of Chinese engineers compiled in 1940 by the National Resources Commission, more than 20,000 Chinese had graduated and were working in China as engineers at the time.[73] The register's entries documenting the academic training and employment history of the engineers show a remarkable range of career moves between private and government sectors and a high level of mobility through work in domestic locations as well as abroad.

Not surprisingly, nationwide professional associations for engineers evolved hand-in-hand with the proliferation of academic institutions and their graduates. The earliest, foreign-dominated engineering societies in China were less oriented toward professional and academic exchange but pursued rather one-sided commercial goals. As a long op-ed article in the *North-China Herald* from December 1912 laid out, the British Engineers' Association in China, founded in April that year and registered with the Board of Trade, came into being "for the purpose of promoting the interests of British engineering firms in China."[74] Although the article's author painted a positive future for engineering projects in China, he considered the protection of British equipment manufacturers and the fight against German and American competition for engineering dominance in China the predominant goal. As announced by the London headquarters, the British Engineers' Association opened its China office on August 1, 1913, in Peking on Wangfujing Street, a premier address close to the imperial palace.[75]

Chinese professional associations began to proliferate in the 1920s and early 1930s when larger numbers of Chinese graduates in professional occupations began to seek organized representation. The Chinese Engineering Society and the Engineer's Society of Peiping (founded in 1912) were amalgamated in 1931 to form the Chinese Institute of Engineers. Naturally, the number of members grew with the expansion of the profession nationwide so that the Chinese Institute of Engineers was able to count 2,500 members in 1934 and 8,292 members in 1944 during the Japanese occupation.[76]

With their meetings, lectures, social events, and publications, engineering associations became a platform for continued knowledge exchange and engagement with international colleagues that went beyond engineering-related issues. The Association of Chinese and American Engineers, with headquarters in Peking, became the most important cross-cultural organization. In 1921, it began to publish its own journal as a forum for Chinese and foreign engineers working in China to discuss ongoing civil engineering projects and facilitate knowledge exchange. Annual conventions with a mix of factory visits, paper presentations, and various social events offered the members informal discussions and the opportunity of "learning to know one another better."[77] As the photograph from the association's 1922 meeting in Shanghai reveals (see figure 5.1), the group of engineers was almost evenly

FIGURE 5.1
Meeting of the Association of Chinese and American Engineers in Shanghai, April 1922 (*Journal of the Association of Chinese & American Engineers* 3, no. 4 (April–May 1922): no page number). Unfortunately, I have not been able to identify the woman at the center of the front row.

split into Chinese and Americans participants, with badges identifying several Chinese members of the local committee.[78]

Slowly, the knowledge exchange between Chinese and foreign engineers began to gain traction, with articles featuring Chinese engineering methods in the association's publications. For example, to close a breach in the Yellow River in 1936, Chinese engineers used so-called "sausages" made out of willow branches, hemp ropes, sorghum stalks, and ropes. In the same year, this ingenious method was published in a detailed paper in the association's journal, noting that foreign engineers would have found this task difficult due to their preoccupation with machine labor vis-à-vis manual labor.[79] The discussions of engineering problems under different socioeconomic conditions in the field—more labor but less machinery in the Chinese case—with unconventional yet successful solutions led to a reverse, outbound knowledge transfer that introduced foreign engineers to different but efficient engineering methods.

Apart from the exchange of professional expertise, engineering associations also offered a platform to shape the social perception of the engineer as a citizen working to promote social progress and modernity. For example, Chinese as well as foreign members of the Association of Chinese and American Engineers became actively involved in philanthropic missions, particularly during the 1920–1921 drought and famine crises in North China. The association formed an action group at its annual meeting in Tianjin in 1920, proposing to the Peking government the establishment of a public works program in combination with famine relief. As a result, with the permission of the Ministry of Communications, two stretches of railroad embankment and about 500 miles of county highways in Shandong were built by some 70,000 local workers, whose famine relief in the form of grain and sometimes cash supported a total of about 500,000 dependents.[80]

When the International Famine Relief Commission (IFRC) was established in Peking in 1923, it appointed O. J. Todd, chair of the engineers' association's special committee on highways, as chief engineer. Apart from managing relief measures, the IFRC organized the engineering staff and mobilized foreign engineers to assist their Chinese counterparts in an important task.[81] The IFRC continued its involvement in programs funding the construction of dykes and roads during flood disasters into the 1930s, and Chinese engineers supervised independent road construction programs initiated by several provinces.[82]

The engagement of China's engineering societies in philanthropic and educational projects during the Republican period helped to establish a vision of the profession and its members that went beyond the mere achievement of technological progress. It also introduced a moral vision of the role of the engineer. Similar to journalists, geographers, lawyers, and doctors, engineers were expected to make a positive contribution to both society and the nation as a whole.[83] In that spirit, the Association of Chinese and American Engineers proudly displayed its guiding professional principles on the title page of its journal (see figure 5.2).[84] With the introduction of June 6 as Engineers' Day in Republican China, the Nationalist government emphasized the economic, social, and political impact of engineers and their profession. Among other professional occupations, only journalists were honored in a similar way by Chiang Kai-shek's government.[85]

Of course, this vision of a professional ethos was not specific to engineering in China; there was ample precedence in the European and American contexts. For example, the Association of German Engineers (VDI) was founded in 1856 with the goal of merging "the putative synthesis of industry

JOURNAL

OF THE

Association of Chinese & American Engineers

The Association of Chinese and American Engineers organized in Peking, November twenty-second, nineteen hundred and nineteen, by Chinese and American Engineers, with the sanction and approval of the Chinese Government, has for its object:

The advancing of engineering knowledge and practice,

The maintaining of high professional standards, and

The fostering of a spirit of co-operation and fellowship among engineers.

FIGURE 5.2

Title page of the *Journal of the Association of Chinese & American Engineers*, 1922 (*Journal of the Association of Chinese & American Engineers* 3, no. 4 (April–May 1922): no page number).

and science with other aspirations for middle-class emancipation into a progressive, patriotic-idealistic vision of the future."[86] In Germany, as in China, the shared identity among members of professional engineering associations through their experiences in institutionalized education and formalized training was reinforced by a positive vision for the future of society and nation.

This vision was put forward most boldly in the "Precepts for Chinese Engineers" articulated by the Chinese Engineering Institute at its annual meeting in Wuhan in 1932. In terms of style and content, the six precepts resembled articles of faith and required that engineers not forsake their responsibilities or neglect their duties, not accept excessive remuneration, not deviate from conduct as expected of people in their profession, neither directly or indirectly damage the reputation of the profession and its duties, not compete with inferior methods for business or positions, and not make false claims about the profession or engage in conduct harmful to the dignity of the profession.[87]

Not surprisingly, these guidelines set the highest expectations for proper professional conduct by engineers based on conscientious, honest, and respectful moral behavior, yet at the same time, they conveyed a general sense of pride and place by equating engineering with a profession with "dignity." In reaction to China's war situation caused by the Japanese invasion, the Society of Chinese Engineers rephrased and expanded the guidelines at its annual meeting in Chengdu in 1941. In a spirit of patriotism and sacrifice for the greater good, engineers were now also encouraged to put "their professional plans at the core of the strategy of the shared ideal of national reconstruction" and to pursue "the objective of industrialization and modernization for the rescue of the nation."[88] The guidelines still emphasized the necessity of impeccable moral conduct and high professional standards, but they now also carried a political dimension by mandating that engineers assist the nation in its current crisis and align their professional goals and identity with the reconstruction efforts for the future of postwar China.

The political context of wars and conflict on a global scale in the first half of the twentieth century led to new expectations of the engineer's roles in shaping society beyond nationally defined borders. Commenting on the World Engineering Congress, which met in Tokyo in October 1929, the *FER* opened its October issue with a long article under the headline "The Engineer as Peacemaker."[89] It should not come as a surprise that the United States registered the largest number of delegates (131). However, China's delegation of fifty-one members sent to the congress was astonishing, surpassing by far engineering powerhouses like Germany (thirty-eight delegates), Britain (thirty-one), and other participating nations' single-digit delegations. We can assume that the relatively close geographical distance of Japan to China to some extent counted for the strong Chinese participation in the Tokyo World Engineering Congress. However, the considerable presence of Chinese engineering professionals might also lead to other interpretations: empowered by the progress of engineering as a discipline and profession under the Nationalist government's sponsorship, Chinese engineers were now ready to participate in conferences on an Asian and global stage and engage with their peers at eye level. Alternatively, the fact that the congress took place in Japan, whose aggressive politics had repeatedly created political trauma for China over decades, might have urged the Chinese government to assemble a large delegation demonstrating that China was catching up with Japan's engineering achievements and no longer intimidated by its neighbor's success.[90]

CONCLUSION

Looking at the global impact and success of Chinese engineers in the early twenty-first century, the relatively short and complex evolutionary trajectory of engineering in China might come as a surprise to readers. Nevertheless, this chapter does not attempt to present China as an exceptional case. In fact, issues related to detrimental political conditions or asynchronous industrial and labor market developments impacted other countries as well, even those where professional engineering developed much earlier.

The case of China demonstrates the paramount importance of institutions in shaping engineering as an academic discipline and profession. Only the establishment of modern universities and schools oriented toward science and engineering allowed for the rapid increase in engineering graduates and their employment in industry, commerce, public infrastructure, and government bureaucracy. China's path to modern engineering also emphasizes the role of the state as an important factor in institution-building, with both negative and positive implications. Although the imperial government was incapable of adequately introducing domestic engineering education, after 1911, the new Chinese nation-state actively promoted the expansion of engineering in various capacities. Of course, engineering as a discipline and its professional norms aligned perfectly with the government's modernization agenda and the desire to express national sovereignty through the production and diffusion of engineering knowledge produced by Chinese engineers. This close relationship became problematic when the Nationalist party increasingly harnessed engineers as ambassadors for their political programs and Chiang Kai-shek's vision of the Republic.[91] However, many examples in this chapter also have shown that despite the goal of becoming less dependent on foreign expertise, global connections continued into the 1940s through interactions with foreign academics, study abroad, and participation in international associations and conferences.

As this chapter argues, the impressive mobility of Chinese engineers in the course of their academic and practical training reflects, to some extent, strategies to compensate for institutional deficiencies in engineering education during its early stage. However, whereas studying abroad around 1900 was the only way for Chinese students to obtain an engineering degree, their visits abroad in the 1920s and 1930s mostly involved postgraduate studies or government-sponsored secondments. Based on these shared educational experiences overseas, it is not surprising that engineers formed strong networks among their Chinese peers but also with foreign colleagues employed in China. Chinese engineers succeeded in incorporating global engineering standards by adapting norms and practices to China's specific social and political setting. In addition, they conducted knowledge exchange via cross-cultural associations and publications that over time led to increased knowledge diffusion and a considerable degree of institutional convergence in the evolution of the engineering profession in China.

The analysis in this chapter demonstrates the engineers' integration into different local, regional, and national labor markets across Asia with distinct political and social environments directing the flow of engineering talent. Different political and economic scenarios in East and Southeast Asia, including semi-colonial ventures in China, state-sponsored projects in Meiji Japan, and industrialization driven by political and economic imperialism in Manchuria, offer a fascinating lens on how engineers performed roles and assumed identities that went beyond their technical expertise. Although most of the engineers encountered in this chapter displayed only modest historical agency as individuals, their participation in large public construction projects or famine relief efforts gives them agency as practicing professionals with a focus on solving engineering problems at the local and regional levels.

However, as symbolic representatives and promoters of modernization, economic prosperity, and political and scientific sovereignty of the Chinese nation-state, they indirectly also demonstrated engagement at the national and global levels. We should not forget that the father of the Republic, Sun Yat-sen (1866–1925), had considered himself a political and social engineer pursuing the goal of construction that would apply to everything from railroads to mines to political organizations and the nation itself.[92] As a result, the institutional trajectory of engineering education and professionalization cemented the role of engineers as China's future technocratic elite that would shape the postwar development of the People's Republic of China and the Republic of China on Taiwan.[93]

6 ENGINEERING KNOWLEDGE AND THE EMERGENCE OF CHINA AS A GLOBAL LEADER IN ANTIMONY PRODUCTION

MARK HENDRICKSON

The American [antimony] market was largely controlled from London prior to the war, but now the Chinese and Japanese brands have become so firmly established, our market is hardly likely to return to the dominance of British firms.
—*Minerals Resources of the United States, 1918*, 35

[Chinese antimony producers] have erected several plants which are the last word in manufacturing methods. The profits they are making enable them to engage the best talent in the world and to install any new device the moment it is known.
—*Engineering and Mining Journal*, June 15, 1918, 1088

During the years leading up to World War I, established global producers of metallic antimony faded, and a new producer prevailed, dominating global production for significant portions of the twentieth and twenty-first centuries.[1] At the dawn of the twentieth century, global leaders France and Italy each produced more than 1,700 metric tons of antimony metal, but by 1909, Italy's production dwindled to almost nothing while France's rose to 5,444 metric tons.[2] One year earlier, China emerged as a global force, producing 9,356 metric tons of the metal, and by 1912, China produced a remarkable 13,313 metric tons of metallic antimony. At the peak of its production in World War I, when demand and prices spiked, China exported some 38,539 metric tons of antimony.[3] To put this into perspective, in 1900, the five largest producing nations in the world combined to produce 4,122 metric tons. In China, more than one hundred antimony mining and processing companies formed in this period, but the Wah Chang Mining & Smelting Company (WCC) of Changsha in Hunan province emerged as the dominant producer in China and, by extension, the world.[4] By World War I, the WCC and the Hunan province had, in a remarkably short period of time, emerged as the world's most important source of not just antimony ore but, importantly and remarkably, refined metallic antimony.

In recent years, leaders in the United States and beyond have looked warily at China's dominance in the production of rare earth elements (REEs) and other critical materials. In the wake of a price spike brought on by Chinese cutbacks in REE exports following a maritime dispute between Japan and China in 2010, James Clapper, director of National Intelligence, testified before the Senate Select Committee on Intelligence that "China holds a commanding monopoly over world REE supplies, controlling about 95 percent of mined production and refining," "China's dominance and policies on pricing and exports," he warned, "are leading other countries to pursue mitigation strategies, but those strategies probably will have only limited impact within the next five years and will almost certainly not end Chinese REE dominance."[5] This chapter looks back to the early twentieth century when China emerged as the global leader in antimony extraction and refining. I cast this emergence as a global rather than national story and one that raised very public concerns internationally about the significance of one nation (the United States in the case of this chapter) becoming dependent on another for its supply of a critical material.

This analysis sheds light on several themes central to this volume. The antimony ores mined and processed in Hunan entered a global marketplace, and the mining engineers who are the central characters in this chapter proved critical to the construction of the infrastructure that facilitated this global movement. Their ability to effect change turned to a large extent on the knowledge and expertise they picked up as they developed a global body of knowledge rooted in their education and training on three continents. Mobility—mobility of people, ideas, practices, technology, and minerals—then, is a critical theme in this chapter. This is not a story of the one-way diffusion of ideas and knowledge. Instead, it is one of interdependencies, global networks, and exchange as Chinese, US, and European experts and leaders learned from one another the exact moment when industrializing nations—specifically and significantly the United States in this case—for the first time became acutely aware of the interconnected nature of a globalizing economy that required seemingly insatiable amounts fuels and minerals that were not equitably distributed in the Earth's crust. More broadly, mining has always been fertile ground for scholars of science, technology, and labor and social movement history, but recently, a number of works have focused attention on specific nonprecious metal and non-fuel-mined commodities. A study of antimony complements the work of historians who have focused on more widely understood and studied elements and materials such as tungsten, REEs, bauxite, manganese, tin, and copper.[6]

The story in this chapter unfolds in four acts, beginning with an analysis of the resource management and human capital investment strategies that laid the foundation for the emergence of the antimony industry in China in this period. The decision to acquire Western technological training in mining engineering and metallurgy—combined with efforts in Hunan to secure control of extractive industries—is central to understanding Chinese dominance of the industry by World War I. These efforts would have been less effective without the invention and refinement of a new processing technique applied to antimonial ores. The second section—with its focus on the Herrenschmidt process and its implementation in works in Hunan—examines the technology put into place by Chinese mining engineers. The third section of the chapter examines the emergence of the Wah Chang Mining and Smelting Company in Changsha and the Wah Chang Company in New York. While the mining company was short-lived, the Wah Chang Company of New York proved important to US economic and military history well into the twentieth century. Its leader, K. C. Li, would become a US citizen and serve in a number of diplomatic, defense industry, and business positions and have a long career that spanned much of the twentieth century. The final section shifts attention to the United States and to efforts by leading mining engineers and economic geologists to make sense of what it meant to be dependent on China for a critical element like antimony.

Rare earth elements and other prized materials like cobalt and lithium receive prominent attention from scholars and policymakers, and while antimony rarely makes front-page news and has gone almost unremarked upon by US historians, it seems to almost always lurk just below the surface. In terms of global consumption, REE and antimony are consumed at roughly similar rates—global consumption of REE is difficult to estimate but generally ranges between 140,000 and 170,00 tons.[7] By comparison, in 2016, global consumption of antimony stood at 188,000 tons.[8] Like REEs, antimony is not actually rare, but also like REEs, it often appears in lists of critical and/or strategic minerals for which there are risks of shortages or supply disruptions. A 2020 study by the European Commission identified antimony as critical based on its economic importance and supply risk.[9] Since the early twentieth century, manufacturers have employed antimony in lead-acid car batteries, shrapnel, and printing, among other applications. Since the 1970s, usage has shifted toward halogenated flame retardants that use antimony trioxide as a synergist.

As we will see, the current concern over the supply of antimony is nothing new.[10] It stretches back to the World War I era when the US Geological Survey's (USGS's) Edson S. Bastin noted that the United States "has always been dependent mainly on foreign sources for its supplies of antimony."[11] More recently, in a 2004 USGS report, W. C. Butterman and J. F. Carlin noted that for "most of the 20th century, the antimony market was one of the more unstable metal markets."[12] In 2017, the USGS warned, "Although antimony occurs in several different deposit types of various ages, ore-grade concentrations of antimony are not common, and economically minable deposits of stibnite [the primary mineral from which most of the mined antimony comes from] are generally small and discontinuous."[13] Growing concern led the Trump administration in December 2017 to issue Executive Order 13817—A Federal Strategy to Ensure Secure and Reliable Supplies of Critical Materials—that charged the secretary of the interior and the secretary of defense with the task of developing a list of critical minerals and mineral groups. Antimony appeared amid the list of thirty-five.[14]

The periodic table is largely divided up between metals and nonmetals, but antimony is one of seven metalloids—along with boron, silicon, germanium, arsenic, tellurium, and polonium. The properties of metals include high electrical and thermal conductivity along with high melting points, strength, malleability, and the ability to stretch under tensile strain. With the exception of carbon—and, more specifically, graphite—nonmetals are poor conductors of electricity, and they are brittle. Metalloids don't have a single defining character, but they do all have properties somewhere between those of metals and nonmetals. Antimony, for instance, has a metallic appearance, but it is brittle and does not effectively conduct electricity. In alloys, it helps resist corrosion and adds strength and hardness.[15] Human use of antimony and antimony sulfides reaches back at least six thousand years. Early applications included the Egyptians' use of stibnite (Sb_2S_3) as an early cosmetic and Chaldeans' application in containers. Gutenberg presses employed it to enhance hardness and fluidity of the printing press, and by the sixteenth century, bellfounders added it in the casting process to enhance the tone of bells. Before the twentieth century, industry employed only modest amounts of antimony, with a limited number of applications. Manufacturers used an of alloy of antimony, copper, and tin to make silver-colored household goods marketed as Britannia metal. The element also found important applications in an alloy of tin, lead, and sometimes copper and arsenic that made up different types of Babbitt metals.[16] Antimony is found in more than one hundred different minerals, but stibnite (Sb_2S_3) is the most common mineral that is mined for its antimony content. The name antimony comes from the Greek words *anti* (opposed) and *monos* (solitude), which allude, correctly, to the fact that the element is rarely if ever found in nature alone, preferring instead chemical bonds with metals such as gold, sulfur, copper, or lead.[17] More recent industrial applications feature antimony in batteries and flame retardants.

RESOURCE MANAGEMENT AND HUMAN CAPITAL DEVELOPMENT

Any successful effort to develop extractive industries in China or any other country requires, among a host of other preconditions, a legal framework that facilitates extraction and processing and the expertise and technology necessary to identify, extract, and refine ores into a product that can be sold. In the case of China and the development of its antimony industry, this process began with Hunan province taking steps to develop and gain a measure of control of its domestic mining industry. Equally important was an early twentieth-century effort to send young potential mining engineers and metallurgists to the United States and Europe to gain training in new techniques and bring them back to China to go to work on local natural resources. As developments later in this chapter make clear, this is not a story of one-way knowledge transfers or diffusion. The knowledge gathered in the West and then refined in exploration, mining, and processing efforts in Hunan quickly circled back to the United States in improved practices and a deeper understanding of antimony resources that helped shape US mineral policy well into the twentieth century. The Chinese mining engineers who had taken their training in the United States and Europe would, in the World War I era, shape how leaders in the United States made sense of the nation's dependence on Chinese antimony.

China's—and, in particular, Hunan province's—rise as the dominant producer of antimony in this period took place against the backdrop of some major changes in the relationship between the Chinese government, Hunan provincial authorities, and foreign mining interests. According to historian Shellen Xiao Wu, Western interests in Chinese mines intensified in the last decades of the nineteenth century, but it was only after the Sino-Japanese War (1894–1895) that "the floodgates of European demands for mining concessions" opened up and a "great underground race took shape in China" that expanded imperialism to the interior and Earth's substrata.[18] In the decade that followed, an assortment of US and European interests vied for concessions in China, but in 1905, the "tide turned against foreign concession hunters." "One by one mining concessions across the country," Wu notes, "once the cause for celebration in boardrooms from London to New York, returned to Chinese ownership."[19]

In Hunan, local policies and developments dovetailed with these larger nationwide efforts. According to Wu, Hunan governor Chen Baozhen surveyed the province and concluded that the mountainous geology, poor soil quality, and abundant mineral resources made certain parts of the province more suitable to mining as opposed to agriculture. In 1896, Chen founded a mining bureau to promote mining activity in the province and raise money for investment. By 1903, the provincial leaders developed a system of mining regulation and worked with Hunan gentry to develop the industry. Wu concludes that "the Hunan gentry developed the mineral wealth of the province successfully without direct foreign investment or loans."[20] Angus W. McDonald Jr. provides further historical precedent for these efforts, writing that between 1885 and 1898 and under Chen, "several antimony mines were opened by members of the gentry."[21]

As Angus McDonald noted, iron, coal, lead, manganese tungsten, antimony, silver, and gold were found in the mountains of Hunan, but between 1900 and 1920, "tin, lead, and—above all—antimony were of prime importance to the economy and province."[22] Chen, according to McDonald, encouraged the establishment of gentry-operated mines both as a means of exerting control over a dangerous group of independent-minded miners and as a way to preempt foreign exploitation of Hunan's mineral birthright.[23] Mining owners operated the mines, in this system, but until 1905, the mining Bureau of Changsha was the only entity that they could sell to.

One family—the Liang family—became involved in antimony mining relatively early and in many respects spearheaded the effort to develop the antimony industry in Hunan. In 1894, Liang Huankui, who was influenced by Chen's emphasis on mining, started work as a correspondent for the Hunan Mining Bureau.[24] It is in this capacity that he, working with the second brother Liang Huanzhang, discovered antimony deposits in the region. Exploration continued, Chen ordered workers to extract the antimony, and he established a provincial mining bureau in Changsha. The mines were not profitable initially, and Chen's replacement, Yu Liansan, invited private businesses and contracted the work out. In 1899, the Liangs, with the financial support of Huang Xiuyuan and Yang Shuchun, founded the Jiutong Company. Jiutong experimented with indigenous smelting methods and tried to apply some Western processing methods, but the mines' remoteness and complications with government policies that restricted how antimony could be sold and transported made the effort unsustainable under current market conditions.[25]

But market conditions change, and in 1906, a dramatic increase in the price of antimony brought on by the Russo-Japanese War (1904–1906) and increased demand for armaments employing antimony allowed the Liang family and Jiutong to accumulate the resources that would allow the firm to modernize.[26] In this same period, Huang Xiuyuan and Yang Shuchun withdrew their shares from the company, which left it solely controlled by the Liang family.[27] In 1908 and on the basis of the Jiutong company, Liang Huankui established the Hunan Wah Chang Smelting Company (*Hunan huachang lianti gongsi*, also known as the Wah Chang Mining and Smelting Company, hereafter WCC) and appointed himself as chairman.[28] Understanding the history of the WCC and its significance to the emergence of China as a mineral powerhouse requires that we first consider the work of a number of young Chinese mining engineers who traveled to the United States and Europe for training in metallurgy and economic geology.

Economic historian Gavin Wright has documented how, in the US case, mineral endowment alone did not guarantee natural resources production.[29] In both the US case and in China, education—and, more specifically, a broad-based and international education often subsidized in different ways by public policies—allowed skilled technicians to identify, extract, liberate, and separate the worthless waste material from the minerals that would transform the twentieth-century world. Although Hunan's antimony mining and processing industry did at least initially develop independent of foreign direct investment, when it came to the knowledge and expertise necessary to refine antimony ore into metallic antimony, the story is quite a bit more complicated. Increased demand in World War I–era war industries, along with a decline in production in France, relatively rich reserves in China, and lower labor costs, helps explain China's standing in the global antimony industry, but these variables tell only part of the story. In fact, the superior processing techniques adopted by Chinese engineers and mining companies as compared with their North American and European counterparts cemented Chinese producers' position.

Explaining this rapid and globally significant change requires special attention to the ways in which the transfer of technical knowledge moved across space and time. This is especially true in the late nineteenth and early twentieth centuries when new technologies, methods, and tools allowed for a truly dramatic expansion in global mineral production that provided the raw materials necessary to fuel and build the second Industrial Revolution. New processing techniques, such as flotation, combined mechanical engineering and surface chemistry in ways that superseded long-held practices that used size and density to separate minerals and waste material. In gold and silver processing, cyanidation allowed for the extraction of extraordinarily low grades of gold ores. In extraction, block caving, open-pit processes, and steam shovels were among the many techniques and technologies that revolutionized extraction methods in ways that increased mineral production and left behind physical and environmental scars on the planet.

Antimony processing in Hunan province benefited from some of these practices, but the major breakthrough that helped China dominate global production came in a French chemist's lab in Paris. Before stibnite ores could be pulled from the earth, transported to processing facilities, and refined into metallic antimony, a different sort of mobility took place—one that involved the transport of ideas and knowledge concerning new processing techniques. Five young Chinese mining engineers were, in no small measure, responsible for bringing this change about. Liang Huanyi (the third youngest brother, often referred to in the Western press as H.Y. or H. I. Liang), Liang Huanting (the fifth of the Liang brothers), Liang Huanzhang (H. C. Liang in the Western press), C. Y. Wang (also known as Chung Yu Wang in the Western press or Wang Chongyou in *pinyin*), and K. C. Li (also known as Kuo Ching Li in the

Western press or Li Guoqin in *pinyin*) drove a transformation of antimony processing in early twentieth-century China and helped to propel China to a global leadership position in the production of antimony and tungsten. All five pursued technical education in the West.

Li, the Liangs', and Wang's study in France was timely in that it allowed them to gain firsthand knowledge of pathbreaking work in antimony processing, but it also occurred at a moment of transition in French antimony mining and processing. France had been one of the few nations that both consumed antimony in industry and produced its own metallic antimony from domestic ore supplies. By the time these young Chinese metallurgists and mining engineers arrived in France, the La Lucetta mine in the province of Mayenne was near exhaustion, and by the early 1920s, production in France of antimony ore had largely ended.[30] But the long history of antimony mining and processing in France meant that some of the leading mining and metallurgical experts in the world still worked in the country.

In her recent work on Chinese students' study in the United States, historian Weili Ye provides a broader context for understanding Wang, the Liangs', and Li's work abroad. According to Ye, in the wake of the Boxer Rebellion, the Qing government embraced Western technical training for Chinese students. Studying abroad was not new, but according to Ye, this early twentieth-century effort amounted to a second wave of a longer foreign study movement. Various types of government sponsorship underwrote this new phase. It came, Ye argues, to be embraced by "both state and society," and it "was marked with a self-conscious willingness on the part of the students to acquire broad learning from the West."[31] China's loss to Japan in the First Sino-Japanese War (1894–1894) gave even greater urgency to these efforts.[32] The story of these metallurgists-in-training dovetails nicely with Ye's broader analysis of the far-reaching significance of these efforts to bring new sources of knowledge to bear on various problems and concerns in China.

Born in Hong Kong in 1879, Wang took a degree in mining from Beiyang University in Tianjin in 1895. In 1901, and with support from the Chinese government, he studied mining and geology at the University of California, Berkeley and Columbia, where he earned an MA in 1904 and studied with James Furman Kemp.[33] By the time Wang arrived in the United States, the University of California and Columbia University emerged as two of the most important institutions of mining engineering in the country, with Columbia having the longer and more established position in the world of mining engineering education. At Columbia, he chose to study with Kemp, who left an indelible impression on many of his former students and a lasting legacy as a teacher and one of the early leaders in the emerging field of economic geology.[34] In 1941 and some fifteen years after Kemp's passing, his former student Thomas Thornton Reed dedicated his book, *The Development of Mineral Industry Education in the United States*, to Kemp, reverently describing Kemp as a mining engineer "who became an economic geologist of international reputation." Kemp, Thornton continued, "not only contributed much to the development of that field of applied science, but, as a teacher, suppled inspiration and enthusiasm to young men in sending them out to play their part in utilizing mineral substances to promote the welfare of mankind." In 1909 and just a few years after completing his studies at Columbia, Wang published his first book, *Antimony*, which was the first English-language book on the subject and one that went through a number of editions and is still cited to this day by experts in the field.[35] He too honored Kemp, writing in the dedication that under Kemp's guidance he had "learned the value of scientific research."[36]

After completing his mining degree at Columbia, Wang stayed for an additional year to do postgraduate work in economic geology. In 1905, he left for Britain, France, and Germany for advanced studies, now as a representative of the Wah Chang Mining and Smelting Company of Changsha. His training in some respects culminated in April and May 1908 with his work at Laboratoire de recherches industrielles pour tous minerais at Le Genests, France, with Henri Herrenschmidt. Between April 5 and May 26, 1908, he worked with Herrenschmidt on a series of experiments on a volatilization process for refining antimony that would earn Herrenschmidt patent number 386,107, which helped to alter the history of antimony production globally.[37] By 1909, he returned to China but soon returned to France with Liang Huanyi to purchase the Chinese patent rights to the Herrenschmidt process. Wang would play a central role in the construction of China's first modern antimony-processing facility in Changsha.[38] Later in his career, he held a number of government and mining posts, many of which were in Canton (Guangzhou). In 1943, Wang returned to the United States, where he served as a technical adviser to China's Ministry of Economic Affairs for three years. At this same time, he reunited with Li at the Wah Chang Corporation of New York, where he served as research director until his death in August 1958.[39] In addition to *Antimony*, he coauthored with K. C. Li one of the more important

English-language books on tungsten that also went through multiple updates and editions, with continued relevance to tungsten scholars today.[40]

Following the Liang brothers' paths is more complicated, but Huanzhang, Huanyi, and Huanting all studied in Europe, along with Huanyi in the United States in the first decade of the twentieth century. It was in this period that they learned of the new volatization process for refining antimony ores developed in Herrenschmidt's lab. Huanyi, after studying mining in the United States from 1904 to 1906 and earning an MA from Columbia, left for London, where he studied mining and processing at the Royal School of Mines. Upon learning about this new process, Huanyi also studied under Herrenschmidt in Paris.[41] In 1915, he served on the Chinese Honorary Commission that traveled through the United States to promote Chinese industry. In this same period, he served as president of the Sue Kow Shan (Shuikoushan) government lead mines and the Wah Chang Mining and Smelting Company, both of Hunan province.[42]

K. C. Li is another Chinese-born mining engineer central to this story, and he is the one who left the deepest imprint on US history. In her memoir, *A Daughter Remembers*, K.C. Li's daughter provides an intimate portrait of her father. He came from a farming family outside of Changsha. When he was twenty-three years old, his older brother took work teaching math at a newly opened Technical High School in Changsha (in other contexts mentioned as Hunan Technical Institute). After several weeks of work, he asked his father if K.C., who would have been about ten years old, could join him at the school as a scholarship student and as a domestic helper.[43] K. C. Li attended and worked at the technical school until 1901, then left China for London's Royal School of Mines (Imperial College) on a Qing government scholarship. After completing his mining engineering degree in London, he returned to Hunan and took on a number of mining and government positions, including secretary of the Hunan Mining Board, general manager of the Kiangwah (Jianghua) Government Tin Mines, general manager of the Hsiao Ku San (Shuikoushan) Government Mines, and codirector of the Hunan Mining Board.[44]

Li would go on to have a long and distinguished career in the United States—and my larger book project looks more closely at his career—where he became better known as the "Tungsten King" for his expertise on the valuable metal. In 1993, he became the 108th individual inducted into the US National Mining Hall of Fame in Leadville, Colorado.

Wang, Li, and the Liangs developed parallel intellectual and training paths, but the Liang family had the connections in China to raise the necessary money to purchase the patent for the Herrenschmidt process and obtain the ore and machinery necessary to make Hunan province a global leader in antimony production. Indeed, Liang Huanyi's pursuit of an education in mining coincided with his family's increasing involvement with mining in Hunan. After the Boxer Rebellion (1899–1901), they took over many of the antimony mines in the province. In a 1915 address before the Commercial Club of San Francisco, Liang described how he worked with Herrenschmidt in Paris, which led to the purchase of the patent for the smelting process for use in China for 100,000 francs.[45] According to US Consul Nelson T. Johnson, when Liang heard about the potential of the Herrenschmidt process to refine low-grade antimony ores, he telegraphed the "Hunanese gentry asking them to raise funds and secure a mine."[46] Yang Du—a national political figure who in 1907 led the Constitutional Government Association[47]—functioned as an intermediary between the Liang family, who lacked the funds, and high-ranking Qing officials Yuan Shikai and Zhang Zhidong, who provided the money for not only the patent but also the machinery necessary to implement modern processing techniques.[48]

THE DEVELOPMENT AND SIGNIFICANCE OF THE HERRENSCHMIDT PROCESS

The major innovation in the Herrenschmidt process turned on how it separated out the antimony in stibnite from the sulfur and gangue materials using volatilization. In his definitive 1988 book on the metallurgy of antimony, Zhao Tiancong described the process as the "first type for volatilization roasting of antimony sulfide materials."[49] Volatilization is a pyrometallurgical technique for processing ores in which a furnace heats up the ore until it vaporizes.

As China emerged as a global leader in antimony production, K. C. Li became one of the leading experts who could explain China's emergence to the mining world and beyond. In September 1917, Li read his widely cited and reprinted paper, "Antimony, Its Metallurgy and Uses," at the annual meeting of the American Institute of Metals in Boston.[50] Previous methods of treating antimony ores—particularly stibnite—relied on a pot system that allowed metallurgists to convert relatively rich antimony deposits into antimony crude via a liquidation process. In the pot system, the upper pot contained the ore and was heated until the sulfur liquidated and trickled into the lower pot, leaving behind a more pure antimony product. This method took advantage of the fact that antimony and sulfur have very different melting points, 630 degrees and 115 degrees Celsius, respectively. This was an energy-intensive system that required not only high-grade ores but also cheap and plentiful sources of energy. If these conditions were met, this system had the advantage of being relatively easy to construct and required less skilled labor than later processes. These early systems did not provide effective means of capturing the volatized material and antimony and other products were effectively burned off into the air.

The basic idea of using volatilization to produce antimony trioxide can be traced at least back to the 1840s, but in 1881, Herrenschmidt began implementing a series of modifications that would significantly improve the efficiency of the process. The initial innovation involved putting low-grade antimony ores in a blast furnace (figure 6.1) for roasting to produce antimony trioxide, which then passed through a series of flues into a condensing chamber with the aid of a fan that pulled the gas through the series of pipes where the gas condensed. The antimony trioxide condensed in the tubes would fall into the bins. Any remaining antimony oxide that made it through the condensing pipes would be captured at the end of the process when it was precipitated out in the water chamber.

The condensation system proved inefficient, and in 1903, Herrenschmidt took out a second patent on a revised process that increased the cooling surface in the condenser, which increased the amount of the antimony oxide that could be deposited. In 1908, he again revised the process in a way that more closely connected the furnace to the condensing pipes by effectively removing the flues. This new process also connected the condensing pipes together with an inverted V piping system. Once through the condensing process, the separated antimony trioxide could then be smelted with a flux to produce metallic antimony.

In addition to the ability to process low-grade antimony ores and prevent losses of antimony oxides that occurred in early processes that relied on liquidation, the advantages of the Herrenschmidt volatilization process over past methods of processing ores were many. For one, antimony minerals contained arsenic oxides, which could be more effectively separated. Other by-products of antimony production, particularly precious metals,

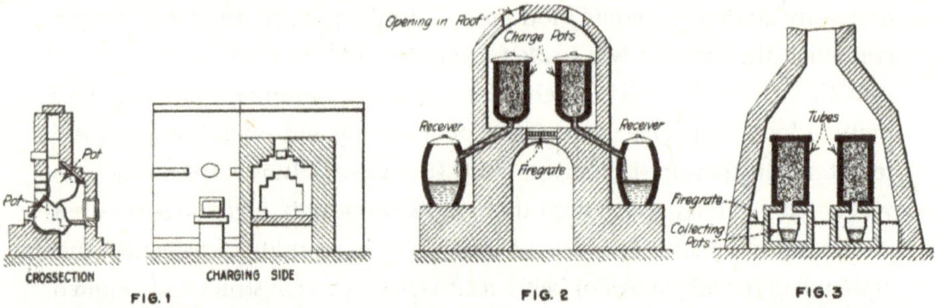

FIGS. 1 TO 3. EARLY TYPES OF LIQUATION FURNACES FOR PRODUCTION OF ANTIMONY CRUDE

FIGURE 6.1
Herrenschmidt apparatus for production and condensation of antimony trioxide. Shown is an illustration of the 1903 Herrenschmidt condensation process, which Hunan producers began using on an industrial scale between 1903 and 1909. *Source*: K. C. Li, "Metallurgy of Antimony," 687.

could be recovered in the residues that remained at the end of the process. And it required less fuel than earlier methods. In the marketplace, the antimony trioxide that came out of the volatilization process could be sent to smelters to produce metallic antimony, but it could also be sold in its current form to paint and enamel producers who used white oxide (antimony trioxide) in their production processes to produce various finished goods.

Completing this process in a way that maximized production, extracted the most antimony oxide possible from low-grade ores, and used the least amount of fuel required highly skilled, experienced, and knowledgeable metallurgists. Fully volatizing the antimony ore depended on the ability of skilled workers to charge the furnace with appropriate proportions of coke and charcoal and to spread out the heat in a way that fully volatized the antimony contents of the ore. Li noted that expert metallurgists could cut down the percentage of fuel "to about 2 to 3% and the yield of the oxide can be increased to 95 per cent."[51]

Other publications drew on Li's work and some provided more insight into the labor conditions in the mine and to the decentralized nature of mining as compared with the greater centralization found in later stages of the refining process. By 1917, the *Engineering and Mining Journal* reported that the antimony industry in China employed around 10,000 people.[52] Three years later, the *Far Eastern Review* gave a close analysis of antimony mining and processing in Hunan. The mining process, as they described it, was highly decentralized. About one hundred companies mined antimony and purchased ore from what would likely be referred to today as small-scale and artisan miners who worked individually or in small groups to extract ore. "Without any system or any idea of the exact whereabouts of the ore," the *Far Eastern Review* ascertained, "the miners merely dig at random," resulting in a mining location that looked "very much like a honeycomb." The *Review* almost certainly slighted the skill required to identify and extract the antimony-bearing stibnite, but the presence of so many individual extraction efforts does speak to the relative abundance of ore and the fact that much of it sat quite close to the surface. The final stages of processing would occur in Changsha with the Herrenschmidt process, but an interesting division of labor occurred closer to the mine site between the one hundred or so "companies" and the small-scale miners who worked in "ore tents." The companies processed the crude antimony in a furnace made up of two pots like the one pictured in figure 6.1. These furnaces employed about 1,000 men and produced five to six thousand tons of crude antimony, but the output was highly irregular and resulted in, at times, around two-thirds of the men standing idle as they awaited more mine feed. The "tents" or smaller-scale miners processed the dust or powder by way of a jigging process that used baskets and then a gravity and water method that further processed the ore that they then sold back to the "companies."[53] Both mined antimony and processed ore required further refining, and it is to the increased refining capacity of Chinese producers that we turn next.

THE EMERGENCE OF THE WAH CHANG CORPORATIONS

This global effort to gather up the leading techniques for refining antimony ore culminated in 1908 with the creation of the WCC, which would soon be the world's dominant producer of metallic antimony. The company, led by Liang Huankui and his younger brother Liang Huanyi, implemented the Herrenschmidt process, but who exactly administered these efforts on the ground remains an open issue. Wang insisted that he took what he learned during his time in the Herrenschmidt lab and implemented it in Hunan antimony-processing facilities, but Liang Huanyi made exactly the same case. It is clear that Wang, Huanzhang, Liang Huanting, and Liang Huanyi all studied in Europe at roughly the same time and learned of the potential of the Herrenschmidt process to refine the antimony ores found in Hunan.

The WCC would quickly become the global leader in antimony production, and while technological innovation played a key role in this process, the policies of President Yuan Shikai early in China's Republican Era (1912–1949) helped ensure the company's global dominance at least for a time. The WCC initially had a monopoly on the use of the Herrenschmidt smelting process, but other antimony producers were able to compete using other processing methods. After the fall of the Qing government in 1911, the monopoly was extended to debar "any other enterprise, whether native or foreign" from smelting, refining, or exporting antimony metal from the province.[54] In his 1918 book on *Mineral Enterprise in China*, vice chairman of the Peking British Chamber of Commerce, William F. Collins, reported that the German and Japanese governments protested these actions as a violation of treaty agreements that forbade this sort of monopoly control. Despite these protests, President Yuan Shikai extended the Wah Chang Company's original concession to 1934.[55]

In fact, the government-sponsored monopoly guaranteed initially by the Ministry of Commerce and President Yuan proved difficult to enforce. In her short history of the WCC, Liang Huankui's daughter Liang Peisu described how WCC leaders initially wanted to concentrate all processing in Changsha, but local miners objected to having to transport ores such a long distance. The highest-quality antimony ores came from the Xinhua mining district west of Changsha. Miners convinced the WCC to build a smelter in nearby Xikuanshan. The diffusion of smelting technology spread well beyond the reach of the WCC, and soon different processing companies were copying and, in some cases, improving on the company's processing techniques, which included the substitution of iron with bricks and the use of wind power. The diffusion of processing technology and production did not supplant the WCC brand, which marketed Chinese antimony internationally well into World War I, but it did frustrate efforts to centralize extraction and processing in a single entity.[56]

Despite these complications in the effort to rationalize production, state polices, international expertise, and diffusion of mining and processing knowledge combined to create a mining and processing industry with a global reach. By 1909, the London Metals Exchange, for the first time, quoted Chinese antimony. By 1916, the *Far Eastern Review* reported that

China supplied about 62 percent of antimony imports to the United States and that 95 percent of Chinese production came from the mines and smelters in Hunan province.[57] The *Mines Registrar* concurred, noting that between 55 and 65 percent of US imports in the World War I era came from China. This trend only intensified after World War I. By the late 1920s, almost all of the metallic antimony consumed in the United States came from China.[58] And the United States was hardly alone in this dependence. Germany, for instance, relied on Chinese imports to supply between 98 and 99 percent of its antimony and tungsten supplies.[59] In 1920, the *Engineering and Mining Journal* noted that the company "virtually controls the production of antimony ore, regulus, and crude" in Hunan and had a "complete monopoly, granted by the Peking government, for the manufacture of" metallic antimony.[60] US observers noted that while Chinese antimony "suffered from a lack of advertising before the war," by 1920, it firmly established itself in the US market, and "its quality has proved equal to the best English grades."[61]

By World War I, with antimony extraction and processing well in place, the Wah Chang company and Liang family worked further downstream to develop and control foreign markets for the metalloid, and it is here where the story gets less clear and more complicated. The family initially sent Liang Huanting to set up a sales office in New York that would allow the WCC to market its own minerals directly to consumers. K. C. Li joined him as sales manager in New York. In an earlier arrangement that expired in 1915, the WCC sold its product to the Chinese Mining Company (a British firm, and this might actually be Cookson) for a fixed price. This was an arrangement that proved disadvantageous to the WCC, and when the agreement expired, it opened up the opportunity for direct sales in New York.[62] The Liang family was well established and connected, but K. C. Li—who would become the most famous of all Chinese-born mining engineers in the United States in his generation—was not.

K. C. Li's rise was in many respects tied to his wife's family and connections to the mining and metallurgy community in China. Luo Buge's—K. C.'s wife—sister married a son of the Liang family. At some point before World War I, K. C. caught the eye of Huankui (the oldest of the Liang brothers), and the two of them seem to have struck up some sort of friendship. K. C. began working for the Liang family. With the outbreak of World War I, the Liang family asked K. C. to go to the United States to sell antimony.[63] K. C. experienced a lot of discrimination and slights while in New York but eventually

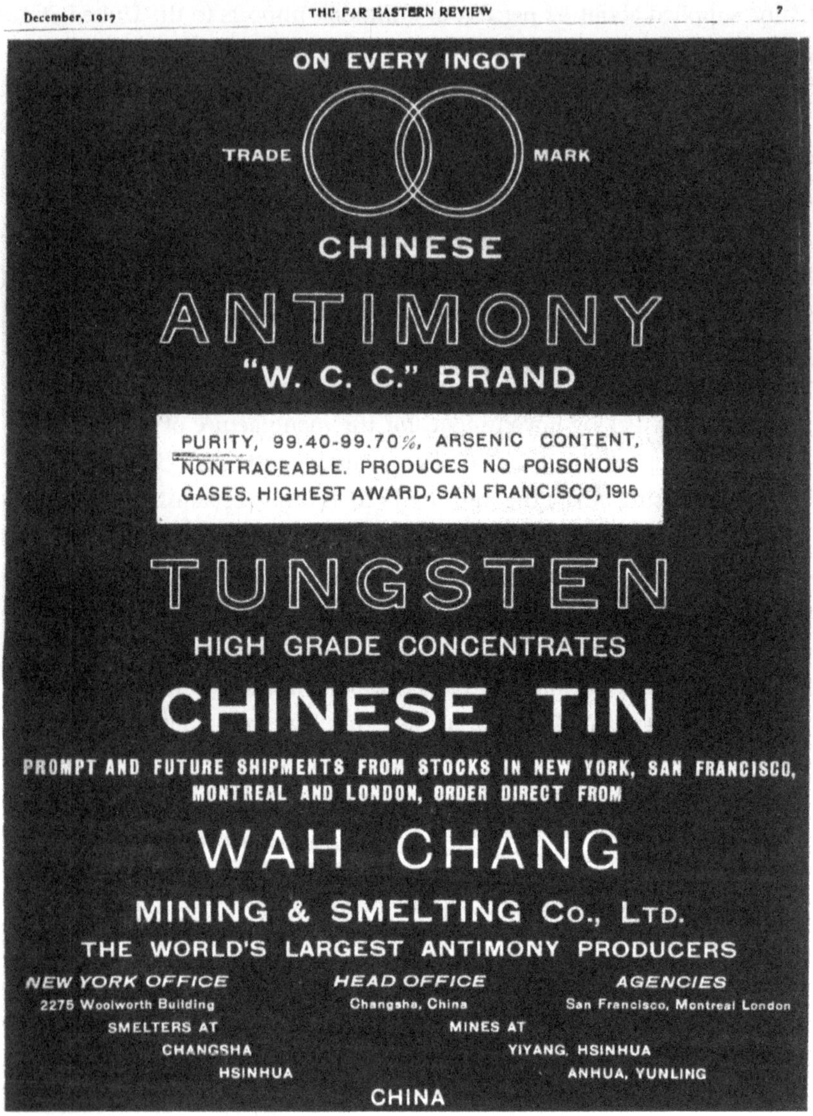

FIGURE 6.2

Advertisement for Chinese antimony and tungsten, 1917. Advertisement from December 1917 edition of *Far Eastern Review* highlighting the purity and availability of antimony as well as tungsten and tin.

won over some of the traders with whom he dealt. In early 1918, the Wah Chang Mining and Smelting company announced that the Wah Chang Trading Corporation of New York would have "the exclusive right to sell antimony, tin, tungsten, zinc, lead and other products of the Wah Chang Mining and Smelting Company."[64] At that point, K. C. was well established in US mining and policy circles, and the Wah Chang Trading Corporation of New York had a strong footing that would allow it to prosper until the 1960s. The fate of the Wah Chang Mining and Smelting Company in China—which lies beyond the scope of this chapter—was less certain.[65]

By World War I, the Wah Chang Mining and Smelting Company claimed in advertisements to be the world's largest antimony producer (figure 6.2).[66] In 1917, the company reorganized and increased it capitalization to $1,500,000. Its principal offices were located in Changsha and New York, but it had agencies in San Francisco, Paris, South America, and London. The company's most important antimony mine was in Yiyang in Hunan province, but the primary antimony smelter was in Changsha. Between 1916 and 1918, the company imported more modern pieces of mining machinery from the United States, including air compressors and drills, that allowed it to double its production of antimony in just two years. At the Changsha processing facilities, the use of forty Herrenschmidt furnaces, twenty-one Reverberatory furnaces, and two large General Electric–made turbine generators allowed the company to produce some 6,320 tons of antimony metal in 1917 alone.[67]

High prices for antimony globally during World War I buoyed the WCC, but so too did its reputation. Across China, creditors opened their doors to the lucrative company and provided it with short-term loans. With these loans and its profits, the WCC issued company script to its employees, which made its way into the economy due in part to a lack of confidence in Hunan's local official currency. The ability to pay script—rather than official currency—to employees further inflated the company's balance sheet.[68] The result, as McDonald describes it, "was a great expansion" in the holdings. In and outside of Hunan province, the company purchased tin, coal, iron, and tungsten and eventually controlled some 500 mines.[69]

World War I laid bare the US dependence on China for antimony, and many people and organizations took note. By the World War I era, China controlled more than 60 percent of the global production of antimony, and the Bureau of Mines estimated that about 90 percent of Chinese production came from the region around Changsha. Control of the production of antimony in the region, the Bureau concluded, rested with the Wah Chang Mining and Smelting company that now had a selling agency of its own in the United States. Chinese antimony, by the early 1920s, "has become firmly established in our markets," the Bureau concluded.[70] Business and professional organizations concurred with the Bureau's assessment. When an owner of a US smelting company that got a foothold in the industry during World War I contemplated the future, he lamented that Chinese producers had "erected several plants which are the last word in manufacturing methods." The ability to "engage the best talent in the world" and to "install any new devices at the moment it is known" further undermined any hopes that the United States could compete in the global marketplace with imported sources.[71] By mid-1917, the *Engineering and Mining Journal* conceded that China was the world's largest producer and "practically controls the market."[72]

CONFRONTING MINERAL SCARCITY IN THE WORLD WAR I ERA: THE CASE OF THE UNITED STATES AND ANTIMONY

During and after World War I, policymakers and the American public worked to size up the new problem of resource scarcity and dependence. In more recent decades, resource dependence has emerged as a major issue in light of US imports of fuel commodities and various minerals such as cobalt, lithium, and rare earth elements, but in the World War I era, this was still a problem just coming into view and not one that impacted all minerals. In fact, between 1900 and 1929, the United States produced about 96 percent of the minerals it used, and when it came to commodities critical to economic development, such as copper, lead, zinc, petroleum, phosphate, and sulfur, the United States actually served as an exporter.[73] Yet, as in the case of antimony, over the course of the early twentieth century, US economic growth became increasingly tied to the import of some raw materials for which the nation had few, if any, domestic mineral reserves.

During the war, the responsibility for coordinating war production fell to the Bernard Baruch–led War Industries Board (WIB). Baruch organized a Raw Materials Division in the late fall of 1917 and asked the mining engineer Pope Yeatman to head its Non-Ferrous Minerals Section, which worked on supply issues for antimony as well as copper, lead, zinc, aluminum, nickel, and mercury.[74] Yeatman had joined Baruch's personal staff weeks before the formation of the section.[75] He came to the WIB as one of the more respected

American mining engineers in the world, having had tremendous success as a close associate of John Hays Hammond in South Africa. When Hammond took work as a consultant for the Guggenheims in the early years of the century, Yeatman played a central role in the development of the El Teniente underground copper mine in Chile as well as some of the open-pit operations in other mines. He was also an important figure in the introduction of flotation in processing low-grade copper-sulfide ores.[76]

The combination of his global reputation as a mining engineer and experience with the WIB gave Yeatman's thoughts on the mineral scarcity problem significant weight in the uncertain times surrounding World War I. For instance, Yeatman's testimony before the US Senate's Committee on Mines and Mining in May 1918 focused in part on antimony and reveals a real-time, wartime effort by policymakers to puzzle through the relatively new and knotty issue of resource dependence. At first blush, the issue seemed straightforward, if a little bit surprising. When asked by Committee Chair and Nevada Senator Charles B. Henderson—who would later head the Reconstruction Finance Committee's Metal Reserve Company—what Yeatman's committee was doing to encourage the production of antimony, he replied that nothing was being done "for the simple reason that at present time there are such large stocks of antimony in this country which were imported at the time the price was very high."[77] As Yeatman's testimony unfolded, he and the senators came back to the question of whether or not the country's almost complete dependence on China for antimony was a problem. Jones queried, "In view of the familiar conditions at the present time, and in view of the stupendous task which we are now performing, do you think it advisable at this time for us to embark upon the business of encouraging the production of antimony in this country?" Yeatman admitted that this very question worried him. He described a recent situation where his nonferrous metals section worked to arrange the purchase of antimony for the Ordnance Department. When they sought out antimony supplies with critical military applications, they found that "very few people, say 8 or 10," had any sizable supply that made the purchase of this material difficult. "Since taking that matter up," Yeatman fretted, "the question of antimony has appeared more serious to me than it had before."

What should be done evaded Yeatman and the senators, and it was and is a complicated issue. Complete dependence on China was a problem, but it was a problem rooted in the price of refined antimony and not a scarcity of the antimonial ores in the ground. Yeatman's former professor Seely Mudd, who served as a mineral expert for the Bureau of Mines and—along with USGS Director George Otis Smith and Bureau of Mines Director Van H. Manning—joined Yeatman at the hearing, chimed in that domestic ores existed in "California, Nevada, and Utah." If you want to promote resource independence by way of domestic production, they noted, you could place a tax on imports on antimony ores and/or metallic antimony, but that could have a lot of unintended consequences. For instance, a tax on antimony ores would hurt domestic processors that imported antimonial ores from Bolivia and Mexico and produced metallic antimony for industry. A tax on imported metallic antimony might work, but the price needed to guarantee domestic production would be many times the cost of imported antimony. And then there was the matter of international trade relations. China imported US-manufactured goods, and as Senator Jones pointed out, "We have got to get paid for that in Chinese products." The government could establish direct contracts with domestic producers or even stockpile antimony, but much of the current demand was driven by military applications. It might take years to establish domestic antimonial production, and by then, the war might very well be over, and the government would left holding a material that it did not need and for which there was no shortage on the horizon. There were opportunity cost issues as well. Using back-of-the-envelope calculations, Senator Jones estimated that it would cost the United States about $4,000,000 to subsidize domestic production. The senator and Yeatman agreed that this money could be spent on "four good big freightships," any one of which could bring in all the antimony (about fifteen to twenty tons) needed by industry for a given year.[78]

World War I increased the visibility of mineral dependence, but as Yeatman's testimony suggests, a clear characterization of the gravity of the problem—if there even was a problem—proved elusive. During World War I, University of Wisconsin economic geologist Charles Leith served as chairman of the Committee on Mineral Imports and Exports of the US Shipping Board and as an adviser to the War Industries Board. By World War I, he was well on his way to becoming the foremost expert in mineral policy. At the close of the war, he had bold hopes as he headed to Versailles with President Wilson, writing before his departure that he hoped that peace negotiations might build on the wartime work of the Inter-Allied Purchasing Commission—the multinational body that had worked successfully during the war to maintain control over prices of critical materials, including nitrates, tin, tungsten, and platinum. Leith hoped to extend the committee's work into peacetime, arguing that the management of global resources needed to be placed alongside political and military considerations at Versailles.[79] Like many others, Leith left Versailles unsatisfied.

Leith continued to think through these issues—increasingly in the company of mining engineers and other experts with long experience in the field—after the disappointment at Versailles. After World War I, the Mining and Metallurgical Society of America and the American Institute of Mining and Metallurgical Engineers joined forces with leading geologists and government officials to produce an investigation of global supply chains for petroleum and eight metals or groups of metals (namely, chrome, graphite, Platinum Group Metals [PGMs], manganese, mercury, tin, tungsten, vanadium, and antimony). The committees were stocked with a who's who of the mining and economic geological community, including Leith, H. Foster Bain (director of the US Bureau of Mines), Yeatman, J. Edward Spurr, and George Otis Smith (director of the USGS), among others (see figure 6.3).

The work of these committees and subcommittees illustrates the global circulation and mobility of expertise and learning in the mining community. The subcommittee on antimony included two names familiar to our analysis in this chapter: C. Y. Wang and K. C. Li. The subcommittee's report on antimony provided a space for deliberation on the significance of mineral dependence, but given the gravity and very public nature of this issue, others outside this deliberative body weighed in as well. For instance, William C. Redfield, who served as the first secretary of commerce (1913–1919), published the provocatively titled *Dependent America* in 1926. Redfield gave special attention to the critical role of antimony in type-casting in publishing, silverware, batteries, telephone cables, and shrapnel, among other common consumer products and military applications. He warned that this was a metal "we cannot supply ourselves."[80]

That the United States was largely dependent on China for antimony caused concern for Redfield, but the collection of mining engineers, policymakers, economic geologists, and government officials brought together on the subcommittee to study antimony came to a very different set of conclusions. Where Redfield looked with alarm and mineral dependence and recommended a robust role for the government in promoting a vision of resource independence, the antimony group largely dismissed the significance of US

MEMBERS OF THE COMMITTEES

Foreign and Domestic Mining Policy

C. K. Leith, *Chairman*
H. Foster Bain
S. H. Ball
Van. H. Banning
George Otis Smith
A. C. Veatch
Pope Yeatman
H. V. Winchell *

Industrial Preparedness

Arthur S. Dwight, *Chairman*
H. Foster Bain
C. K. Leith
H. S. Mulliken
George Otis Smith
J. Edward Spurr
Pope Yeatman

SUB-COMMITTEES

Chrome

Albert Burch, *Chairman*
E. F. Burchard
Frank Probert

Graphite

Benj. L. Miller, *Chairman*
Edson S. Bastin
Henry A. Wentworth

Manganese

C. Minot Weld, *Chairman*
J. W. Furness
D. F. Hewett
Robert Linton
John A. Mathews
J. V. W. Reynders
Bradley Stoughton
R. A. F. Penrose, Jr.

Mercury

George J. Young, *Chairman*
Murray Innes
F. L. Ransome

Petroleum

A. C. Veatch, *Chairman*
A. W. Ambrose
Roderic Crandall
Van. H. Manning
Chester W. Washburne
David White

Platinum Metals

A. F. Keene, *Chairman*
J. M. Hill
J. A. Schloss

Tin

G. Temple Bridgman, *Chairman*
H. H. Alexander
A. R. Ledoux *

Tungsten

Fred W. Bradley, *Chairman*
O. H. Hershey
A. G. McKenna
A. V. Udell

Vanadium

W. Spencer Hutchinson, *Chairman*
Harold Boericke
F. L. Hess

Antimony

Guy C. Riddell, *Chairman*
H. Foster Bain
J. W. Furness
F. C. Schrader
K. C. Li
C. Y. Wang

Publication Committee

Donald M. Liddell, *Secretary*, M. M. S. A.
H. Foster Bain, *Secretary*, A. I. M. E.
F. F. Sharpless, *Secretary Emeritus*, A. I. M. E.

* Deceased.

FIGURE 6.3

Committee to Investigate Mineral Supply Chains, highlighting the participation of Chinese engineers. *Source*: Committee membership list from the last page of *International Control of Minerals* (Baltimore: The Lord Baltimore Press, 1925).

dependence. They found that China produced 90 percent of the global supply of antimony, and the US produced next to nothing, leaving it entirely dependent on foreign supply chains. Yet, the report concluded that this was nothing to worry about and that the present "*laissez faire*" policy—as they termed it—on antimony production and imports worked just fine.[81] Interestingly, the antimony report was the longest of any in the volume, which raises questions about the need to spend so much time and effort on something that was not a problem.

Part of the explanation for the length of the report and its confidence in current supply chains can be found in the global perspective that committee members brought to the topic. The subcommittee that focused on antimony brought together national and international experts with experience that spanned institutions, industries, continents, and nationalities and exemplifies that sort of mobility of ideas, people, and knowledge critical to the history of mineral extraction and policy. Though Yeatman had led the section of the War Industries Board focused on antimony, he served on both of the two major committees and left the study of antimony in the capable hands of Guy C. Riddell, who chaired the committee along with H. Foster Bain, J. W. Furness, F. C. Schrader, Li, and Wang. Wang and Li have been discussed earlier, but the other members of the committee each brought unique perspectives to the study that merit consideration.

Guy C. Riddell, who chaired the committee, would have been intimately familiar with the issue of antimony supply chains and public policy. At the time of the investigation, he served as the head of the new Federal Mineral Division with the Bureau of Foreign and Domestic Commerce. Before taking on that position, he worked as a mining engineer and metallurgist for the Wah Chang Trading Corporation, which, of course, was the major source of US imports of antimony.[82] He became a specialist of sorts on critical and scarce metals and would later serve as the chairman of the AIME Committee on Rare Metals and Minerals, and he worked in Australia and Russia. His career also included a stint with the US Tariff Commission, where he served as an expert on metallurgical and mining problems.[83]

Bain had considerable experience in government and industry and brought additional hands-on experience in China to the table. Born and raised in Indiana, he studied economic geology and mining engineering at Johns Hopkins University and the University of Chicago, where he earned a PhD. Between 1916 and 1925, Bain joined John Wellington Finch's investigation of the mineral resources of China organized by the New York Orient Mines Company (a subsidiary of the American International Corporation). In the mid-1920s, Bain, Leith, and others made the case that while China might have antimony and tungsten deposits of global importance, it lacked geological supplies of iron and other metals necessary to achieve industrialization on par with more developed nations.[84] He served as the editor of the *Mining and Scientific Press* in San Francisco before moving to London to take on the same role with the *Mining Magazine*. During World War I, he worked with Herbert Hoover on Belgium Relief. After the war, he conducted a number of investigations in Africa and China. He followed Hoover into government service and would work for the USGS and the Philippines Bureau of Mines. As the report went to press, Bain transitioned from the US Bureau of Mines, where he served as director from 1921 to 1924, to the American Institute of Mining and Metallurgical Engineers, where he served as secretary from 1925 to 1931. He would also succeed Leith as the sixth president of the Society of Economic Geologists in 1926. He remained in the Philippines during World War II. While there, Japanese forces captured and interned him at Santo Tomas University prison camp for two years.[85]

Furness and Schrader rounded out the committee. J. W. Furness, who served as the head of mineral statistics for the Department of Commerce and Bureau of Mines, brought additional government expertise and experience.[86] Schrader joined the USGS in 1896 and spent nearly forty years with the Survey. His work was largely confined to the continental United States and Alaska, but between World War I and 1930, he was the Survey resident expert in antimony, writing in most years the Survey's annual entry on antimony for its annual *Mineral Resources* publication.[87]

On the face of it, the supply chain for antimony would seem problematic for US consumers in the decade following World War I. Global production of the metalloid stood at 20,000 tons, and US industry consumed about half of that production, but US mines produced only a negligible amount of metallic antimony. Potentially even more problematic, the world relied on one nation, China, for nearly 90 percent of global demand.[88] But, of course, the committee did not see it that way. In addition to the nature and experiences of the committee members, there were several other reasons that explain what would seem a cavalier attitude toward antimony. First, the committee pointed out, though China produced almost all of the antimony consumed in the world, it used relatively little and was thus dependent on foreign consumers for its market. What is more, until the 1920s, only the highest grade of easily accessed antimony ores had been extracted in China, which left lower-grade ores available in the future. A close look at Chinese production revealed an industry that had plenty of room to increase its efficiency, particularly in the processing stages, and little in the way of an internal market for the finished product—usually metallic antimony.[89]

Second, although wartime demand for antimony surged and prices followed, the war also demonstrated an elasticity in supply. During the war, China dramatically expanded its production and, at one point, produced twice as much antimony as the entire world demanded annually in peacetime. Evidence of elasticity could be found in North and South American production as well. A sustained price of twelve to thirteen cents, the committee estimated, would lead to an increase in extraction in Mexico and Bolivia. The committee reminded readers that during the war, Bolivia had briefly produced some 60 percent of global production. An even larger price increase would stimulate production at home. When the price and demand surged and prices rose from 5½ cents in 1914 to 45 cents per pound in 1916, domestic production increased from almost nothing before the war to a peak production of 1,760 tons in 1915.

These dramatic price and production increases were, of course, merely spikes. By 1918, prices fell dramatically, and the United States returned to its nearly complete reliance on foreign supply chains, but this had everything to do with the market price of antimony and nothing to do with an exhaustion of domestic or regional supplies. The committee estimated that 20 cents per pound was the threshold to restart domestic production but that only a price floor of 40 cents per pound would be sufficient to stimulate sufficient domestic production to satisfy US industry's demand.[90] The committee could not imagine a scenario in which prices could rise that high, leading it to conclude that "it was hard to conceive of any condition in time of peace or duress" that "would enable the United States to supply it needs from domestic sources."[91] Given China's comparative advantage, the committee concluded that importation of antimony was a necessity and that the United States "could not foster an infant industry to our economic advantage."[92]

Third and most importantly, Wang and Li's backgrounds made them uniquely qualified to size up China's antimony reserves and resources and its ability and willingness to continue to serve as a, if not the, global source of production. In fact, all of the global antimony knowledge that went into this report rested at least in some measure on their knowledge base. Wang knew as much about the metalloid as anyone in the world, and Li, along with Riddell, understood as well as anyone the antimony trade. Li and Wang had extensive experience and knowledge of the geological and extraction process in China, and it showed. Fully one-third of the antimony report dealt specifically with China, and it provided close scrutiny of the geology, economics, and processing of the industry. Measured in page numbers, the six pages devoted to China were more than any other producer of any of the other ten metals and fuels addressed in the larger report.

In a report that oozed with confidence concerning global supplies, there was one small note of concern. In the conclusion, the investigators noted that the United States was separated from all supplies by water "whose maintenance may at any time be cut off."[93] Though they did not address it in any detail, much of the supply of Chinese antimony came through the city of Changsha in Hunan province before heading north to the Yangtze River, where it was transported to Shanghai before entering into the global marketplace. Just over ten years later, when Japan took control of this critical waterway, this vulnerability became extremely relevant to global consumers of Chinese antimony. This fresh round of supply chain crises facilitated another in a long line of inquiries rooted in World War I and continuing until today that wrestle with the constantly shifting costs and benefits of living in an interconnected world. The bedrock of which is firmly rooted in the effort to identify, extract, and process materials buried sometimes miles below the Earth's surface.

CONCLUSION

An analysis of mining engineers' role in antimony production and in the global expansion of engineering in the early twentieth century reveals, to borrow an idea and phrasing from Emily Rosenberg, "how transnational social and cultural currents circulated across and beyond national states . . . drew the world together in new ways."[94] A focus on antimony allows us to see this process unfold across time and space as ideas, experts, ores, and elements make their way around the world in a set of highly interdependent processes that extended the boundaries of global capitalism and mining engineering knowledge. As economic historian Gavin Wright documented in the US case, mineral endowment alone does not guarantee natural resource production.[95] In both the US case and in China, engineering education—and, more specifically, a broad-based and international education often subsidized in different ways by state policies—led to profound changes in global supply chains. Connections developed by these agents of globalization in professional, business, and education organizations and institutions played a critical role in this process, but so too did provincial leaders, government officials, and local gentry who helped this process through a variety of often statist means that facilitated not only mining and processing but also the very sort of connections described by Rosenberg. Global shifts in supply chains and supply and demand needed to be interpreted, and here too, mining engineers played an important role whether in explaining processes to a global body of mining engineers in professional meetings and publications or in developing knowledge that informed public debate over the significance of resource dependence.

7 FRENCH MINING CAPITALISM ABROAD: MOBILITY OF ENGINEERS, TECHNOLOGY TRANSFER, AND INVESTMENTS FLOWS

MARCO BERTILORENZI

This chapter investigates the role of French mining engineers in spreading French industrial investment abroad. It provides insights into the complex relationships that existed between the mobility of highly skilled workers and foreign direct investments (FDIs) between the late nineteenth century and the 1960s.[1] It expands our understanding of the international reach of French mining engineers, focusing on a specific typology, the civil mining engineers (*Ingénieurs civils des Mines*, or ICMs) who graduated at the École des Mines of Saint-Étienne, by adopting a prosopography approach to explore their mobility and impact outside France. This topic is interesting from two standpoints that are usually disregarded by scholarly studies about French engineers: their global reach, which made French mining engineers a tool for the global mobility of techniques and managerial capabilities, and the relationship between this mobility of people and the flow of investment capital abroad.

A BRIEF HISTORY OF MINING ENGINEERING EDUCATION IN FRANCE

The historiography on French mining engineers usually adopts a national focus because of the specific national goals that underpinned the formation of mining engineers in France.[2] Since the French Revolution, within the sphere of higher education in France, the mining sector has represented a key field of specific policies aimed at enhancing the exploitation of national resources.[3] After the *Code minier* came into effect in 1810 to regulate state ownership of national resources and define the norms governing concessions to private firms, a system was created that included two different types of mining engineers: on the one hand, state engineers who belonged to the Corps des Mines and, on the other hand, the ICMs. The main task of engineers belonging to the Corps des Mines was to survey and control French mines, while the ICMs were hired by the private companies. Both kinds of engineers were created to improve the exploitation of national resources.[4]

The engineers of the Corps des Mines were trained at the École Polytechnique and then went on to further specialize in their studies at the École des Mines of Paris. This corps of engineers was originally drafted to provide the French state with the power of control over the process of exploitation of national resources, within the framework of public concessions to private companies. Thanks to the *Code minier*, the corps of state engineers became an important governance tool of France's industrialization process.[5] The involvement of the French state in the exploitation of national resources was not limited to the state engineers: in fact, the law passed in 1810 also gave impetus to the creation of another type of engineer, the ICM, and made the state responsible for educating engineers working for private enterprises to increase the productivity and the rationality of the mining and metalworking industries. During the nineteenth century, ICMs were usually trained at the École des Mines of Saint-Étienne, established in 1816 in the center of one of the most important coal districts of France, and at the École des Mines de Paris, which was initially established in 1796 to train state engineers before the *Code minier* came into effect. The Paris École des Mines gradually opened its doors to ICMs after the Napoleonic period at the same time as the establishment of École des Mines de Saint-Étienne.[6] After the Great War, another school for ICMs was opened in Nancy, which focused on metallurgy as its main specialization.[7] The aim of the ICM training was to provide high-level technical expertise to private companies exploiting French national resources. Even though ICMs worked for private companies, they were trained by the state according to the same principles that inspired the creation of the state engineers.[8] As Anne-Françoise Garçon has pointed out, the training of engineers for the mining industry in France was an example of a dialectic nexus between private and public interests, in which enhancing national resource exploitation was of mutual interest.[9] The training of ICMs in the state schools in Paris, Saint-Étienne, and Nancy entailed the creation of a highly skilled group of engineers with both technical and managerial skills. State engineers of the Corp des Mines could expand their professional activities to the private sector, becoming part of the economic élite, like managers or general directors of the French mining companies.[10] In contrast, ICMs ranked somewhere in between and often occupied top corporate positions, although examples of ICMs who became "Grand Patrons" are not lacking.[11] Like many other engineers, ICMs were assigned to both technical and managerial duties.[12] However, one of the specificities of ICMs was that these engineers were a depository of both technical and managerial skills, and their careers were usually characterized by positions that incorporated both these aspects, often without a clear distinction between them.[13] ICMs were highly trained and provided with a top-level knowledge of the mining technical system: from geology to mechanics, chemistry and metallurgy. ICMs were also taught specific skills in the fields of management and economics, both very useful when working with modern and large-scale mining corporations like those of France, which in the second half of the nineteenth century progressively adopted modern structures and strategies of vertical integration and diversification. For instance, Henri Fayol, one of the fathers of modern management, was an ICM trained at Saint-Étienne.[14] Fayol, in his 1916 masterpiece, *General and Industrial Management*, stressed that the work of ICMs consisted of both managerial and technical duties: the higher the position of engineers in a company's organizational structure, the more managerial their duties and vice versa. But while technique was not completely excluded from management activities, management was also part of the duties of the lowest-level engineers because of the specific technical environment associated with the mining industry.[15]

MINING ENGINEERS, MOBILITY, AND FOREIGN INVESTMENTS

Historians have typically argued that French foreign investments in the nineteenth and twentieth centuries were primarily driven by colonization and imperialism, by corporate strategies, and by French macroeconomic financial strength.[16] The research in this chapter follows a different approach: it investigates the international mobility of mining engineers and their links with the global mining industry, exploring how the mobility of people has been interconnected with the flow of capital. The specific technical, managerial, and administrative knowledge of engineers proved essential to create the framework for the flow of investments in the mining sector. Their capabilities were locally focused, because of their training in the French mining schools; however, their field of activities engaged a global scale, with activities abroad that spanned the French colonial empire as well as other countries. In short, the mobility of ICMs created economic and technical ties between France and host countries. The mobility of engineers followed personal, professional, entrepreneurial, and conjunctural considerations over time, which directly affected the way in which French companies invested, the means of technology transfer, and even the flow of managerial or administrative ideas. The mobility of ICMs can be understood not only as a by-product of the strategy of private enterprises but as an independent driving factor in spreading capital abroad and in establishing rapport between countries or regions.[17]

The original goal of rationalizing *national* resource exploitation through the training of ICMs does in fact obscure the actual role that mining engineers had in exploiting resources *abroad*. Despite the national emphasis behind the creation of ICMs and the ways in which each school was anchored to regional mining sites (for example, Saint-Étienne was intended to provide local directors to the French mining companies in the Loire region), these engineers progressively became important actors in the overseas flow of French international investments. Many French ICMs contributed to the spread of French companies in global mining capitalism as consultants for firms or as prospectors. Notable examples of mobile ICMs from Saint-Étienne can be found already in the first half of the nineteenth century. These included Jean-Baptiste Boussingault (1802–1887), traveling in the Americas, or Louis Simonin (1830–1886), who traveled around the world, to gather information and provide specific expertise to French companies and entrepreneurs intent on investing abroad. ICMs underpinned the growing flow of French foreign direct investments, which, especially after the 1880s, marked the expansion of French business across places in eastern and southern Europe, Africa, Asia, and Latin America. In the context of globalization and the imperial race for resources, the technical and managerial expertise of ICMs came to represent a decisive advantage for French investments abroad. ICMs helped companies acquire multinational status and supported venture capital in mining activities, underpinning many freestanding companies as they expanded their investments abroad.[18] Part of the French technical and economic élite, ICMs also contributed to the process of French colonization, providing expertise and know-how to the *mise en valeur* of the growing French colonial empire. Specific studies have already underlined the role of ICMs in the process of resource exploitation in North Africa,[19] the Far East,[20] and southern Europe,[21] providing important case studies that illustrate the relationship between engineers' mobility, on one hand, and reginal and firm-level investment patterns, on the other.

However, despite their importance as agents of French investment abroad, the role of ICMs in the creation of global mining capitalism has never been thoroughly investigated in a systematic way. This chapter exploits a new data set of ICMs to provide a prosopography of ICM mobilities, focusing on the group of engineers who graduated at the École des Mines of Saint-Étienne, one of the most important schools for training civil mining engineers, and linking their global mobility to French FDIs. To explore the global diffusion of ICMs, a new relational data set was created with the comprehensive data from all the ICMs trained at the École des Mines of Saint-Étienne between 1890 and 1962. In this period, ICMs received a similar training. In 1890, the École des Mines of Saint-Étienne became a top-level engineering school within the French education system. The specific expertise of ICMs was characterized by a strong connection to the mining and metalworking industries and their related fields.[22] Until 1962, the focus of ICM training remained more or less unchanged; afterward, the sunset of national coal production triggered a gradual diversification toward other fields, such as computer science and chemistry.[23] As a consequence, ICMs trained between 1890 and 1962 can be considered depositories of a common set of skills and expertise. The database comprises a total of 2,595 ICMs. For each graduating engineer, the data set includes the employer, the individual's position within the company, and the location in which they operated through ten benchmarks between 1893 and 1973. As we will see, out of about 2,600 ICMs, about 600 had an international career, which represents about 23 percent of all trained ICMs, whose mobility can be analyzed through the data set.

The data set makes it possible to analyze the evolution of the careers of ICMs with international mobility and to link those career patterns with the dynamics of French FDI, the foreign strategies of companies, and French expansionism abroad. The study adopts a largely quantitative and focuses on a prosopographical analysis of the Saint-Étienne ICMs. Moreover, this data set has been processed by a georeferenced information system (GIS), which mapped the international mobility of ICMs at each benchmark.[24] This methodology provides a more accurate and comprehensive understanding of the part played by French investment in a kind of global race for resources, within which the ICMs were specific agents. The evidence demonstrates that engineers belonged to a "peregrine profession," and their mobility also shaped patterns of international investment and created bridges between territories with different levels of economic and industrial development.[25] Their mobility was linked to technological networks that were locally focused. French capabilities, embodied in the technical and administrative/managerial expertise of ICMs, contributed to the emergence of new territories as resources providers, expanding the reach of French mining capitalism. Thus, by mapping engineers' mobilities, we can access information on companies, branches of activities, investments dynamics, exploitation of mining regions and natural resources, and technology transfer. After presenting the data set and some general considerations on the occupational structure of ICMs in the following section, the chapter examines the spatial mobility of ICMs and the business structure of this mobility.

THE DATABASE AND THE METHODOLOGY: SAINT-ÉTIENNE'S ICMS

Instead of using data from companies or financial actors, which usually characterize the study of multinational companies and FDI, this chapter focuses on the professional trajectory of individuals, of personal careers, in order to explore FDI. An original data set created by the author contains information on each ICM who graduated at the École des Mines of Saint-Étienne between 1890 and 1962, using the archive of the École to gather personal information and to track their records as students (including evaluations, specialization, grades, and ranking). Since it belongs to the system of Grandes écoles, the École des Mines of Saint-Étienne organized its graduates in "Promotions" (usually referred to as "Promo," which can be translated as "cohort"), which for the Saint-Étienne mining school are identified by the graduating year. Since these students were part of a prestigious institution, after graduation, they were also organized into a Société Amicale des anciens élèves de l'École des Mines of Saint-Étienne (hereafter "Amicale"), a powerful and long-lasting alumni association that gathered information about professional activities over time and, since the late nineteenth century, has published a yearbook called the Annuaire de la Société Amicale des anciens élèves de l'École des Mines of Saint-Étienne (hereafter "Annuaire"). The Annuaire contains important information on careers that ICMs provided to the Amicale on a voluntary basis: each year, former students of Saint-Étienne sent information about their activities to the Amicale, which was afterward published in the Annuaire.

By combining archive documents with information from the Annuaire, it is possible to follow the careers of 2,595 ICMs who graduated from Saint-Étienne between 1890 and 1962. For each ICM, the Annuaire contains information about the employer (firm), their position in the corporate hierarchy, the place of activity (country and town), and sometimes also more specific information about the company, such as the name of the mine in which the ICM worked or the specific division/utility to which he belonged. A series of benchmarks have been established about every ten years (1893, 1903, 1913, 1924, 1933, 1939, 1950, 1957, 1963, 1973) to follow the career evolution of each ICM in terms of employer, hierarchical position, and localization. The benchmarks tend to avoid critical years, such as wars or periods of strong financial crisis, which might introduce bias to the analysis of the data set.

Table 7.1 presents the occupational distribution of all Saint-Étienne graduates. Although ICM training focused primarily on mining, only just over half of the ICMs who trained at the Saint-Étienne between the 1890s and 1960s and who are tracked in the data set found occupations in the mining sector, as we see in the fourth column. Other fields of employment for graduates include metallurgy (15 percent), other heavy industries (11 percent), engineering and consulting (8 percent), energy and public utilities (5 percent), army and public administration (4 percent), transport and infrastructures (1 percent), and others (1 percent).

We can explore the evolution of the occupational fields of ICMs in greater detail by splitting the group of ICMs into three subgroups: one formed by the ICMs trained between 1890 and 1914, another represented by interwar "Promotions," and the last formed by the ICMs trained after World War II, from 1946 to 1963. These three periods correspond to three different phases of global economic development, from the globalization of the *belle époque* period to interwar deglobalization and the postwar era of economic growth. If we look at the data on the three subgroups of ICMs trained in 1890–1914, 1915–1945, and 1946–1962, noted in the first three columns of table 7.1, we see that the occupations of ICMs underwent significant change over time.

Before World War I, the first group (1890–1914) saw the largest proportion of graduates with an occupation in the mining industry and related fields,

Table 7.1
Occupational fields of ICMs, 1890–1962.

Sector	1890–1914	1915–1945	1946–1962	Full period
Extraction	77%	57%	44%	55%
Metallurgy	12%	12%	14%	15%
Heavy industry	2%	5%	11%	11%
Engineering/consulting	4%	11%	15%	8%
Energy	2%	5%	11%	5%
Army, administration, church, etc.	1%	8%	2%	4%
Transports/utilities	2%	2%	0%	2%
Others	1%	1%	2%	1%
Total number	691	854	348	2,117

Percentages indicate the proportion of all Saint-Étienne graduates who found jobs in each field, for each time period; $N = 2{,}117$. *Source*: calculated from the Saint-Étienne alumni database, constructed by M. Bertilorenzi. See text for further discussion.

with about 76 percent of ICMs employed in mining activities. During the interwar era, this percentage fell to 56 percent of ICMs active in the mining sector, while after World War II, only 44 percent found a job in the mining industry. Metallurgical jobs ranked second as an occupational sector for ICMs of the first period (1890–1914), with a modest 12 percent, which remained constant through the other two periods. Across all three periods, the main change in terms of occupation was an increase in engineering consulting activities, which passed from 4 percent before World War I to 11 percent in the interwar period and 15 percent in the post–World War II period, together with jobs in the energy sector, which comprised only 2 percent before 1914 and rose to 5 percent in the interwar period and to 11 percent thereafter. After World War I, ICMs progressively found greater employment in the oil industry, while after World War II, a new field of activity gradually expanded in the atomic and nuclear industries. At the same time, other heavy industries, especially chemicals and petrochemicals, which had accounted for only 2 percent of positions before World War I, grew to 5 percent in the interwar period and 12 percent afterward.

The technical foundation of ICM training in the French schools recognized that ICMs played a critical role within an expansive technological system. Mining in itself was only a part of a broader technological picture. Initially, their training included metallurgy, steam, and mechanics but over time evolved to include new fields of expertise and technologies. In fact, training and education in the schools evolved to follow the needs of industry because in the École des Mines of Saint-Étienne, businessmen were directly involved in the definition of programs within the school advisory board and council.[26] As we noted with regard to occupation distribution above, over the century covered here, ICMs progressively moved from being "pure" mining engineers typically working in a coal-steel technological system, including some diversification toward nonferrous metal and other minerals than iron and coal. Over time, their work became more closely related to new technologies involving different energy sources than coal, including geological surveys and petrochemical industries for the oil and nuclear industries.[27]

ICM SPATIAL AND BUSINESS MOBILITIES

Thanks to the information in the data set, we can analyze that part of the ICM group that experienced international mobility over time, using these data to better understand the transfer of technology, know-how, and capital supported by this kind of international mobility. The spatial location of ICMs was georeferenced using ArcGIS software, which made it possible to identify the changing geographical position of French engineers over time, not simply at the national level but also at subnational, regional, and even city levels. Out of the full group of 2,595 ICMs, 605 recorded at least one international mobility at one of the benchmark moments, and 106 ICMs experienced international mobility in two or more benchmarks. Some of these 106 experienced a substantially nomadic career, moving their activities over three continents in less than twenty years. Combining all the data from 1890 to 1962, we see in column 4 of table 7.2 that these 605 ICMs included 34 percent who moved to Africa, 29 percent to Western Europe, 19 percent to Eastern Europe, 11 percent to Asia and the Far East, 4 percent to Latin America, and 3 percent to North America.

Over time, the destination of ICM mobility evolved with changes in the broader political and economic context, following both globalizing/deglobalizing macroeconomic trends and the specific history of French FDI and colonial activities. In the database, colonial mobility has been considered as international mobility, despite the fact that colonial mobility occurred within the borders of the imperial territories. The French colonial empire was

Table 7.2
International mobility of ICMs, 1890–1962.

Area	1890–1914	1915–1945	1946–1962	Full period
Africa	22%	39%	48%	34%
Western Europe	35%	24%	28%	29%
Eastern Europe	29%	17%	2%	19%
Asia	10%	13%	8%	11%
Latin America	2%	5%	7%	4%
United States and Canada	1%	3%	6%	3%
Total number	242	262	99	603

Percentages of the subset of internationally mobile ICMs who moved to specific regions; $N = 603$. *Source*: same as table 7.1.

complete by the time of World War I, but its *mise en valeur* from an economic standpoint underwent substantial acceleration during the interwar period in a context of deglobalization and also after World War II when additional investments were made by the French government to more closely integrate the colonial territories into French economic development.[28] At the École des Mines of Saint-Étienne, a specific postgraduate course was established in 1936 for the purpose of enhancing ICM capacity with specific administrative knowledge of colonial rules.[29] Despite this emphasis on colonial careers, the French colonial empire was only one among the varied destinations of ICM mobilities.

In fact, the French colonial empire was not the first target of ICM mobilities. Eastern Europe dominated international job locations at the end of the nineteenth century, when ICMs underpinned the dramatic expansion of French mining and metallurgical interests in the Russian and Ottoman territories. At the town level, these mobilities were not concentrated in the political capitals of these countries but in specific mining regions. In the case of the Russian empire, for instance, many ICMs settled in the Ukrainian or Polish regions, where territories next to towns, such as Donetz and Dąbrowa, were the target of French investment in mining activities and steel production. Another wave of mobility was concentrated in the early twentieth century in Western Europe. This target emerged again in the 1920s as an exit strategy for the divestments that were made from the ex-Russian empire because of the Bolshevik revolution. In broad terms, Africa started to appear as a significant destination after World War I, but it was not the main target of ICM mobility before 1945. In the aftermath of the Great Depression, the general trend of international mobility toward European and African destinations was reduced. This can be deemed an outcome of the decline of French FDI in these countries but also reflected other destinations emerging as new opportunities for ICMs. For instance, Asia and the Far East, as well as Arab countries, increasingly appeared as good opportunities. For the ICMs, these mobilities also represented a technology change from the coal-mineral mining industry, which dominated the Belle époque period, to new energy industries such as crude oil.

During the 1930s, Africa and other French colonial destinations became more crucial than before, concentrating a good portion of ICM mobility. This evolution is consistent with the ongoing *mise en valeur* of French colonial administration, which also had a propulsive effect on both private FDI by French companies in the colonies as well as the personal mobility of ICMs, who moved toward colonies to work as consultants and to facilitate investments. Many ICMs moved within the administrative staff of the French colonies. However, Africa only emerged as a larger target for mobility after World War II, when French administration began more vigorous policies of industrialization in the colonies, efforts that required a growing number of engineers.[30] Other destinations like Latin America and North America were marginal throughout all three periods: these mobilities were always linked with specific technical opportunities opened to leading French companies. For example, the Schlumberger firm found opportunities for investment in Argentina and the United States, including the oil fields in Texas.

Despite the links to mining group strategies and to the French colonial framework, however, the mobility of ICMs does not appear to have been driven simply by corporate investments and French colonial policies. International mobility was an outcome of a complex set of opportunities, which spanned technical, mining, and even personal contexts. From an overall perspective, some "promotions" (cohorts) were more mobile than others, reflecting specific conditions for ICMs in the job market when they graduated. In some annual promotions, for instance, more than half of graduated ICMs fled France to land positions abroad, while others saw less than 10 percent of the total graduates undertake the same path (see figure 7.1).

The reasons for this fluctuation are complex, combining both macro and micro levels of analysis. On the one hand, the École des Mines decided year by year how many ICMs to train, and when ICMs obtained their degree, the situation of the job market did not necessarily match the forecast made three years prior when the managing council of the school had decided upon the number of ICMs to be trained. Thus, in certain situations, France lacked ICMs for the national mining industry and factories; in other years, some ICMs could not find domestic jobs, and the surplus had to migrate to find good opportunities, typically moving back to France when the domestic job market was hiring again. On the other hand, specific conditions and opportunities influenced the mobility paths of ICMs, like the opening of new mining activities in any one country, the discovery of new resources in another, and the ability of the ICMs to respond to such opportunities following their entrepreneurial and career-building aspirations. From figure 7.1, we can see a rising trend before about 1915, which was followed by a declining trend afterward. The general trend of ICM mobility underwent a gradual reduction through the middle of the twentieth century, despite a number of peak periods shortly after World War I. If we split the period in two and consider the trends before and after World War I, there is a clear upward tendency toward international mobility before 1915 and a severe reduction afterward. The interwar period sees some decline but still represents a period of substantial mobility, even though it was reduced compared with the situation before 1915. The greatest decline comes after 1945. This was not the outcome of deglobalization or of Bretton Woods settlements, but it was more clearly the outcome of the industrial policies of the French government following World War II. These pushed investment and productivity increase in France's coal and steel industries, including the Charbonnage de France, the national coal company created in 1946, which hired a growing number of ICMs to increase productivity within the so-called *Bataille du charbon*. In other words, domestic pressures and opportunities reduced the likelihood that French-trained engineers would move abroad.

However, charting the international mobility of ICMs by dividing them into the three macro groups (pre–World War I, interwar period, and post–World War II graduates) implies that macroeconomic conditions were decisive in shaping the mobility of mining engineers. The first three columns of table 7.2 illustrate how the geographic destination of ICMs changed substantially across the three time periods. First, the period with the highest international mobility was the interwar period, characterized by a level of mobility greater than prewar globalization. As in the previous period, the interwar era saw highly skilled workers like the ICMs facing a challenging domestic

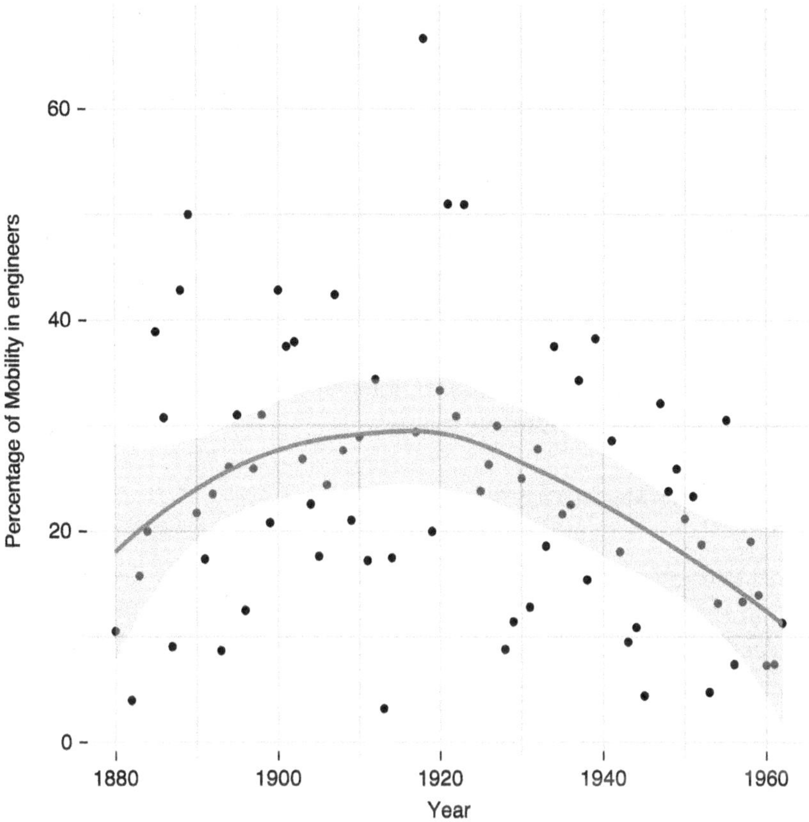

FIGURE 7.1
International mobility of ICMs, percentage per annual promotion. Annual percentages indicate the percentage of graduates from that year who found occupation abroad. *Source*: same as table 7.1.

job market. Divestments and crises, like the depression of 1880–1890, which heavily affected the coal and steel industry, and that of the 1930s had similar impacts, prompting ICMs to look for remunerative jobs abroad when the French job market did not offer them many opportunities. Part of engineers' mobility toward the Russian empire, which ended with the Soviet revolution, found a partial replacement in the same general region, in countries that had good diplomatic connections with France such as Poland or Romania, explaining the persistence of Eastern Europe in the data.

Second, the period after World War II saw a significant decline in international mobility compared to the preceding period. Eastern Europe almost completely disappeared as a target of ICM mobility, and the appeal of Africa and Western Europe also waned. Africa's drop in appeal can be explained by the decolonization process, while that of Western Europe was driven by the generally buoyant job market that ICMs found in France, where the state was investing heavily in key ICM business sectors (mining, steel, and energy, which were nationalized and boosted). The period after World War II was one in which, generally speaking, ICMs were able to refocus their careers within the national context, which became more and more attractive for them. As we will see in the next section, major changes in engineers' international mobility patterns varied according to the field of activity: while traditional sectors became increasingly national, international mobility was especially present in new and innovative activities.

ICM MOBILITY AND INDUSTRIAL STRUCTURE OF FDI

This section examines ICM mobility from the perspective of the industrial sector. We utilize the name of employing companies as well as the job function or office declared by the ICMs in the Annuaires. It is impossible to clearly distinguish between managerial and technical duties, as the recent study of Solares and Beatty shows, due to the specific features of French ICMs. Many followed the classic career path from technical to managerial positions.[31] However, analyzing ICM functions from the data set highlights another relevant tendency. Many ICMs switched from working for large companies, with functions linked to the direct exploitation of resources or factories, to being self-employed experts, working as consultants in technical fields. As we will see, this is an important development with regard to the jobs that attracted ICMs abroad throughout the period. This suggests a gradual evolution from technical to consulting or auditing functions, not necessarily linked to more administrative and managerial occupations, compared to the engineers who continued to work for large mining corporations.

As we see in table 7.3, about half (49.5 percent) of all ICMs who worked internationally were employed by a company whose core business was in the mining industry. This confirms that ICMs were part of France's expanding role in global mining and resource capitalism during the long twentieth century and that French companies abroad required skilled engineers to exploit resources in countries where there were no local skills available to carry out mine prospecting activities. Metallurgy accounted for 8 percent

Table 7.3
Occupational fields of internationally mobile ICMs, 1890–1962.

Sector	1890–1914	1915–1945	1946–1962	Full period
Extraction	64%	43%	33%	50%
Army, administrative, church, etc.	6%	20%	16%	14%
Engineering/consulting	6%	15%	13%	11%
Metallurgy	13%	4%	5%	8%
Energy	1%	7%	24%	7%
Heavy industry	4%	4%	8%	5%
Others	3%	5%	2%	4%
Transports/utilities	2%	3%	0%	2%
Total number	242	261	101	604

Percentages indicate the proportion of the subset of internationally mobile ICMs; *N* = 604. *Source*: same as table 7.1.

of ICM placements, other heavy industries 5 percent, general engineering and consulting 11 percent, energy 7 percent, public administration 14 percent, transports 2 percent, and others (mainly commercial activities and agricultural-related businesses) 4 percent.

This division only partially follows the occupational structure of all Saint-Étienne graduates documented in column 4 of table 7.1: some fields of activity were more mobile than others, such as engineering/consulting (11 percent compared to 8 percent of the total) and administrative activities (14 percent compared to 5 percent of the total). While in France, a career in public administration was not easy or remunerative because of the presence of state engineers who had specific functions provided by the law of 1810, in the colonial context, ICMs could be hired by the colonial administration with surveying functions and good wages. Mining was smaller than the total (49.5 percent compared to 54 percent), but more ICMs worked in a related field as consultants, while energy had a larger share (7 percent compared to 5 percent) because some specific energy sectors, such as oil and nuclear, involved essential resources that were exploited by French companies abroad. In the case of transport, which was only a marginal field of occupation for ICMs, international mobility was twice as big as that of the general occupational structure (2 percent compared to 1 percent).

To understand these dynamics, we explore in columns 1–3 of table 7.3 the sectorial structure of ICM employability in the three subperiods adopted for the other statistical observations from our data set (1890–1914, 1915–1945, and 1946–1962). Before 1914, the mining industry was by far the largest sector in which ICMs worked abroad, but its relative weight, with 155 ICMs out of 262, was less critical than that of mining in the general occupation of ICMs (which was over 70 percent for this period). Engineering and consulting were more relevant for internationally mobile engineers, about 11 percent compared to 8 percent for the general occupation. These data suggest that only some ICMs were sent abroad directly by a firm. At the same time, ICMs personally moved to a foreign country to provide services to companies intent on investing in particular locations. By becoming a local consultant for foreign "French-speaking" businesses, ICMs became an important agent of technology, know-how, and financial transfer independent of (or complementary to) French FDI flows. Companies were an essential dynamic of transfer but were not the only one, and the mobility of ICMs was not simply the outcome of the deliberate strategy of corporate investments abroad. FDI and personal mobility constituted a complex interaction, complementary but also independent, each representing an opportunity for the other (and vice versa).

Another relevant field of ICM employability was in public administration. ICMs worked for the French colonial administration, which was considered a good occupation, often better than anything they could find in France (e.g., the position of assistant to public engineers belonging to the Corps des Mines, with poor wages and bad career perspectives).[32] During the interwar period, ICMs found positions in internationally located public administration positions at significantly higher rates than before or after. About 20 percent of all internationally located engineers worked abroad in such jobs, either in the German-occupied zones or in the French colonial empire. French ICMs were part of the diffusion of not only managerial and technical know-how but also administrative approaches. Some of these were French nationals, but not all. For instance, in the 1940s, the École des Mines of Saint-Étienne trained a small group of Persian engineers who graduated as ICMs and, after World War II, were hired by the Iranian Ministry of Economics and shaped that country's national approach to its oil resources. It is also interesting to note that, in the early 1930s, the École des Mines of Saint-Étienne organized a postgraduate course on "colonial mining and administration" (*école coloniale*), the purpose of which was to train students intending to pursue an international career. The specific approach of the French authorities to colonial administration, often called "direct rule" in broad political terms, can be seen in the education and occupational path of ICMs as public servants. The central regulations that France applied to its colonies were often underpinned by the work of ICMs, who supported the regulation of resource exploitation by the French authorities. In that case, ICMs were trained with a specific focus on colonial business and administration, which could lead to obtaining very good positions in the colonial administration and specific government-controlled companies, such as the Bureau de recherches géologiques et géophysiques (established in 1941) and the Bureau de recherches géologiques et minières, established just after World War II.

As table 7.3 shows, the engineering/consulting field also grew during the interwar period, marking a deeper tendency toward external mobility driven by rising corporate FDI. In contrast, after World War II, new fields of industrial employment saw the only significant growth in international mobility, seen here in the booming energy sector. Oil and nuclear power surpassed traditional mining and metallurgy abroad. New industries provided more opportunities outside France, while in their home country, ICMs were more easily able to find jobs in the mining and steel sectors than in the past.

CONCLUSION

The contribution of ICMs to the spread of French mining capitalism abroad is complex. To say that it was driven by the flow of French FDIs is to oversimplify. Engineers were not simply agents of capital. Historically, ICMs' contributions to the mobility of technology and capital were related to three other aspects. First, engineers' international mobility reflected the personal strategies of individuals who moved independently from corporate strategies and set up consulting activities abroad. This entrepreneurial aspect is crucial in looking at the careers of ICMs who were not simply passive company agents or bureaucratic workers but individuals characterized by personal aspiration and motivation, leading them to frequently change jobs, contexts, and countries. Second, with its administrative and political goals, the French state also represented a major impetus for the international mobility of ICMs, who were long accustomed to working for public services in colonial contexts. However, this should not lead us to believe that ICMs simply followed the big waves of French colonization and public policies. The growing role of states and colonies ran parallel to the rapid growth of FDIs during the first wave of globalization at the end of the nineteenth century and continued afterward, shaping economic ties between regions that also lasted through the interwar deglobalization era.

ICMs, in other words, were part of a complex nexus between public and private interests because of the very nature of French resource policies that assigned a specific role to both companies and governmental offices. This reflected the attitude of the French state abroad, which tried to establish similar administrative policies that passed through the adoption of specific economic agents trained to work within this administrative approach.

Third, technology itself represented an essential factor that shaped patterns of international mobility and the transfer of skills. All branches of activity were highly mobile during the investigated period, but traditional sectors were less mobile than new ones, and the relative mobility of engineers working in different sectors changed over time. In any subperiod, sectors of innovation and growth witnessed higher levels of international mobility. The contribution of ICMs was marked by the spreading of new technologies abroad, especially when the mobility related to older technologies was reduced by other factors, such as World War II. The emergence of new venues for international mobility, linked to oil and uranium after World War II, reflected an important role of ICMs in the country's changing energy regime, which relied on imported materials that progressively substituted for coal.

8 FROM EMINENT ENGINEER TO AUTHORITY ON ECONOMIC PLANNING: SIR M. VISVESVARAYA IN INTERWAR INDIA

APARAJITH RAMNATH

The term *engineer* is most commonly understood to mean (in the words of the *Oxford English Dictionary*) "a person who uses specialized knowledge or skills to design, build, and maintain complicated equipment, systems, processes, etc."[1] But since the nineteenth century, when engineering began to emerge as a modern profession around the globe, engineers have done much more than "design, build and maintain." In contexts ranging from the public works departments of colonial India to large industrial corporations in the United States, engineers have been engaged in diverse activities, including management, budgeting, and public administration. In some settings, they have functioned as representatives of the state; in others, as coproducers (along with industrial tycoons) of modern capitalism.[2]

The subject of this chapter is the Indian engineer Sir M. Visvesvaraya (1861–1962), whose career provides a vivid example of the multifaceted roles played by engineers across the world between the mid-nineteenth and mid-twentieth centuries. Visvesvaraya remains, to this day, the most iconic exponent of his profession in India. He is associated with spectacular engineering projects and technical institutions, and his birthday is celebrated all over India as Engineers' Day. But he is equally famous for his technocratic interventions as a statesman and public intellectual. In the 1910s, not long after taking early retirement from the Bombay Public Works Department, he served as dewan (prime minister) of the princely state of Mysore, where he spearheaded an ambitious program of modernization. In the interwar years, he built a pan-Indian reputation as an evangelist for industrialization, a writer on economic development, and a promoter of economic planning. Although his role has been overshadowed in the public imagination by that of Jawaharlal Nehru and the statistician P. C. Mahalanobis (who worked with Nehru in the 1950s),[3] Visvesvaraya's contemporaries considered him among the earliest Indians to popularize the concept of economic planning.[4]

Visvesvaraya was a centenarian who was professionally active over seven decades, from the 1880s to the 1950s. In other words, his working life spanned virtually the entire period in which, as the present volume emphasizes, the engineering profession rose to prominence on the global stage. His career, in fact, epitomizes several of the themes animating the volume. He lived through and contributed to the widening institutional base of engineering as a profession in late-colonial and independent India. His work was ostensibly done within a national context, but his reference points and influences were much wider. His career encompassed several forms of mobility. He worked not just in the princely state of Mysore, where he was born, but across the length and breadth of British India. In the latter half of his career, he moved with remarkable success into administrative and political roles. He was deeply invested not only in the project of constructing infrastructure across India but also in that of building the incipient nation.[5]

In this chapter, I am concerned with a particular question: How did Visvesvaraya the civil engineer evolve into Visvesvaraya the economic planner? I also deal with a number of related questions. What were the origins of his interest in planning? In what ways and to what extent was his approach to planning influenced by his training and experience as an engineer?[6] How did he, a professional engineer by training, convince the public that he possessed the expertise to pronounce on economic planning and national governance?[7] Finally—and this is especially relevant given Visvesvaraya's many long study tours abroad—what role did global trends in engineering and technocratic thinking play in the fashioning of his worldview?[8]

Answering these questions requires a detailed and critical exploration of the various stages of Visvesvaraya's life and career. In this chapter, I draw on research I have undertaken for my ongoing project, a book-length biographical study of Visvesvaraya. The sources for that project include Visvesvaraya's private papers and published writings, the memoirs of his contemporaries, government archival collections and official correspondence, and contemporary press reports.

I argue that Visvesvaraya's experiences during his early years as an engineer profoundly shaped his interest in and approach to economic planning. In this he was part of a global trend: across the world, from France to Mexico, engineers in the twentieth century moved from designing public infrastructure to contributing to national plans.[9] As a result of his professional background, Visvesvaraya had a belief in *progress*, a materialist outlook, and unbounded confidence in humankind's ability to process natural resources and raise its standard of living. As an irrigation and sanitary engineer, he grappled with problems related to drought, famine, and disease. These concerns took him into the realm of macroeconomic policy, especially through his friendships with some of the leading nationalist writers on economic matters in the nineteenth century, who were interested in similar problems. Later, Visvesvaraya's record as Dewan of Mysore and his working relationships with influential figures across the political spectrum were crucial in lending credibility to his views on planning. So were his travels abroad and his study of world affairs, which gave him the stature of a modernizer bringing to India the best the world had to offer.

PRACTITIONER OF ENGINEERING, STUDENT OF ECONOMICS

Mokshagundam Visvesvaraya was born in 1861 in Muddenahalli, a village 36 miles north of Bangalore, in the princely state of Mysore.[10] He belonged to one of the earliest generations that came of age after the colonial government decided to play an active part in education at various levels. In this period, the state focused on building institutions on Western lines, pushing traditional Sanskrit- and Arabic-based learning to the margins.[11]

Throughout his youth, Visvesvaraya studied in missionary and government-run institutions, where he was taught largely by British professors. His mother tongue was Telugu, and his early education would have been in Kannada, the main language of Mysore state, but he thought and wrote (and probably read) almost exclusively in English throughout his adult life.[12] He began his schooling in Chikkaballapur, the town nearest Muddenahalli, and later joined the Wesleyan Mission High School in Bangalore.[13] He then earned a BA degree (specializing in mathematics and natural philosophy) at the Central College in Bangalore, run by the Mysore government but affiliated with Madras University. This was followed by a Licentiate in Civil

FIGURE 8.1
The Poona College of Science, established in 1854 to train engineers to work for the colonial Public Works Department; in 1911, it became the College of Engineering Pune and was India's first college to offer a formal engineering degree. *Source*: Wikimedia Commons.

Engineering at the Poona College of Science, which had been established to produce junior engineers and subordinate officers for the Bombay Presidency's Public Works Department (figure 8.1).[14]

The institutions Visvesvaraya attended had a distinct disciplinarian flavor. As a later alumnus of the Wesleyan Mission High School put it, "There is nothing soft about these Mission Schools. Law presides over sentiment. . . . There is a certain firmness and stability in their moral construction."[15] In addition, they had thoroughly Anglo-centric curricula. For his BA at the Central College, Visvesvaraya had papers not only in various branches of mathematics and physics but also in history, philosophy, and English language and literature. In those years, the history and philosophy curricula were dominated by British and European themes and authors, while studying English meant delving into the work of writers such as Shakespeare, Locke, and Tennyson.[16] The two colleges Visvesvaraya attended were presided over by an Englishman (Charles Waters in Bangalore) and an Irishman (Theodore Cooke in Poona), respectively. While these men were more liberal and sympathetic to their Indian wards than the average European in India, they epitomized the ideal of the Victorian gentleman as intellectually well rounded, energetic, and fond of the outdoors. Waters was an enthusiastic member of the Boat Club in Bangalore. Cooke tutored the college band in Poona, encouraged field trips and hands-on learning in the workshop, and gave scientific lectures for the public in his free time.[17]

From his education in the 1870s and 1880s, Visvesvaraya seems to have absorbed at least one quintessentially Victorian trait: a devotion to the idea of "self-help," a phrase that appears often in his speeches.[18] Samuel Smiles's book of that title is among Visvesvaraya's personal effects displayed at a small museum at his birthplace.[19] Visvesvaraya believed in local initiative and completely internalized the engineering ideal in the sense of molding nature to serve human needs and wants.[20]

In 1884, Visvesvaraya graduated at the top of his engineering class in the Poona College of Science, which meant he automatically received the offer of a job as assistant engineer in the Bombay Public Works Department (PWD). He worked in the PWD for the next twenty-four years, stationed in the city of Poona for more than half that time. He was credited with the design and/or execution of several important projects. Among these were water-supply works at Sukkur in Sind; a drainage scheme in Aden; the patenting of automatic sluice gates, which he designed for the dam at Khadakvasla near Poona; and the introduction of a new system of irrigation known as the "block system" in the Deccan region.[21]

Visvesvaraya's career as an engineer developed in an era in which "native" Indians had to battle negative stereotypes to prove their worth in the colonial engineering services.[22] Visvesvaraya's response was to adopt a thoroughly British style of dress, comportment, and bureaucratic functioning.[23] As I have shown elsewhere, engineers of the Public Works Department were expected to be "gentlemen generalists," and their duties were as much administrative and managerial as they were technical.[24] In his role as engineer in charge of water-supply, drainage, and irrigation works, Visvesvaraya frequently interacted with the public, becoming the interface between them and the Bombay government. In this role, he became a master of drafting reports and memos, full of facts and figures, designed to convince his colonial superiors to undertake new works, and of creating propaganda to win the cooperation and confidence of the public in his technical-administrative schemes.[25]

From the late 1890s onward, Visvesvaraya showed a keenness to pursue reforms in the system of irrigation. This interest went beyond technical tinkering and encompassed what were essentially administrative innovations. His famous "block system" of irrigation involved marking off consolidated blocks of land in each village along an irrigation canal, with cultivators sharing portions of land within the block by rotation. The blocks would receive enough canal water to encourage the cultivation of sugarcane (a high-revenue crop) alongside other crops; farms outside the blocks would largely depend on rain water. This was in fact a form of social engineering, for it required the cultivators to change their traditional ways of parceling land and water among themselves and was designed to increase the production of sugarcane as opposed to more traditional crops.[26]

These efforts came in the context of Visvesvaraya's firsthand experience of famine and plague, both of which afflicted large parts of the Deccan in the 1890s. He tried to devise methods by which the Bombay presidency could be made less dependent on nature in all its unpredictability. He was a strong advocate of what were known as "productive" irrigation projects (large reservoir-and-canal projects designed to help produce heavily watered commercial crops yielding high revenues) as opposed to purely "protective" ones (which were designed to carry canal water over large distances to enable the maximum number of farmers to grow subsistence crops in times of drought). He reasoned that when food was in low supply, the revenue generated by the sale of commercial crops would enable a province to import food grain from other parts of the country.[27] In essence, he had internalized the emerging colonial philosophy of irrigation, which was based on a commercial view of cultivation. Revenue maximization was at its heart, and integrated markets across the subcontinent were assumed.[28]

Thus, Visvesvaraya's experience in the Bombay service caused him to think deeply about the agricultural population's total dependence on the monsoons and how the threat of famine loomed large in their lives. He became convinced that this very dependence on agriculture was perpetuated by the colonial government's lack of encouragement to Indian industry and its failure to create alternative modes of employment.[29]

Deeply preoccupied with very similar concerns was another prominent denizen of Poona, the renowned judge and social reformer M. G. Ranade. Visvesvaraya struck up a friendship with Ranade and his protégé, Gopal Krishna Gokhale, spending a lot of time with the two men. With them he worked to establish the Deccan Club, a space in Poona for Indians and Europeans to meet socially. Visvesvaraya was also exposed to the work of a citizens' organization that discussed matters of public policy, the Poona Sarvajanik Sabha, in which both Ranade and Gokhale played central roles. On one occasion, Visvesvaraya even contributed an article to the society's journal.[30]

Ranade, who was Visvesvaraya's senior by a couple of decades, became a sort of mentor to him.[31] Ranade wrote extensively on economic issues. He belonged broadly to the school of thought established by Dadabhai Naoroji, who made famous the "drain" theory of colonial economics. According to this reading, the British were plundering raw materials from India to make industrial goods that were sold back to India, impoverishing the latter in the process. Ranade argued that Indian industries were being destroyed by competition from cheaper British manufactures and insisted that the colonial government must take steps for the protection and promotion of Indian industries.[32]

Ranade also prefigured Visvesvaraya's conception of a national economy driven by industry. As historian Manu Goswami has shown, Ranade, Naoroji, and others were instrumental in constructing the idea of the Indian economy as a coherent and clearly defined entity. They were deeply influenced by the German economist Friedrich List, who saw "the spatially bounded nation . . . as the sovereign subject of economic development."[33] As a corollary, they questioned the validity of "classical economic theory," shaped as it was by "free trade and laissez-faire ideologies." The British propounded these principles while flouting them in practice; there was no good reason why they should be imposed on India.[34] These arguments surely shaped Visvesvaraya's support for protectionist policies in the interwar years.[35]

The engine that would drive this entity called the Indian economy was industrialization and occupational specialization. In 1893, an article in the journal of the Sarvajanik Sabha (probably from Ranade's pen) held that "the real standing of a country in the scale of civilization . . . is measured by the practical pursuits of its people, the number and magnitudes of their callings, their industries, their enterprises, their skill, their ambition and their performances and by the extent to which they employ natural forces as aids to production."[36] Forty years later, Visvesvaraya wrote, "With the growth of civilization and multiplication of human wants, the occupations associated with industries and manufactures have increased in importance and are found to be more remunerative than those of agriculture; and industrialism has come to be regarded as a necessity and more or less as synonymous with civilization."[37]

Also prominent in Poona circles was the fiery politician Bal Gangadhar Tilak. Although his temperament made Visvesvaraya gravitate toward the moderate and constitutionalist Ranade and Gokhale rather than the more politically radical and socially conservative Tilak,[38] he would surely have been attracted by Tilak's enthusiastic championing of industry and technological education through the newspapers *Mahratta* and *Kesari*. There were also several industrial exhibitions and conferences in 1880s and 1890s Poona.[39]

Thus, Visvesvaraya's interest in the concept of a national Indian economy, as well as the steps needed to rejuvenate it, was ignited during his Poona years. Although his professional duties kept him busy, he made an effort to keep himself informed on these issues. In 1898, he traveled of his own accord to Japan, a country that Indians were beginning to look up to after its post-1868 modernization drive. He spent three months there as a pilgrim of modernity. He records in his memoirs that after his return, he had a discussion with Ranade on the subject of Japan.[40] This visit and a subsequent trip he made to Japan two decades later left a deep impression on him. He was enraptured by what he saw as a concerted nationwide effort to build a strong, modern, self-sufficient economy. He made note of Japan's policies, ranging from its investment in technical and university education to its encouragement of village self-government; he returned to these unfailingly in his later writings.[41] He may well have noted also that the Japanese Ministry of Agriculture and Commerce had issued in 1884 a policy document, *Kogyo iken*, which stressed, as historian Sydney Crawcour puts it, that "government direction, guidance, and protection" would be crucial for industrializing Japan and strengthening its economy. Crawcour argues that—like Ranade and his ilk—*Kogyo iken*'s author, Maeda Masana, was also probably influenced by the work of List.[42]

MYSORE: FIRST STEPS TOWARD PLANNING

In 1908, after a long and successful career in Bombay, Visvesvaraya decided to take early retirement from the PWD. He went on a long tour of Europe and North America, once again to observe the latest trends in the industrially advanced countries of the time. In 1909, he accepted the invitation of the Maharaja of Mysore to take over as chief engineer of his home state.[43] Among various schemes he started at this time was the gigantic Krishnarajasagara dam and reservoir project on the river Cauvery at Kannambadi, upstream of the existing hydroelectric station at Sivasamudram. This was a multipurpose project, contributing to irrigation as well as electricity generation. It would increase the flow of water at Sivasamudram, enabling the government to produce electricity in greater quantities. Some of this power could be sold to the private companies operating the gold fields in the state's Kolar region and the remaining supplied to new industrial ventures. It also aimed to irrigate thousands of acres of farmland in the state, enabling the cultivation of commercial crops such as sugarcane. Notably, in trying to convince his government to fund this costly project, he deployed what we would today call Keynesian arguments. He was strongly of the opinion that the state should borrow from the public if necessary to finance the project, which would bring in revenues and help stimulate industries in Mysore.[44]

In 1912, Visvesvaraya was appointed Dewan (prime minister) of Mysore, a post that was usually occupied by civil servants who were generalists. This was a momentous development, as he was the first engineer to be appointed dewan in Mysore—indeed, probably in any princely state.[45] In this role, he pushed an energetic program of industrialization and modernization in the state. His was clearly a top-down approach, characterized by Bjorn Hettne as a "state-capitalist strategy": he aimed to use the state as a catalyst for industrialization by creating the conditions for the emergence of a capitalist class.[46]

Visvesvaraya had been laying the ground for these transformations even before he became dewan. In 1911, when he was still chief engineer, he convinced the Maharaja to establish a body called the Mysore Economic Conference. It was made up of committees devoted to agriculture, education, and industries and commerce. The committees functioned year-round and met once a year at sessions of the Economic Conference, presided over by the dewan—which, from 1912, meant Visvesvaraya himself. In addition to government officials, the committees also included representatives of agriculture and business. At the core of their duties was compiling statistics on the relevant sectors, so that they could suggest fresh initiatives and policies to the corresponding departments of the government.[47]

The Economic Conference was a key tool in Visvesvaraya's project of modernization. Its deliberations were directly or indirectly responsible for establishing key institutions during his dewanship, including the Bank of Mysore, Mysore University, soap and sandalwood oil factories, and an iron works in

Bhadravati.[48] Significantly, the emphasis on data and expert opinion made the Economic Conference a precursor of planning bodies that Visvesvaraya was to suggest later at a national level.[49]

THINKING ABOUT THE NATIONAL ECONOMY

Following his retirement from the Mysore dewanship in 1918, Visvesvaraya divided his time between his homes in Bangalore and Bombay city, diving into a public career as consulting engineer, committee man, and public intellectual. He was now appointed to a series of committees and commissions that were not always related to his expertise as an engineer. In the 1920s and 1930s, he chaired or participated in bodies that looked into ways of collecting better data for the formulation of economic policy, examined the controversial Back Bay Reclamation project in Bombay, investigated the causes of a communal flare-up in Bangalore, made recommendations on technical education in the Bombay presidency, and drew up schemes of retrenchment for the cash-strapped city corporations of Bombay and Karachi.[50] A number of factors may have contributed to his being chosen for these roles. First, he had made a very favorable impression on his colonial bosses during his time in the Bombay PWD. Second, he was not formally a part of any political party and was thus seen as a neutral professional. Third, he nevertheless had a cordial relationship with moderate politicians in the Congress (especially Madan Mohan Malaviya and M. R. Jayakar), who were in positions with some power after 1919, when a measure of provincial autonomy was introduced.[51] Fourth, his status as a former Dewan of Mysore meant he came with substantial administrative experience.

Visvesvaraya was clearly gaining visibility at the national level. He now broadened his canvas. In these years, he began to articulate his view of the policies that would be required to strengthen the Indian economy as a whole. He published a series of tracts on this subject, including *Reconstructing India* (1920), *Planned Economy for India* (1934), and *Nation Building* (1937).[52] He was qualified to write these manifestos, he suggested, by virtue of his years as an official and a statesman, and especially his foreign travels, undertaken "for the purpose of obtaining first-hand knowledge of world conditions and problems."[53]

The first of these publications, *Reconstructing India* (1920), was written in the aftermath of World War I and in the context of the Montagu–Chelmsford reforms of 1919, which were designed to give Indians a larger (but still severely limited) role in the governance of their country. Visvesvaraya had stepped down as Mysore dewan in 1918 and toured Japan, Canada, and the United States in 1919. He then camped in London to put down his thoughts, which came out in the form of a book.[54]

As the title suggests, *Reconstructing India* was framed as a program of "reconstruction" following the war. But Visvesvaraya was really interested in a fundamental change in the way the government approached Indian economic problems, war or no war. In a telling passage, he acknowledged that unlike the direct belligerents, "India . . . has had no devastated cities to reconstruct, no ruined homesteads to restore, and no sunken ships to replace." Nevertheless, "reconstruction . . . is more urgently needed in India than in other countries, because political, social and economic developments have been insufficiently considered for many years past, and . . . the standard of living has reached a level far below the minimum recognized in civilized communities as necessary for decent existence."[55] In other words, concerted economic policies, formulated at the national level, were required to reconstruct India after a long period of neglect.

Although the book did not refer explicitly to "planning" or to formal "five-year plans," Visvesvaraya did occasionally use the phrase "national plans" in a more general sense. To prevent "unproductive effort" in the country, he wrote, "A definite move should be made towards building up an Indian nation by outlining national plans and programs in the political, economic and social spheres."[56] Indians must not stand around waiting for market conditions to allow the growth of industries: the experience of countries like Japan showed that "industries and trade do not grow of themselves, but have to be willed, planned and systematically developed."[57]

Visvesvaraya's proposed program was divided into the broad spheres of "political and administrative reconstruction," "economic reconstruction," and "social reconstruction."[58] Accordingly, he suggested a staggering range of measures. He advocated compulsory education at the primary level. He declared that India must have many more universities—beginning with twenty new ones within five years—along with facilities for technical and commercial education. Students and delegations were to be sent abroad on educational tours. Basic industries and other large-scale industrial ventures were to be established with government support and granted economic protection from foreign competition. Agricultural production should be modernized and made scientific. A network of banks was needed to provide finance to industrialists and cultivators. The goal was to make "India a nation economically strong, socially accomplished, and politically a self-governing unit of the British Commonwealth."[59] Once again, as in Mysore, the approach was one of fostering capitalist growth through an interventionist state.[60]

Through these years, Visvesvaraya developed an almost fanatical devotion to quantitative data and statistics. He wrote about it in the press[61] and harped on it in all his publications. In his speeches and writings, he obsessively deployed comparative statistics to drive home his points. When he spoke in 1912 about the need for Mysore to have a university, he noted that Canada, which had only around 1.25 times the population of Mysore, had eighteen universities. He also quoted the ratio of universities to citizens in the United Kingdom and Germany.[62] When he spoke about basic education in India in the 1930s, he compared the literacy rate in the country to that in Great Britain, France, Germany, the United States, Japan, Canada, and Australia.[63] The compilation of statistics, we recall, had been one of the key activities underpinning the Mysore Economic Conference during his years as dewan.

His many interventions on such questions won for Visvesvaraya a reputation as an economic thinker. In 1925, the Government of India asked him to chair a body that came to be known as the Indian Economic Enquiry Committee. (The other members were Pandit Hari Kishan Kaul, a retired civil servant and former commissioner of Lahore, and A. R. Burnett-Hurst, professor of commerce at the Muir Central College in Allahabad University.[64]) This came in the context of debates on the level of taxation the Indian populace could bear. To throw light on this question, the Central Legislative Assembly—which at this time contained a substantial proportion of Indian members, both nominated and elected—had asked the government to undertake an economic survey. (Whether they had any influence in his selection is not known, but the legislature at this time had several members known to Visvesvaraya, including Congress leader Madan Mohan Malaviya and industrialists like Purshotamdas Thakurdas and Lallubhai Samaldas.) The colonial government's Finance Department responded by setting up the Enquiry Committee. However, they stopped short of asking the committee to report on the economic well-being of the people; instead, the committee was to specify *how* such a study might be carried out and what additional sources of data might be required for the exercise. Indian legislators and industrialists considered this a dodge, but the establishment of the committee, even with this narrower mandate, remained an important step.

Visvesvaraya and his colleagues on the committee gathered in Delhi in February 1925. There they met legislators and held discussions before traveling to various provinces where they interviewed witnesses and spoke to government officials. In April and May, this work took them to Allahabad, Calcutta, Rangoon (Burma then being administered along with British India), Madras, Poona, Bombay, and Simla.[65] In total, 108 witnesses submitted written responses; 81 gave their views in person. They included professors and lecturers of economics, government officials, legislators, business journalists, and representatives of chambers of commerce.[66]

The committee investigated a crucial theoretical question: what were the appropriate measures of a country's economic status? After extensive discussion of this and related questions with scores of witnesses, they concluded that the key measures were "income (including production), consumption (including cost of living), and wealth."[67] While the government was already collecting some demographic data and statistics related to agricultural and industrial production, there were no systems in place to calculate parameters such as "income, wealth, cost of living, indebtedness, wages and prices." The committee recommended "intensive enquiries . . . in limited areas in every district" to measure these parameters. They also suggested "a detailed quinquennial census of industrial production," "a comprehensive quinquennial wage census," and the drawing up of "cost of living index numbers . . . for the principal industrial centers." They urged that "all statistical work should . . . be co-ordinated and centralized, the aim being to provide a common purpose and give the statistics an economic trend by means of a central thinking office." A bureaucratic machinery was to be created to give effect to these steps, including Central and Provincial Statistical Bureaus.[68]

These recommendations largely embodied Visvesvaraya and Kaul's thinking. Burnett-Hurst, the British member of the committee, wrote a "Note of Dissent." "As a statistician and economist" (which his colleagues were not), he disagreed with "some of the principles and most of the details of [the report's] recommendations." He found his colleagues' suggestions too ambitious and criticized them for aiming to model the collection of statistics on the practices prevailing in the British Dominions (such as Canada and Australia). India was much larger, and the costs of collecting data would be exponentially greater. The population was predominantly rural, making it difficult to obtain data; the people were largely illiterate; there were too many "scattered unorganized undertakings" when it came to the industrial sector. Burnett-Hurst also disagreed with the measures his colleagues had chosen. There would be little use in the government calculating the national income, which in any case was "very largely an academic exercise which should continue to be left to statisticians, economists and other non-officials." Individual incomes were not as relevant to measure in India as family incomes. Moreover, "the money value of the income of a family is no true index of its economic condition." What the government should focus on was obtaining "a full and detailed description of the standard of living of different classes of the people," which could be done "by collecting and analyzing a large number of family budgets."[69]

Despite Burnett-Hurst's critique, Visvesvaraya's involvement in the Economic Enquiry Committee would have given him a detailed understanding of the kinds of debates economists and statisticians were conducting at the time. In any case, the committee's exertions had established the existence of "a wide-spread desire in the country for an economic enquiry." Moreover, it reinforced the principle that the purpose of such an undertaking would be to "facilitate the shaping of the economic policies and the solution of current economic problems with a view to meet existing deficiencies, improve resources and increase the country's prosperity generally."[70] This was, in effect, an early expression of the need for economic planning.[71]

GLOBAL TRENDS: PLANNING AND ITS VARIETIES

Visvesvaraya was a keen follower of economic policies across the globe, and no doubt the *Zeitgeist* in the 1920s and 1930s further strengthened his belief in what was now being termed *planning*, just as it influenced other prominent Indians like Meghnad Saha.[72] Around the world, interwar debates on economic policies centered on the concept of planning, although there was no consensus on what exactly it meant. Proponents of planning came from different ideological backgrounds and interpreted it in various ways. Lionel Robbins, an economics professor at the London School of Economics, noted in 1937 that "'planning' is the grand panacea of our age. But unfortunately its meaning is highly ambiguous. In popular discussion it stands for almost any policy which it is wished to present as desirable. . . . When the average citizen, be he Nazi or Communist or Summer School Liberal, warms to the statement that 'What the world needs is planning,' what he really feels is that the world needs that which is satisfactory."[73]

Planning was as much a feature of the Nazi German economy as it was of the Soviet economy.[74] Even within a single country, different kinds of planning were conceived of and debated. Daniel Ritschel refers to varieties such as "National Planning," "Socialist Planning," "Capitalist Planning," and "Progressive Planning" in 1930s Britain.[75] Marcia Balisciano, writing of debates at a similar time in the United States, posits that there were four different conceptions of planning: "Social Management Planning," "Technical-Industrial Planning," "Business Economy Planning," and "Macroeconomic Planning."[76] In India too, different schools of thought were to emerge, as Medha Kudaisya argues. Communist thinkers wanted a form of planning with complete nationalization of industries; Nehru and others were not averse to some form of private capital; the colonial government too formulated a version of planning. Only Gandhi, who believed in a decentralized, village-based economy, opposed the idea of planning. But Visvesvaraya, who vigorously promoted planning before most of these actors, fell into none of these categories. For him, industrialization was to be at the core of planning,[77] but as Benjamin Zachariah puts it, he was "less committed to questions of social organization in terms of capitalism or socialism."[78]

Visvesvaraya's conception of planning, in other words, was fundamentally technocratic. The most obvious model for this could have come from the United States, where mechanical engineers like Morris Cooke and Henry Gantt had evinced a strong interest in social engineering. Among their counterparts in the profession of mining engineering was Herbert Hoover, who went on to have a successful career in politics, becoming president of the United States.[79] Balisciano argues that among the schools of planning being discussed in interwar America was "technical-industrial planning," which was strongly influenced by the emergence of the industrial doctrine of scientific management.[80]

Visvesvaraya, a man of action, rarely spent time discussing the genealogy of his beliefs. He probably saw them as ahistorical laws of society. So we do not know if he read Veblen or what he thought of Taylorism. What we do have is a brief mention in his memoirs of a meeting with Hoover. Traveling in the United States after World War I, Visvesvaraya reached Washington, DC and went to see Hoover, already an influential man in the American capital. He later recalled:

> I had a long conversation with Mr. Hoover on various aspects of national development . . . I had learnt he was taking a great deal of interest in the development of American industries. He explained to me how rapidly industries were being developed in his country. I asked him that he thought was wrong in India and why my countrymen were so backward. He replied, "You people have no hustle in you," which meant, of course, that Indians were slow, sleepy and easy-going.[81]

Hoover had no firsthand experience of India.[82] But his analysis appealed to Visvesvaraya, who often repeated the Orientalist trope about Indians being laid back because of the warm climate in their part of the world.[83] To the Indian engineer, industry—in both senses of the word—was the answer to economic problems.

VISVESVARAYA AS PLANNER

It was in the 1930s, with the Round Table Conferences in London leading up toward the drafting of a new Indian constitution, that Visvesvaraya first gave systematic expression to his ideas on planning in a book title *Planned Economy for India*. The book was first published in December 1934, and his intervention owed not a little to his frustration with "the prolonged delay in constitution-framing and the absence of any move to grapple with the unemployment problem."[84]

Although this was probably the first attempt by an Indian to draw up a nationwide plan, Visvesvaraya's intervention did not take place in a vacuum. As Raghabendra Chattopadhyay has shown, the colonial government had proposed its own steps toward economic planning in 1930, when they set up an Economic Advisory Council. Ideas of state intervention were in the air as a result of the Great Depression and Roosevelt's New Deal in America, and the colonial government calculated that an initiative of this kind might stem the political unrest they were contending with at the time. In the next four years, various provincial governments established similar advisory bodies to help steer them out of the depression. Prominent individuals, including the business tycoon G. D. Birla, began to talk of planning.[85] These developments must have encouraged Visvesvaraya to organize his thoughts on planning as they had evolved over the preceding decade.

In *Planned Economy for India*, Visvesvaraya revisited many of the themes he had discussed in his 1920 book. But now he made explicit his endorsement of formal economic planning. Planning was now the norm rather than the exception around the world, he argued. Most countries had historically used "some sort of plan, but plans were usually kept secret" until the Soviet Union came out publicly with its first plan in 1928. Other countries had brought out plans (Italy, Turkey, China, Denmark), and yet others were considering planning in some form (Mexico, Germany, Sweden). In addition, many Western industrialized countries had economic councils. India too must have these two measures, namely, "economic planning and economic organization."

Visvesvaraya suggested the following institutional arrangements. At the national level, there would be an Economic Council made up of "expert economists and leading business men [*sic*]" from across India. The council, which was to assess the needs of the Indian economy and suggest policy measures to increase the nation's wealth, would have a small Standing Committee in Delhi. The central government would create a Development Department, overseen by a cabinet minister. This department would have a Development Committee made up of civil servants "approved by the Economic Council and the Legislature," which would work with the Economic Council's Standing Committee to recommend policies and ensure that planning goals were being met. Analogous institutions (an Economic Council and a Development Department) would be set up at the provincial level, while provinces would prepare five-year plans.

Thus, there was a sort of expert body, the Central Economic Council, that helped formulate plans, and an official body, the Central Development Department, responsible for the success of the plans. The actual implementation of the plans, however, would occur at the level of districts and cities and be taken care of by the regular administrative departments of the provincial governments.[86] This was clearly a technocratic setup, for none of the personnel involved in planning—except the development minister—was democratically elected (though the legislature would have some say in the choice of civil servants in the Development Committee).

Visvesvaraya then proceeded to sketch the lineaments of a ten-year plan for the country as a whole. (He suggested a ten-year plan because the economic data available were insufficient to form the more accurate estimates required for a shorter plan.[87]) The aim of Visvesvaraya's prototype plan was "to double the income of the country within ten years." Targets were set under various heads—industries, agriculture, public works and infrastructure, commerce, finance/banking, (combating) unemployment, and education/ training.[88] A sum of Rs 700 crore (Rs 7 billion) would need to be invested in fresh industrial ventures. Increasingly mechanized agriculture and cooperative farming were to help raise agricultural income by 25 percent, while moving 20 percent of the agricultural workforce to industrial occupations. Roadways would be nearly doubled in extent, as would the installed electric power capacity. Rail coverage would be increased by around 29 percent.[89] The literacy rate—languishing at 8 percent—would be brought up to 50 percent. Targets were also set, *inter alia*, for the production of coal, iron and steel, automobiles, shipping capacity, the capacity of the cotton textile industry, and the number of students in universities.[90]

Some of this would be undertaken directly by the government and some of it by capitalists, whom the government would help with financing. Naturally, the government would need funds for this. For this purpose, they should take out "loans to the extent necessary either from the local public or from foreign countries." By appealing to their sense of public duty, people hoarding wealth should be encouraged to invest it in productive undertakings.[91]

In order for economic planning to be successful, Visvesvaraya stressed, India must have political autonomy. In his seventies now, Visvesvaraya had become much more direct in his criticism of the colonial setup, at least as far as the economic relationship between India and Britain was concerned. "Great Britain wants India as a market for her manufactured products; so there is no incentive to the growth of industries in this country." The consequence of "Dependency Rule," he argued, "has been to lower the country in almost every field of human activity, to foster special interests and to impede the well-being of the vast masses of the population." To tackle this state of affairs, India needed a new constitution that gave it the same level of autonomy as the British Empire's settler dominions. Indians, too, must reform their attitude. Measures were needed "to change the outlook of the people, modernise their business practices and habits, prepare them for independent thinking and collective self-help and bring them into the company of the progressive nations of the world."[92]

This last point, the need for educating the people and modernizing their mindsets, was a key component of Visvesvaraya's worldview as an engineer. It became an increasingly important component of his conception of planning.

In 1937, elections were held under the provisions of the reformed constitution as defined by the Government of India Act 1935. The Indian National Congress came to power in a majority of the provinces of British India, where previously colonial governors had held the bulk of executive power.[93] Visvesvaraya now addressed himself to these Congress governments, producing a short pamphlet, entitled *Nation Building*, that purported to advise them on running a planned economy. He reiterated the institutional setup he had outlined in *Planned Economy*, focusing this time on how the provincial governments should go about creating their five-year plans.

As before, Visvesvaraya was clear that the purpose of these plans was to fight poverty and raise the standard of living of the masses. Now he made an estimate of how much people's incomes would need to be raised. He concluded, based on some back-of-the-envelope calculations, that the average per capita income was Rs. 60 per year. One of the aims of economic planning should be to boost this figure by 50 percent within a decade.[94] In addition to his usual advocacy of heavy industries, Visvesvaraya stressed the importance of cottage and small industries in rural areas and the need for "real Self-Government in villages."[95] This may have been at least partly a gesture to his audience of Congress members, at least some of whom had Gandhian views on the economy.

But nation-building was about more than this: it also required "reforms and reconstruction in other spheres of national life—political, social, intellectual and spiritual."[96] Education was key to create an efficient workforce. Women must be empowered to participate in the formal economy. Indian towns must have "better libraries, hotels, inns, travel bureaux, chambers of commerce, banks, business houses, co-operative stores": all the accoutrements of a commercially vibrant society. He even went to the extent of suggesting the advantages of "a standard dress, a uniform language, besides certain well-recognised international habits and practices in matters pertaining to business, society, travel and self-defence."[97]

VISVESVARAYA, THE CONGRESS, AND THE INDUSTRIALIST CLASS

Visvesvaraya's overtures in *Nation Building*, in addition to his earlier work, made him an obvious candidate for the Congress's ambitious National Planning Committee (NPC), established in 1938 by Subhas Chandra Bose, the party's president, with Jawaharlal Nehru as its chairman. Its establishment was a clear indication that the Congress was preparing for a time in the not-too-distant future when they would form an autonomous all-India government. In its first session, held in Bombay in December 1938, the NPC worked on preparing a questionnaire to be sent out, *inter alia*, to the governments of the various provinces and princely states. In the second session, held the following June, they constituted twenty-seven subcommittees to collect and interpret data under various heads.

Packed with industrialists, scientists, and academics, the NPC was unabashedly expert-driven and urban-centric. Despite opposition from some sections of the Congress, the Nehru-led committee insisted on the importance of promoting large-scale industries alongside cottage industries. In these senses, it was very much in tune with Visvesvaraya's approach. But he soon grew impatient with what he saw as its ponderous progress. In the first session, he asked that the NPC instruct each provincial government to establish at least one significant industrial venture forthwith. His colleagues, however, felt that the committee's role was to advise, not to order. At the second session, Visvesvaraya made a similar demand, only for Nehru to stress the importance of collecting and analyzing data thoroughly before proposing action. Visvesvaraya resigned from the NPC sometime after the second session.

In September 1939, war broke out in Europe. Soon India was swept up in the currents of what became World War II. Soon after, the Congress governments in the provinces quit, objecting to Britain's dragging of India into the conflict. The NPC continued to function for another year, by which time nineteen of its subcommittees tabled their reports. But with the war on and Nehru in and out of prison (for opposing India's involvement in the war), the committee's work went into limbo for the next few years. It reconvened after the war, and a full set of the subcommittee's reports was published a few years later.[98]

Meanwhile, Visvesvaraya kept up his interest in planning, especially its industrial aspect. During the war years, he became the president of a new industrialists' body, the All India Manufacturers' Organization (AIMO), based in Bombay. Under the aegis of the AIMO, he brought out a number of manifestoes on planning and industrial policy.[99] In 1946, he took a team of AIMO members on a study tour of Britain, Europe, and North America.[100]

In 1944–1945, a high-profile plan was released by leading Indian industrialists (including G. D. Birla, J. R. D. Tata, and Kasturbhai Lalbhai) and a few economists. These industrialists were more prominent than those who formed the kernel of the AIMO. Known as the Bombay Plan, it was undertaken in clear anticipation of an independent Indian government at the end of World War II. In it, the industrial class saw itself as working closely with the state to double the per capita income of the Indian population in a fifteen-year period. The state would provide essential infrastructural services and establish heavy industries, while private business would step up the production of goods.[101]

This was in tune with Visvesvaraya's thinking. He was largely in favor of capitalism and what he saw as the civilizing influence of a business-friendly environment, but he was not a free-market liberal of the classical type either, for he clearly believed in large-scale public investment by a strong and decisive state. In the 1930s, he had stated that he wanted his proposed plan for India to "avoid *communistic* tendencies; its basic policy should encourage collective effort without interfering with individual initiative."[102]

It is not clear whether Visvesvaraya was approached by the Bombay Planners to be a part of their endeavor. He was certainly well acquainted with some of India's big industrialists and had strong ties with the Tatas (he served as a director in their steel company for several years).[103] In any case, he was in sympathy with their effort and endorsed it heartily. Some months after the Bombay Plan was released, Visvesvaraya published his own views in a pamphlet titled *Reconstruction in Post-War India: A Plan of Development All Round.*[104]

Incensed by the colonial government's lack of encouragement to Indian industry during World War II, he contrasted the Bombay Plan favorably with the government's preparations for postwar reconstruction.[105] The government had set up committees, but these had only come up with "piecemeal proposals . . . ; proposals mostly unproductive which have no relation to one another and which give no idea of the aggregate financial liability involved."[106] What India needed instead was planned, coordinated policies. His own suggestions were compatible with those of the Bombay Plan.[107] The Bombay Plan had been drawn up for a period of fifteen years. Visvesvaraya suggested that "from the Fifteen-Year Plan, a plan for five years can be deduced and put forward in complete detail." Both the longer and the shorter plans should be refined after seeking expert and public opinion.[108]

Historian Medha Kudaisya argues that the Bombay Plan established the template for the Nehruvian Five-Year Plans and the post-Independence "mixed economy." Among the important ideas they emphasized were the

FIGURE 8.2
Visvesvaraya with Jawaharlal Nehru, probably in 1955. *Source*: Wikimedia Commons.

need to improve substantially the per capita income of the population and the concept of "deficit financing" (the idea that it was legitimate and desirable for the state to incur debt to stimulate economically productive activity).[109] As the preceding discussion shows, Visvesvaraya's 1934 book, *Planned Economy for India*, anticipated the Bombay Plan on both these issues.

By the time India gained Independence, Visvesvaraya was in his mid-eighties and coming toward the end of a long and active public career. He had spent more than three decades in the vanguard of efforts to promote planning in one form or another. Now, due perhaps to his age and his earlier falling out with the National Planning Committee, he was relegated to the sidelines of the planning exercise as it developed under the new national government. Although he and Nehru continued to share a cordial relationship,[110] Visvesvaraya never became a member of the Planning Commission. When the Nehru government focused its energies on public-sector industries under its Second Five-Year Plan (launched in 1956), it was the statistician P. C. Mahalanobis who steered the effort.[111]

Yet, as scholars have argued, Visvesvaraya's body of work had an important influence on planning as it developed under Nehru (figure 8.2). Both believed the state should take the lead in establishing key industries (which, in turn, they believed, would help promote the growth of other industrial sectors). Both established a range of institutions to contribute to economic development. These included Mysore University and the Bank of Mysore in Visvesvaraya's case, as well as the Indian Institutes of Technology (IITs) and the State Bank of India (the latter by nationalizing the Imperial Bank of India) in Nehru's case. The Mysore Economic Conference was a forerunner to the Planning Commission. Visvesvaraya emphasized the need to establish small industries in the villages to generate employment; Nehru established a "community development programme" with similar objectives.[112]

CONCLUSION

Although his career appears to fit neatly into three phases (Bombay engineer, Mysore dewan, and public intellectual), Visvesvaraya's transformation from civil engineer to pundit of planning was neither sudden nor a fluke. It occurred over several decades, and each phase informed the next.

I have argued that the seeds of Visvesvaraya's interest in *planning* were planted in the late nineteenth century when he was a young engineer, even though nobody then used that precise term. Formed in these years, his close friendships with the leading Indian theorists of national economics informed his worldview for the rest of his life.

Visvesvaraya was well primed to imbibe the lessons of Ranade, Gokhale, and their ilk. As an engineer, he agreed with their advocacy of industries and technological education. Indeed, he ratcheted it up further, coming to see industries, urban as well as rural, as the beating heart of what would be a developed India. He also had a Victorian belief in progress and a strongly materialist conception of what it entailed. While he spoke about the cultural and social sides of nation-building, his canvas was fundamentally economic. He was appalled by the poverty and illiteracy he saw around him and made it his life's mission to promote measures to raise the standard of living of the average Indian. This essentially meant more and better production. It required a modernized workforce that would shed the baggage of superstition and outmoded tradition and adopt methods that had proved successful elsewhere in the world.

This emphasis on modernization fit well with his training in colonial educational institutions and his professional socialization in the public works bureaucracy, which was premised on the state's dominion over nature and natural resources. His belief in the importance of an interventionist state was further strengthened by his travels to and lifelong admiration for Japan. His experience of famine while he was an irrigation engineer also made him think deeply about the more efficient use of water. But his solutions went beyond the technical; they also included administrative innovations designed to encourage cultivators to participate more directly in the market economy.

When he became chief engineer and then prime minister of the state of Mysore, he had the opportunity to implement some of his ideas on molding society. He brought to the job his engineering worldview. Like an irrigation engineer who must draw up plans (designs), make estimates, and control variables, Visvesvaraya the dewan surveyed the scene (by amassing statistics), set targets, and put in place structures (institutions) that he believed would produce a transformation in Mysore's populace. He had an extraordinary belief in the power of committees—probably a result of his long service in the colonial bureaucracy. And although he took some steps toward democratizing the polity in his princely state, he was unabashed in placing experts and industrial magnates at the center of his plans for reform.

Visvesvaraya's standing as an ex-dewan of Mysore had much to do with his gaining acceptance as an expert on economic matters. It gave him the experience of negotiating with individuals in powerful positions, from viceroys down to maharajas. It also (along with his stellar record in Bombay) made him a figure the colonial powers were willing to engage with, even if they didn't always trust him absolutely.

On the other hand, his friendship with leading nationalists (albeit of the Moderate variety) and his professional successes at a time when "native" Indian engineers faced many hurdles made him an acceptable figurehead for the cause of national development. His pioneering of the Mysore Economic Conference added to his credentials, as did the many visible achievements—institutions and engineering projects—during his time in Mysore (although his critics maintained these were exercises in vanity). He also entered his career as a public intellectual at a time when constitutional reforms were resulting in a greater degree of political representation at both provincial and central levels.

The legitimation strategy Visvesvaraya himself most used was to refer to his foreign travels and his knowledge of global affairs. He invested time and money to undertake personal study tours of the more prosperous industrial nations (he began traveling abroad at a time when this was taboo to many of his compatriots). He took notes assiduously and collected statistics obsessively, quoting them ad nauseam in his speeches and writings.

In many ways, then, Visvesvaraya's career as a planner would have been impossible without his professional experiences as an engineer. His career in engineering gave him a reputation as a doer, not just a thinker. It gave him the opportunity for foreign travel, endowed him with a materialist philosophy of progress, and led him, for better or worse, to see the industrialized West as the only worthwhile model for the rest of the world. In turn, his own administrative initiatives, combined with his efforts at promoting the importance of planning, were key influences on younger leaders who set out to engineer a new India.

9 THE GLOBALIZATION OF AMERICAN ENGINEERS: US INTERNATIONAL DEVELOPMENT POLICY AND THE ARMOUR RESEARCH FOUNDATION

AURORA GÓMEZ-GALVARRIATO

Engineering developed along with the acceleration of globalization during the last third of the nineteenth century because it was essential to forging the empires industrialized nations built during that period. Thus, engineers became global very early on. They were vital in locating and making available the best and cheapest raw materials that industrial firms required, and mining, agricultural, transportation, construction, and manufacturing companies hired them worldwide. They were also crucial in aiding their governments' efforts to consolidate imperial power.[1] However, technical experts were not mere tools of empire. The authority, prestige, and connections they enjoyed "invested them with power to shape international relations, trade and resource management."[2] As Marsden and Smith point out, engineers were "empire-builders: active agents of political and economic empire."[3]

American engineers became increasingly global through the twentieth century, in tandem with the United States' growing international hegemony. They were crucial in designing and implementing the less formal type of economic imperialism the United States developed, in which soft power played an increasingly larger role than direct territorial control.[4] Engineers' global presence resulted from the expansion of US companies beyond the nation's borders and the government's policies to strengthen the nation's overseas military, economic, and political power.

Nonprofit organizations, such as the Armour Research Foundation (ARF), played a key role in this process since they were the first to find venues of cooperation with foreign countries, charting paths that the government later followed. Their involvement in projects with underdeveloped nations gained clout during World War II and its aftermath as state support for scientific and technological research grew more vigorous. These initiatives began in Latin America, the first frontier of American international expansion. Once decolonization opened new regions to American ventures, this knowledge was applied in Asia and other regions, expanding the global frontier.

The role of technical experts in US imperial expansion began early. In 1899, only one year after the end of the Spanish–American War, they founded in Cuba the Harvard Botanic Station for Tropical Research and Sugar Cane Investigation. This "great open-air laboratory" supplied a training space and a source of samples for US experts to produce agricultural and botanical knowledge on the tropics.[5] Its research on the selection and hybridization of sugar cane and its pests and diseases was vital for US sugar companies in the Caribbean and other tropical regions.

After World War I, as the United States evolved into a world power, its foreign policy increasingly required the services of scientists and engineers to foster its strategies; "bilateral technical aid missions designed to offer American agricultural and economic consulting to many Asian and South American countries proliferated through the interwar years, focused on Latin American countries and the Philippines."[6] During the 1920s and 1930s, several private philanthropic organizations, such as the Rockefeller Foundation, carried out agricultural research and educational programs in Latin America.[7] On the eve of World War II, Nelson Rockefeller, head of the Rockefeller Foundation, convinced President Roosevelt of the need to pursue a policy that committed US government resources to Latin American economic development tied to the strategic interests of the United States. In 1939, Roosevelt appointed Rockefeller to head the newly established Office of International Affairs, which sent diplomatic missions throughout Latin America to examine potential locations for proposed internal improvement programs. In April 1940, following the escalation of World War II in Europe, Roosevelt created the Office of the Coordinator of Inter-American Affairs (OIAA), chaired by Nelson Rockefeller. It carried out projects in Latin America to facilitate rubber production and extract minerals necessary for wartime production in the United States.[8] Also in 1940, Congress appropriated

$500 million in capital funds through the Export–Import Bank to stimulate Latin American industries, and US investment in Latin American nations increased by around $400 million.[9]

After the war ended, President Harry S. Truman's inaugural address in January 1949 portrayed the new direction American foreign policy would take. "The fourth point of his speech proclaimed a 'bold, new program' to redress the economic backwardness and political instability that plagued underdeveloped regions of the world."[10] The "old imperialism," which Truman defined as "exploitation for foreign profit," should be replaced with a constructive project designed to develop the productive capabilities of foreign nations to foster global prosperity and stability. These ideas soon coalesced into the "Point Four Program," which proposed that American experts share their technical knowledge with underdeveloped countries "to create a domestic environment in which collaborative efforts of private capital and local labor could facilitate economic growth and raise standards of living."[11] At the same time, Truman transformed the OIAA into a government-owned corporation, the Institute of Inter-American Affairs (IIAA), to coordinate all assistance programs to Latin American nations under Rockefeller's supervision.[12] While managed by government officials, IIAA and Point Four programs relied heavily upon private investment.[13] In September 1950, the Act for International Development institutionalized these policies as part of US efforts to outbid Communist influence in underdeveloped nations.[14] Instead of large-scale government funding, the act called for coordination among private companies, philanthropic foundations, and supranational organizations to finance development.[15]

Although much has been written on these policies from the perspective of international politics, the role played by technical experts and the organizations that coordinated them remains to be explored. This chapter begins to fill this gap by studying the role played by the ARF and the scientists, engineers, and other technicians who participated in the diffusion of practical technological knowledge to several underdeveloped nations as part of these government programs. As we will see, technical experts were not solely the executors of governments' strategy; they were crucial agents in shaping it through a global network of technical cooperation and in gaining the support of American and local governments and companies to sponsor them.

THE ARMOUR RESEARCH FOUNDATION

The ARF was established in April 1936 as a nonprofit industrial research organization "founded to render scientific and experimental service to the industry."[16] Like other industrial laboratories in the United States during that period, it was "set apart from production facilities, staffed by people trained in science and advanced engineering who work toward deeper understanding of corporate-related science and technology, and who are organized and administered to keep long-term company needs."[17]

Initially, it was affiliated with the Armour Institute of Technology (established in 1892), a private university in Chicago specializing in "practical education." In September 1940, the Armour Institute merged with the Lewis Institute (founded in 1896) to create the Illinois Institute of Technology (IIT).[18] It was conceived to provide the research requirements for industry in the Midwest of the United States, where few companies had facilities for extensive research. Its founders believed that the Institute was better suited than educational institutions to provide these services since "Armour research men are full-time scientists, not teachers" and "when they attack a project for a commercial sponsor, their findings belong to that sponsor, and they remain secret until he chooses to release them. Patent rights are likewise owned by the sponsoring company."[19] The ARF focused on industrial research "aimed at creating or improving specific products, services, methods or processes; reducing costs, or extending technological information."[20] The Institute was unique in its operational process, called "the Armour Plan for Industrial Research." It assumed that every industrial problem required a particular study within a specific scientific field but also demanded engagement with other fields, so the resources of all sections were coordinated within the Institute and geared for cooperative effort.

The ARF started as a relatively small organization with appropriations of $30,000.[21] During its first years, it had only three employees, its premises were limited to three rooms, and it carried out only two industrial projects.[22] However, as World War II increased research demands, the ARF expanded rapidly. In 1942, more than half of the fifty-seven projects on which the ARF was working were directly connected with war efforts, sponsored by the US government or by leading war contractors.[23] Two years later, the ARF employed 300 persons, occupied five buildings, and had a budget of more than 1.5 million dollars.[24] The ARF was involved in projects such as radar, the atomic bomb, and the magnetic recorder.[25] It invented a device to record sound on fine steel wire manufactured exclusively for the US Navy and established the National Registry of Rare Chemicals, a complete catalog of the rarest chemical compounds.[26] It also participated in several international projects as a critical agent of the US government's foreign strategy.

Through the 1950s, many of its projects continued to be related to defense needs, such as developing a new guided missile warhead and "two new weapons of immeasurable value to the armed forces."[27] By 1953, it had become the world's second-largest independent research organization and managed a research portfolio of $12,662,600. Its staff surpassed 1,150, including 725 full-time engineers and scientists who specialized in practically all the fields of science and technology.[28] In 1958, its gross research volume reached $14,648,000, and the ARF carried out 782 research projects, 407 for private industry and 375 for the government.[29] In 1963, the ARF was renamed the IIT Research Institute.

WAR EFFORTS AND THE BEGINNING OF ARF INTERNATIONAL PROJECTS

The ARF's international ventures began in 1941 when Harold Vagtborg, director of the ARF and an expert in municipal and sanitary engineering, was appointed by the National Research Council (NRC) as a leading committee member of an industrial exploration tour of South America.[30] The tour organizers were the NRC, Nelson A. Rockefeller, and Jesse H. Jones, US secretary of commerce. Its purpose was "to assist in the speeding up of industrialization of the more progressive South American countries" and to locate industrial raw materials to support the imminent war.[31]

The tour participants comprised six technologists, research experts, and twenty-one executives of American companies such as Colgate-Palmolive-Peet Company, Goodyear Tire and Rubber Company, International Business Machines, Standard Oil, and the United Fruit Company. The group traveled for seven weeks, with stops at various industrial centers in Colombia, Peru, Chile, Brazil, and Argentina.[32] Its committee prepared a report on the industrial possibilities and limitations they observed for presentation to the Inter-American Development Commission and the Department of Commerce.

Vagtborg's observations show he was looking for ways to advance American ascendancy. "British craft line the harbors of South American ports, greatly outnumbering American and other national's vessels," he remarked. He also observed a strong German presence. There were "sections of or groups in South American republics that are heavily Nazi in sympathy, Chile, for instance," he explained. Nonetheless, he believed that most South American republics saw the United States as a friendly political force, looked forward to increased trade relations, and were "fervent in hope for heavy capital and technological investment of American firms."[33]

The trip allowed him to identify major Latin American problems the ARF could intervene to help alleviate. One of them was a dire need for technology and technical education. Moreover, he noticed that "wherever there has been technological or scientific advance in South America, German influence has had much to do with it." For instance, the Industrial Institute of Santa María, near Santiago, one of the few schools that offered adequate engineering and scientific education, imported almost its entire faculty and all equipment from Germany.[34] The United States could offer channels for Latin American students to study in its universities. Still, a better "solution of the problem was to staff South American schools with a reasonable number of North Americans willing to live for periods of several years at a time abroad."[35] Another major problem he encountered was that "with Brazil as an exception, no extensive scientific or engineering application in the industry has been made on the continent" and that the introduction of large-scale mass-production methods had still to get underway.[36] Vagtborg considered that US investments could help overcome the lack of proper communications and transportation that was a great handicap for the region and "heavy industries which South America lacks so badly can be best supplied by North American interests."[37] He believed that the United States' greater involvement would be beneficial not only to American interests but also to the South American population. "Few of us know the considerable resources for creating a better life, based on a higher standard of living, that South America has within its power to develop," he argued.[38]

The tour opened the way for the ARF to engage in several projects. In April 1941, as part of the tour, NRC committee members met several key Argentine government officials: Raul Prebisch, then general manager of the Argentine Central Bank,[39] and several representatives of the Corporación para la Promoción del Intercambio S.A. (CPI).[40] They sought to explore ways to increase the application of new technologies in Argentine industries to improve their potential export products.[41] In the following months, the CPI, through its New York office, examined the capabilities of several industrial research organizations and selected the ARF.[42]

By then, Argentina's relationship with the United States was strained since it had declared neutrality in the war and failed to negotiate a treaty with the United States to export beef and cereals.[43] Moreover, in retaliation for opposing American interests in the Pan-American conference of Rio de Janeiro, the United States imposed an embargo on weapons, stalled the concession of credits from the Export–Import Bank, and canceled the supply of oil and machinery.[44] Despite these problems, in February 1942, the CPI commissioned the ARF to undertake a broad and comprehensive study of Argentina's industrial and agricultural possibilities. The field party consisted of Dr. Francis W. Godwin, a chemical engineer and assistant director of the ARF, who headed the group; the biochemist Dr. John A. Shellenberger; and the economist Dr. John A. Hopkins, a professor at Iowa State College.[45] The survey was also supported by personnel at the ARF in Chicago, such as Dr. Henry G. Fisk, head of the mineral branch of the Chicago laboratory, who studied the samples collected by the field crew.[46]

The survey's purpose was to take American technology to Argentina, to discern new methods of industrialization possible to Argentina, and to seek opportunities to tap new resources since "Argentina may prove to be a source of some of the critical chemicals needed by America's war effort."[47] The group traveled through seven Argentine provinces to study twenty-six industries, several centers of technological education, agricultural experiment stations, and chambers of commerce. The party found several essential chemical and mineral products in practical quantities of which the United States had a shortage, such as beryllium in the oxide form, butyl alcohol, wood distillation products, and asbestos.[48]

In August 1943, the ARF submitted a report of its thorough exploration.[49] It established criteria to foster industrialization in those sectors that benefited from an abundant and cheap supply of raw materials. It also highlighted the main weaknesses of the Argentine industry, such as the lack of uniform standards and difficulties in shipping and payment conditions, offering solutions to confront them. It underscored companies' extreme lack of technical staff in research departments or laboratories as a significant problem. The report recommended the creation of a national institute of industrial research.[50] The Argentine government did not carry out ARF proposals since, in June 1942, a military coup supported by nationalistic forces ousted the government that had sponsored the survey. The new government pursued policies that promoted industrialization based on import substitution rather than on producing competitive goods for world markets, as the ARF suggested, and reinforced Argentina's neutrality position in World War II. The United States retaliated by freezing Argentine bank deposits in the United States.

Despite its aborted ending, the Argentine project bore some fruits. After spending nine months in Argentina with his family, Dr. Godwin became very knowledgeable and fond of the country and became its advocate. Through 1943, he gave several lectures on Argentina with motion pictures where he explained that despite the government's positions, most Argentinians were strongly sympathetic to the Allied cause.[51] In May, he gave a talk in the offices of Nelson Rockefeller that may have had some influence on Rockefeller's interest in improving the US relationship with Argentina after he became responsible for Latin American affairs in the US Department of State.[52] The US and Argentina subsequently reestablished a diplomatic relationship, and in March 1945, Argentina declared war on the Axis.[53] Moreover, the ARF Argentine experience became the seed that began what was later called "El Plan Armour," meant to provide "technological assistance to enable nations, such as those in Latin America, to import U.S. technology with a minimum depletion of dollar reserves."[54]

MEXICO AND THE CREATION OF THE ARF'S INTERNATIONAL DIVISION

"El Plan Armour was brought to Mexico in 1944 at the request and sponsorship of the Banco de Mexico."[55] Raul Prebisch must have talked about it with several high-ranking officials from Mexico's central bank when he held conferences in that institution.[56] In May 1944, Eduardo Villaseñor, director general of the Bank of Mexico, conferred with Vagtborg and Maurice Holland "to ascertain ways and means in which the U.S. and Mexican industrial research activities can cooperate in assisting Mexican development."[57]

In October, Villaseñor asked the ARF to undertake a nationwide technological study of Mexico. The objective was to "make a technological audit of Mexican industries, both existing and potential with special emphasis on basic fields: hides, leather, solid fuels and by-products, fibers and their industrialization and all types of forest products as well as other industries related to these fields."[58] The report sought "to aid in pointing ways in which Mexican products can be made more generally useful in world markets and in aiding the United Nations' war needs until the end of the war with Japan." Since the end of the war was near, it also was interested "in the general postwar development of the Republic of Mexico and in increasing their home technological development, with a view to a more balanced hemispherical industrial economy." A vital element of the project was training local scientists and engineers so they could propel future development independently by taking them to Chicago "to learn the latest research methods in order that they may further augment Mexican technology in starting new projects after their return to Mexico."[59] The program began in November under the leadership of Dr. Godwin. "The ARF sent a carefully selected group of its staff who made their home in Mexico and made everything possible to absorb a keen understanding of the problems under study."[60] In addition to the field staff sent to Mexico by the ARF, much research was done in Chicago.[61]

As the audit was being prepared, it became clear that Mexico needed an institute for industrial research since it was crucial for developing practical technological capacities in the country and was too costly for Mexican firms to hire consulting abroad. However, given the recent failure in Argentina to create a similar proposal, Godwin realized that it was necessary to convince key Mexican decision-makers of the importance of such an institute. To pursue this objective, he organized a conference in Chicago attended by several of them so they could get firsthand experience with industrial research.[62]

The "Mexican-American Conference on Industrial Research" took place in Chicago from September 30 to October 6, 1945, to show "industrialists of Mexico what research has done for the United States and how it can best be used to pay off in Mexico."[63] It was attended by thirty-six Mexican leaders in industry, government, industry, and universities, including the director of the Bank of Mexico, the minister of economics, and the presidents of Mexico's most influential companies.[64] The group visited several organizations near Chicago, where they were shown fundamental and applied research in action and participated in several discussion panels. Dr. Henry T. Heald, president of IIT, said,

> We want to tell them how research in this country is financed and managed to give them an understanding of the far-reaching benefits to be obtained from the application of science and engineering to industry. We want them to become familiar

> with our systems for training scientists and engineers, the motivations we have found for developing creative ability, and the facilities which we have found necessary for productive research.[65]

While the conference did not induce Mexico's businesspeople to provide resources for the research institute, Villaseñor became so convinced of the importance of the project that the Bank of Mexico sponsored it.[66] The report on the audit of Mexican industries submitted in 1946 included a detailed proposal for creating the Mexican research institute that considered the views expressed at the conference and several reunions held afterward. The model it envisioned was based on the way ARF operated but tailored to fit Mexico's circumstances.[67] In 1947, the Bank of Mexico asked the ARF to initiate its establishment.

In May, the ARF created its International Division, with headquarters in Mexico City, to carry out the project. Dr. Godwin was appointed its director and the chemist, Dr. Milton E. Nelson, its assistant director. Godwin moved to Mexico City while Nelson worked on the project from Chicago. During its first years, its efforts focused on establishing a new complete laboratory and a pilot plant for the practical solution of industrial problems. It was a modern structure occupying 1,500 square feet aimed at "giving Mexican enterprises the same level of scientific aid that enjoyed the industries in the United States."[68] Its objective was "to meet the need—already voiced by officials and business leaders of a dozen countries—for a well-staffed, central, non-governmental, non-profit, technical research agency, with personnel entirely familiar with the technical and industrial picture in the United States."[69] It was the first unit of its kind in Latin America and expected to make a real contribution to the development of the Mexican industry.[70]

At the same time, the group conducted several projects jointly with Mexican engineers and scientists. They included the industrialization of *cascalote* tanning, the domestic extraction of quinine and other antimalaria drugs from the cinchona tree of Chiapas to produce supplies crucial to the war effort, the production of oils for leather treatment with improved methods, and the manufacture of semi-coke from Mexican coals that could save the Mexican forestry by providing a satisfactory fuel for domestic use.[71] It also developed a new wax obtained by a solvent extraction process from the bagasse or waste pulp of sisal fiber from the Yucatan peninsula in the Laboratorios Nacionales de Fomento Industrial that ARF helped establish.[72] However, its most important project was the development of a mass-produced corn tortilla flour acceptable to the Mexican public, which transformed the production of Mexico's most important staple.[73] This innovation had a huge economic and social impact by "free[ing] Mexican women from the daily physical effort required to prepare nixtamal and the corn dough necessary for the tortillas" (see figure 9.1).[74] Private Mexican industrial firms also began posing a steady stream of research problems focused on improving the production of glass, textiles, vitamins, matches, salt, fluorspar, several vegetable oils, and corn products.[75]

According to the plan, the ARF initially managed the laboratory, but gradually, its control was handed to Mexican scientists and engineers trained by the ARF. As Dr. Haldon A. Leedy, director of ARF, explained, it was not easy to produce an effective industrial researcher.[76] "Academic training is effective in acquainting the prospective researcher with scientific principles, but it is only the first step." The other crucial ingredient was on-the-job training, working under a good researcher and with other experienced researchers. This was the only way to enable "the young researcher to acquire and develop the many and varied qualifications he must possess," including "the ability to sell ideas to his superiors, particularly to management; good business judgment, realizing the necessity to make research pay for itself."[77] Working in the ARF laboratories was thus a fundamental part of the training of Mexican technologists. The ARF arranged several fellowships to get this experience in its laboratories in Chicago and Mexico.

From the ARF's perspective, the Mexican endeavor was successful in endowing Mexico with "its own industrial research organization," which was considered the culmination of the development program.[78] On June 1, 1950, the Instituto Mexicano de Investigaciones Industriales (IMIT) was formally established. The IMIT was placed under the administrative direction of Mexican officers. Still, until 1953, the ARF remained in charge of technical matters, maintaining Dr. Nelson as director of laboratories, with control of the ARF's international operations. Thomas P. Collier, a civil engineer, became the manager of the International Division in Chicago.[79]

Although the International Division was at first focused on carrying out the Mexican project, it was from the beginning conceived to serve many developing countries and head the Armour Plan for International Technical Assistance.[80] Its functions were to (1) study natural resources, industry, agriculture, nutrition, and related technical-economic problems of other countries as a major phase of industrial plans and improvement of living

FIGURE 9.1
Advertisement for MINSA, nixtamalized corn flour for tortillas, developed in the laboratories of IMIT in collaboration with the ARF. *Source*: *El Nacional*, September 9, 1958, 14.

standards; (2) conduct laboratory and field research for the solution of specific problems and creation of new industries; (3) aid in the training of additional native technologists; and (4) help in the establishment of industrial laboratories and modern research facilities in other countries.[81]

The International Division enabled the ARF to become a pioneer in exporting American technology and know-how, "a basic factor in the similar development of other countries and in the growth of international trade."[82] It selected its specialists based on their technical skills, linguistic talent, and other attributes of foreign service.[83] Most of them were engineers from universities in the Midwest; some were immigrants, and a few were recruited from abroad. They all had substantial expertise in practical technical applications they learned while working in diverse industries and ARF projects through long careers within the ARF.

The ARF conducted "El Plan Armour" on a nonprofit basis and made special arrangements "to reduce the operating costs and conserve dollar requirements by accepting a considerable percentage of domestic currency from countries facing a shortage of dollars."[84] Thus, the International Division did not represent a relevant funding source for the ARF. In 1948–1949, for example, it raised only $147,000 of the $4,250,000 the ARF made during that year.[85] In 1948, it undertook new projects in Haiti, where it undertook a study to improve sisal operations, and Puerto Rico, where it carried out field tests on laboratory developments in the decortication of coconut fibers.[86] Although Puerto Rico was a US territory, it was handled by the ARF International Division, headquartered in Mexico City, perhaps because the island faced similar problems to those of the Latin American and Caribbean countries with which the division had gained expertise.

THE ARF AND TRUMAN'S POINT FOUR PROGRAM

In 1949, when President Truman promoted the Point Four program, the ARF was well positioned to play a crucial role. The ARF had been undertaking projects based on the same philosophy and method Point Four proposed, sharing its technical knowledge with underdeveloped countries to improve their living standards and redress their socioeconomic environment. The ARF also embraced the ethos of "self-help" that Point Four proposed by working collaboratively with local officials, entrepreneurs, and technicians and training them. Knowledge of the ARF projects might have influenced Truman's address.[87]

Although Congress authorized only modest funding for Point Four, it paved the way for private investment and loans from other organizations, such as the Export–Import Bank, the United Nations, and the Organization of American States, to support its proposed projects. In 1953, the administration of Dwight D. Eisenhower transferred all Department of State technical assistance and foreign aid programs to the new Foreign Operations Administration (FOA).

While the programs initiated under Point Four remained intact, the term "Point Four" became increasingly rare after that year. Eisenhower's administration moved away from small-scale assistance, advocating more extensive government funding for economic development and military aid. The committed government funding for development projects increased dramatically. In the summer of 1954, the Mutual Security Act eliminated the FOA, but the programs created, implemented, and developed under Point Four remained operative. In 1955, the International Cooperation Administration (ICA) replaced the functions of the FOA and became a leading source of international development funding. Technical assistance programs continued into the 1960s.[88]

Mexico's successful experience became a showcase to the ARF. Godwin explained that it had "demonstrated its effectiveness by contributing to the technological needs of the nation at a relatively small cost, assisting in training Mexican technological personnel, and helping Mexico to put its research on a self-sustaining basis."[89] Shortly after Truman's address, requests for ARF to participate in international projects increased. Given the "continually expanding programs of research and development in several Central and South American countries," in March 1949, the ARF created a new post of manager of Mexican Operations, headed by Walter J. Armstrong. Henceforward, the International Division focused on projects in other countries but coordinated between their Chicago offices, led by Thomas P. Collier, and in Mexico City, led by Milton E. Nelson.[90]

In April, the International Division surveyed Latin America's needs for technical assistance at the request of the United Nations Economic Commission for Latin America (ECLA).[91] It stressed the need for more domestically trained industrial and agricultural technologists and proposed ways in which technical agencies in the United States, both private and governmental, could render assistance in the various fields of technology. The report was used in preparing materials for the ECLA Havana conference that took place in May 1949.[92]

Trying to replicate the Mexican experience, the ARF organized a tour for members of the Club of Consuls of Latin America to visit its laboratories in Chicago in June 1949, representing fifteen Latin American countries and Spain.[93] They were offered a luncheon with Dr. Leedy, who discussed "ways in which research can assist Latin Americans in furthering the policy of international technical assistance voiced in President Truman's inaugural address."[94] Then, between October and November 1949, Leedy and Godwin made a one-month trip to Puerto Rico, Brazil, Uruguay, Argentina, Chile, Peru, Colombia, Venezuela, and the Dominican Republic. They held meetings with several financial, industrial, and governmental officials to stimulate interest in scientific research for their technological and industrial advancement. They also explored the possibilities of establishing fellowships for qualified advanced Latin American students to work and learn in the Chicago laboratories of the ARF.[95]

Through 1950, the ARF's International Division carried out a project in Cuba, sponsored by the International Bank for Reconstruction and Development (IBRD), to make a comprehensive survey of its economy and make specific proposals on investment possibilities and methods of raising agricultural and industrial production. Armstrong and Godwin acted as advisers on the survey's industrial phases, coordinating a fifteen-man fact-finding group that spent three months on the island.[96] During the following year, the ARF undertook a project to determine the possible sites and advisability of founding a Construction Materials Development and Training Center in South America, requested by the Department of State's Technical Cooperation Administration, sponsor of the Point Four survey. It was meant to give small rural businessmen an opportunity to learn the latest construction techniques using locally available materials. If they found such a center practical, the ARF would start a three-year project to organize and operate the center, but it was not undertaken.[97]

In June 1951, Thomas P. Collier, manager of the International Division, left for Mexico City to discuss expanded activities of the international division with Dr. Milton E. Nelson, ARF representative there as well as director of laboratories at the Mexican Institute of Technological Research. That year, the ARF began projects beyond Latin America, hiring Paul B. W. Gollong as a research engineer in charge of Far Eastern affairs.[98] By the end of that year, Gollong made "a 35,000 miles globe-girdling research trip" to several Asiatic nations. The trip was the initial phase of a program carried on by the ARF for the federal government to facilitate industrial development in India and

Pakistan, which had recently gained independence, and to assess the opportunities open to the United States.

After returning from the trip, Gollong explained to several potential American investors that "among the chief needs of such nations as India and Pakistan" were "tactful and technologically-qualified men" and that providing them would result in a greater market for American goods.[99] Both nations had a great need to develop metals, cellulose, paper, and other industries, and their modernization, with their consequent expansion, would result in the exportation of American conveyors, tamping devices, shakers, control instruments, and many other items. Also, they would require large amounts of farm machinery to achieve larger food production. Gollong's views must have convinced several key agents to support the ARF to begin several projects as part of the Point Four program.

In February 1952, the International Division assigned John F. Schnur, a foundry technologist, to a project in India to conduct studies of foundry practices and make recommendations for modernized techniques.[100] Schnur taught foundry practices, made recommendations for modernized techniques, and gave lectures "aimed to keep the foundrymen of India well posted with technical advancements and to assist them in changing century-old practices for better and more efficient methods."[101] He visited and worked in about thirty different foundries throughout India, spending between one week and two months at each plant. He traveled around 40,000 miles in that country, some of them by elephant-back, looking for molding sand since the lack of a good, naturally bonded sand was one of the problems that hindered India's foundries. Schnur recommended the best possible sand mixture for each plant, advanced plans to establish sand conditioning plants, helped the foundries improve their pattern making, and taught them modern molding methods. By the end of 1953, Schnur explained that by adopting current practices, Indian foundries increased their production by 400 percent at no additional expense in new equipment. He returned to India in February 1954 to establish a training center for foundrymen at the Indian Institute of Technology in Kharagpur, about 75 miles from Calcutta. He considered it would be more beneficial than spot assistance to individual companies and strike at the root of India's industrial development problem, the lack of skilled engineers and technicians.[102] By September 1954, the foundry training center was a reality.

In November, a new project was undertaken, led by James Dewey Long, an agricultural technologist at UC Davis, working with the Damodar Valley

Corporation (DVC) in India to develop small industrial companies in the 150-mile-long valley, which contained a population of some 6,500,000 people and was rich in several minerals.[103] The project was part of the US Foreign Operations Administration program and also sponsored by the Indian government and those of West Bengal and Bihar. "The Damodar valley project is a good example of the type of program being conducted by India to prove that a democratic government can better the well-being of its people," said Long.[104] It aimed to encourage private industry by upgrading traditional crafts and industries by improving techniques, designs, materials, and equipment, both manual and mechanized, and providing additional market outlets. It also initiated new industries and the establishment of cold storage facilities for locally grown fruits and vegetables to prevent glutting the markets at harvest time.[105]

The ARF also conducted several industrial development projects in Pakistan.[106] In August 1953, Frank R. Mahurin, ceramic technologist of ARF, went to Malakwal to advise and start up a modern fire brick plant that had been recently established.[107] Under a Point Four industrial project, in September, S. Frederick Magis, steel technologist at ARF, surveyed Pakistan's steel plant operations in Karachi, lent technical assistance to speed up production, considered using local resources, and trained plant personnel.[108] At the same time, Howard D. Goodman, an automotive technologist at ARF, was assigned to standardize Pakistan's bus and truck designs, set up mass production systems, and train operators to teach handling, service, and maintenance.[109] Magis observed that slow but steady progress was being made to help Pakistan surmount the main obstacles impeding the development of its steel.[110] After visiting these projects, Leedy noted that ARF's know-how was helping Pakistan, "a young country struggling to build a stable economy."[111]

SPREADING THE TECHNOLOGICAL AUDITING-INDUSTRIAL RESEARCH INSTITUTE MODEL

The Mexican experience encouraged the ARF to replicate the model in other countries, a priority clearly reflected in the cartoon image published in the *ARF News* in 1951 (figure 9.2). In December 1949, the ARF began an exhaustive study of the technological conditions of El Salvador upon the request of Manuel Enrique Hinds, minister of economy. According to Godwin, technological auditing was "the logical first step of a country's development program since it includes specific needs and suggestions which a government

The Foundation of the Future

The *ARF News* artist was feeling somewhat bilious again during the early part of the month, and, one afternoon, when at his lowest ebb, he agreed to draw a page of cartoons depicting what he believes will be the Foundation of the future.

The editor asked that the cartoons represent each of the departments as well as the administration.

To answer the request, the artist, who has talents far exceeding those expected from a man of his craft, took out his crystal ball and began to gaze at it hopefully.

Through the aid of some very black magic—or call it what you will—the artist came up with these 10 cartoons. When last seen, he still was at his crystal ball, obviously impressed by his ability as a seer.

FIGURE 9.2

The Foundation of the Future. The caption reads: "The ARF News artist was feeling somewhat bilious again during the early part of the month, and, one afternoon, when at his lowest ebb, he agreed to draw a page of cartoons depicting what he believes will be the Foundation of the future. . . ." Note the inclusion of international technological audits in the lower left. *Source*: *ARF News*, February 1951, 8.

can use in making subsequent decisions on research investments."[112] Under the leadership of Dr. Nelson, Walter F. Huppke, a consulting chemical engineer, several Salvadorean engineers conducted an intense four-month technical survey to evaluate existing industries and identify the technical education that would be of most value to the country.[113] The ARF submitted its report to the government in May 1950.[114] It remarked on the need for improved irrigation, trade schools, and more engineering courses in the local university. It also urged the formation of a Salvadorean industrial research and development institute.[115]

One year later, the Salvadorean government had already acted on several of its recommendations. The International Bank approved a $12.5 million loan, and the Salvadorean government destined another $5.2 million to build a hydroelectric plant on the Rio Lempa. The government set up a vocational school with assistance from Truman's Point Four program. In addition, it created an industrial bank that included in its mandate the establishment of a scientific laboratory to provide research services to manufacturing and agricultural industries. The Ministry of the Economy and the Central Reserve Bank encouraged programs for expanding the cotton textile industry and reforestation of the country using the *aceituno* tree. In addition, it passed legislation to repeal laws that prohibited importing shoe machinery that enabled the establishment of the country's first mechanical shoe factory. The government also passed bills granting free entry for fishing vessels and special concessions to fishing and canning industries to encourage large-scale fishing enterprises. Finally, salt began being produced by solar evaporation instead of burning large quantities of wood.[116]

The ARF carried out one of its most successful projects in Burma (Myanmar since 1989). It began during the spring of 1953 when Dr. Christopher E. Barthel Jr., ARF's assistant director, was sent on a three-year project to Rangoon to reorganize, expand, and strengthen the State Industrial Research Institute of Burma. He made a detailed analysis of the existing organization and facilities of the Burmese Institute. He supervised its reorganization and enlargement by erecting a modern research facility on a 35-acre tract in Rangoon.[117] Patterned after the ARF, its purpose was to conduct confidential research and development projects for industry, carry out government-sponsored research, offer technical advice and assistance for the modernization of enterprises, and pursue fundamental research problems.[118] The ARF sent several specialists for six months to help establish research programs in such fields as ceramics, metallurgy (ore processing), fiber technology, and extraction analysis of vegetable oils.[119]

By then, some leading figures in the international division in Mexico started to focus on other projects, and in 1954, the ARF ended all internal influence on the Mexican IMIT.[120] In November 1953, Dr. Nelson, who had recently returned from Mexico City, went to Rangoon to oversee the establishment of an applied chemistry section at the institute.[121] After serving three years as director-general of the Union of Burma Applied Research Institute at Rangoon, Dr. Barthel returned to Chicago in May 1956 as assistant director at the ARF.[122] However, "in view of the service he rendered towards the development of scientific and engineering applied research program in Burma," the Burmese government appointed him a member of the Burma Research Council to advise on scientific and engineering matters. Niels C. Beck, acting manager of program development at ARF, became the new director of the Burmese Institute.[123]

ARF assistance was key in the development of Burma's atomic energy program. In 1955, the Union of Burma Atomic Energy Center in Rangoon was formed as a division of the Union of Burma Applied Research Institute.[124] In May 1957, Dr. Richard F. Humphreys, a physicist who had been instrumental in establishing the first private nuclear reactor for industrial research at ARF, went to Burma to advise on this project. Some years later, in July 1959, Clinton A. Stone, supervisor of nuclear physics research at the ARF, went to Burma for several weeks to aid the Burmese government in establishing the nuclear research laboratory, the first step of a complete nuclear program.[125] The ARF also assisted other projects. In January 1954, J. Franklin Miller, a steel mill specialist, went to Burma to study and evaluate bids for the construction of a steel mill built by the government.[126] In October 1958, Joseph E. Fleury, a paper chemistry specialist, joined the institute's staff to advise on programs for making paper from local materials such as bamboo and rice straw.[127] By 1959, the ARF considered that the Union of Burma Applied Research Institute had been "a model of what American technology can do for an underdeveloped country." It offered "Burmese scientists new methods with which to uncover and use the natural wealth of their nation," solved many of Burma's most pressing technological problems, and had been prominent in stimulating economic and industrial growth.[128]

The ARF also helped establish an industrial research institute in Colombia in 1955. With the sponsorship of the Agricultural, Industrial, and Mining

Credit Bank of Colombia, the ARF provided the technical assistance to create the Instituto de Investigaciones Tecnológicas in Bogotá, the first laboratory in the country for technological research.[129] The following year, John Farmer, an ARF mechanical engineer, went to Bogota as a resident engineer of the research center. In September 1957, Dr. Nelson and Dr. C. Roland McCully joined him. Nelson guided the Colombian director of the research institute on administrative techniques, while McCully served as research operations adviser and instructed the Colombian staff in industrial research techniques. In addition, the ARF participated in research on termite resistance of materials for prefabricated houses and field studies on fertilizers and insecticides conducted at experimental stations.[130]

The ARF also aided in establishing the Venezuela Institute of Technical Industrial Research (INVESTI) in Caracas in 1959 and directed research activities in building and construction materials using local natural resources. The Creole Foundation, a nonprofit organization established by Creole Petroleum Corp., the largest oil producer in Venezuela, gave the ARF $75,000 for the project, which was also sponsored by the Venezuelan government. Its objective was to carry out "new fields of research leading to new industrial activities and thereby lessen the country's dependence on petroleum" and to train Venezuelan scientists and engineers to undertake this work in the future. Alexander Populidy, a ceramic engineer at ARF, was the project director.[131]

Other projects were less successful at establishing industrial research institutes. In September 1953, Thomas P. Collier, Irving D. Canton (chief of Latin American operations), and Layton C. Kinney (assistant manager of ARF's chemical engineering department) traveled to Uruguay to make a preliminary technological audit of the country "with a view toward creating in Uruguay an independent industrial research institute to serve the techno-economic needs of that country."[132] The project, financed by Uruguayan capital, was successful at providing critical recommendations for developing the wool textiles industry, but the industrial research institute was never established.[133]

After Uruguay, Collier traveled to Hawaii, along with Gollong and Maurice J. Day, assistant director for program development, on a mission to study the possibility of undertaking a technological audit of Hawaii and find ways to help the island transform into a more diversified economy.[134] Under a grant from the Industrial Research Advisory Council of the Territory of Hawaii, the team visited forty industrial operations, met and talked to 300 business and government leaders, and held fifty individual and group conferences.[135] After an intensive two-week study in January 1954, they concluded that it was necessary to support small- and medium-size industries "to make the Hawaiian economy more self-productive and self-sufficient" and advised establishing an industrial research center. However, these plans were not carried out.[136] As in the case of Puerto Rico, Hawaii was approached by the ARF through similar procedures as those applied to foreign countries despite being a US territory. The ARF also attempted unsuccessfully to establish a research institute in Kabul, Afghanistan. In the spring of 1960, Beck undertook a one-month assignment to explore this possibility. The project was part of a University of Kabul program supported by a grant from the Asia Foundation, a California not-for-profit organization with offices in fourteen Asian countries.[137]

GLOBALIZATION AND THE EXPORT OF AMERICAN TECHNOLOGY

ARF engineers and scientists, who until World War II had not traveled beyond the American Midwest, became frequent world travelers and critical agents of US global expansion. Through the Point Four program, they undertook several research projects fostered initially by war needs and later by postwar US foreign policy priorities. "That the world is shrinking is becoming more apparent every day. ARF's scientists no longer consider Colombia, Burma, and Pakistan as exotic places on the map. To them, these countries are project sites," noted Paul Gollong in 1957. "Exporting American technological know-how to underdeveloped areas of the world has become a specialty with Armour Research Foundation of Illinois Institute of Technology," and its projects were, in its view, "helping to strengthen the free world and at the same time create good will toward this country."[138]

Besides foreign policy and development objectives, the ARF was aware of its projects' economic benefits to the United States. In March 1952, Collier explained to several Illinois engineers that America had twin reasons for encouraging technological advancement worldwide. "First, no less than 73 percent of our imports of strategic and critical materials are shipped to us from under-developed countries." Exporting technical know-how by this country would ultimately lead to "greater production, better quality and reduced prices" of the vital materials we import from other nations. The second reason was that the United States retained its position as a scientific and technological leader. "Many countries which once looked to Europe for their technical leadership now have turned to this country," a turn resulting from "her resourcefulness and ingenuity in building up the fantastic production record of World War Two."[139] In 1957, he noted, "Highly developed countries, as well as underdeveloped areas, turn to the Foundation when they need technological assistance to bolster their national economy; ARF scientists and engineers travel to all parts of the globe to provide this assistance."[140]

Since 1942, when the ARF launched its first overseas activities, the institute had "provided research assistance ranging from industrial development and increased production to modernization and creation of industries, finding uses for raw materials, training personnel, and scientific and business counsel in setting up research centers." By 1959, the ARF had conducted projects for more than twenty-five foreign countries, including nine nations in Latin America.[141] Besides those countries whose projects have been described in the previous pages, it carried out projects in Brazil, Peru, Chile, Costa Rica, the Dominican Republic, Aruba, Jamaica, the Virgin Islands, Lebanon, and Saudi Arabia. It also carried out projects in several developed nations such as Austria, France, Germany, Italy, the Netherlands, Canada, and Sweden. In addition to exporting American know-how, in 1956, the ARF established a European Office in The Hague, Netherlands, to provide scientific and technical liaisons for some twenty American firms to import European technological development ideas for American industry.[142] The European Technical Observation Group provided specialized technical research to European industries, helped keep American science and industry informed on the latest developments abroad, and advised American industries on the feasibility of entering the European common market.[143]

CONCLUSION

The effort and resources that ARF engineers and scientists placed into many development projects worldwide during the 1940s and 1950s are staggering. It contributed to giving shape and content to US foreign policy and global influence during that period. Some were more successful than others, but many have had a profound impact on fostering economic growth and improving living standards. The IMIT, for example, following the principles and methods it learned from the ARF at its inception, until 1994 kept generating technological solutions and providing technical assistance to the country's private and public industries, which were crucial for Mexico's development. The Mexican experience had a pivotal role in shaping ARF's international strategies, and its office in Mexico played a role in some of the ARF's initial global projects. Unfortunately, the sources available for this study, such as ITT news releases, provide the image the ARF wanted to project but not how its projects were perceived by the people they were meant to benefit. It would take an in-depth study of each ARF international project, exploring local sources, to fully assess their domestic short- and long-term consequences as well as their impact on the consolidation of US hegemonic power. Yet, this study suggests that they must have contributed to the direction Truman initially envisioned.

Studying US development policy during this period from the perspective of the research organizations and the engineers and scientists who worked in them shows that despite scant funding, Point Four did not remain "in the realm of abstraction rather than material fact," nor was it "too petty" to produce meaningful results to match Truman's rhetoric, as Gabriel and Joyce Kolko argued.[144] Moreover, it reveals the fundamental role that engineers and scientists played in shaping the postwar modern world, not only through their inventions and innovations but also by actively spreading them throughout the globe. It shows that early on, scientists and engineers realized that technical expertise was not enough to achieve successful international projects; it was also necessary to understand the cultural, political, and social settings for technological development. In addition, it was crucial to support the formation of local experts and cooperate with them through broad international networks. Participating in global projects changed American scientists' and engineers' views of the world and of their work and profession. This study also suggests that despite the need to highlight American government and business objectives in sponsoring ARF projects, and besides their own economic motivations, many scientists and engineers were also driven by the belief that they were undertaking meaningful jobs for the benefit of humanity.

10 CONCLUSION: MOBILITIES AND THE EMERGENCE OF A GLOBAL PROFESSION

EDWARD BEATTY AND ISRAEL G. SOLARES

On May 14, 1964, thousands of people began congregating along the Nile to follow the path of an old ship as it navigated the widest river in the world. Aboard were Gamal Abd al-Nasser, president of Egypt, and Nikita Khrushchev, secretary-general of the USSR, who waved enthusiastically to the crowd. Later, both addressed the thousands of attendees: officials, diplomats, engineers, construction workers, and others. During his speech, the Soviet premier conferred the Order of Lenin on Nasser, who accepted it "in the name of the people of the United Arab Republic."[1] Khrushchev's comments that day were largely directed at the presidents of Algeria, Iraq, and Yemen who attended the ceremony, as Nasser had been vying for leadership among Arab nationalists. The Soviets, meanwhile, were vying for increased influence in the Middle East. After the speeches, Nasser and Khrushchev together pushed a button to dynamite a channel to the Nile, diverting the river's flow to the massive rock- and clay-filled embankment of the new Aswan High Dam. Though still under construction, the structure was designed to enhance water control and economic planning in the Nile Valley. It drew engineering expertise from at least three continents and culminated seven decades of transnational technical (and political) interactions and influences.

Although construction had begun with the technical and financial support of the Soviet Union four years earlier, the dam had come to represent the industrial and nationalist aspirations of the Arab world. It would take six more years to complete, but the crowds gathered that day were already cheering. As the leaders observed cargo ships transporting rocks and sand to mid-river to block the flow of water, off-duty laborers on both sides walked the ground that now cut through the wide waters of the Nile. A Norwegian photographer, Tor Eigeland, captured the moment when an Egyptian worker on one side of the river and a Russian engineer on the other extended their arms across the world's largest waterway to shake hands with one another. As in the picture, Nasser and Khrushchev emphasized the project's transnational alliance of technicians and workers. Nasser had used the massive construction project to bolster his country's technical capacity by hosting Soviet engineers and providing students in Egypt's engineering and technical schools with a final year of job-site training at the dam site, as well as guaranteed jobs on the project after graduation. Many went on to obtain higher degrees in engineering after, some in the USSR and some in Egypt. Khrushchev sent dozens of Soviet engineers and technicians to work on the project, noting in his speech that "we consider they were here not only as technical experts but also as good Socialists."[2]

This highly symbolic event highlights many of the themes in this volume. It shows the expanding technical capacity of the Soviet Union and the global reach of Soviet engineers, as Cold War motivations replaced older imperial aspirations. Soviet leaders had seen technological advance as key to the survival and success of the Communist state since its inception, a virtual "love affair with science and technology."[3] They invested heavily in developing technical capacity, including engineering schools and a large network of research institutes that employed well over twenty thousand engineers and technicians by the end of the 1930s.[4] By the time Khrushchev traveled to Egypt, the number of trained engineers in the Soviet Union had reached over 700,000, making it the country's largest profession, surpassing in sheer numbers those of any other country.[5] Their influence in Egypt, however, was not a simple extension of national technical capacity. British engineers had built the first dam at Aswan at the beginning of the twentieth century. By midcentury, an Egyptian-Greek engineer, Adrian Daninos (educated in France, who had earlier worked with an international group of hydrologists along the lower reaches of the Nile basin), proposed to Egypt's monarchic authorities the construction of a massive new dam just upstream of the original. Danino's plan was initially rejected by the government of King Farouk, who favored a British system of river management that privileged water storage further upstream in Sudan and Ethiopia. Nevertheless, soon after Nassar and the Free Officers Movement took power in 1952, Daninos's project was approved.[6]

Who would provide the financial and technical assistance for the project was not initially clear, and Nasser negotiated on both sides of the Iron Curtain. After frustrating negotiations with the United States and the World Bank resulted in lost opportunities for US support (and US influence), the Soviet Union quickly agreed to a $1.2 billion loan and a technical partnership to complete the project, even as engineers from the US Bureau of Reclamation continued to work along the Sudanese section of the Blue Nile, far upstream.[7] In contrast to the British-led construction of the first Aswan dam a half-century earlier, the new High Dam project was neither a wholly national nor an imperial affair. Negotiations with competing global powers yielded a central Soviet role, adding one more global workplace for highly mobile Soviet technicians and new opportunities for geopolitical alliances. At the same time, the dam provided an opportunity for a generation of Egyptian technical workers and engineers to develop their skills on a major project with global visibility and powerfully nationalist overtones, a symbol of their political and economic independence from Anglo-American influence.

In this book, several scholars have presented cases in which interactions and influences across spatial scales offer a new narrative for a global history of engineering. We argue that this history was heavily influenced by the indigenization of technical practices and knowledge, which played out in an interconnected global arena of technical formation, negotiation, and cooperation. The technical connection between the USSR and the Arab countries, for example, was not limited to direct control of the dam's construction or other similar projects in Iraq, Syria, North Yemen, and Algeria. Students from the Arab countries comprised about 40 percent of all international students from non-Communist countries studying in the USSR.[8] The Soviets sent Russian professors to universities in the Middle East and participated in creating technical institutes across the region. This cooperation was not uncontested, nor did the technical experts formed under Soviet sponsorship necessarily follow the national or imperial interests at play. A number of Iraqi engineers trained to develop the Iraqi energy sector in the 1960s ended up working for Western-owned companies in Iraq, the Emirates, Saudi Arabia, and elsewhere.[9] Egyptian technical experts who were trained in Egypt, like Kawthar al-Subki, head of the chemistry lab at the Aswan High Dam (and the only woman on the project), frequently resented the privileged place of Soviet specialists.[10]

The story of the Aswan High Dam, like the research presented in each chapter of this volume, reaffirms that histories of engineering that are organizationally bounded are insufficient to fully understand the broader pattern of global influences and interactions that have forged the profession. Engineers worked for and within universities, firms, nations, and empires and were responsible for the interests of those institutions. Many engineers also, however, moved across organizational boundaries.

This final chapter reasserts the argument we laid out in the Introduction: the modern profession of engineering was born global in the late nineteenth century, and the work of engineers involves the interaction of global influences with local problem-solving. Cross-boundary, transnational, and often global interactions were part of the tangible reality in the formation of engineers and of the modern profession. We argue that engineers' institutions and practices emerged from and indeed were dependent on a globalized, transnational movement of knowledge, institutional models, and individuals. This concluding chapter synthesizes how the preceding eight chapters illustrate the globally situated interactions and influences that shaped engineering knowledge, institutions, and careers during this period. Only by pulling back from more narrowly focused stories in which institutions, organizations, and nation-states have been foregrounded can we effectively see the cross-border mobility of knowledge, institutions, and individuals and the essential role of global interactions in the emergence of modern engineering.[11]

THE GLOBAL FROM WITHIN

One by one, the chapters of this volume present case studies of engineering and of engineers at work in particular places around the globe. In India, we have seen M. Visvesvaraya earn a degree in civil engineering in 1884 and begin his career as an assistant engineer in the Bombay Public Works Department. He rose rapidly through the ranks and moved from technical to managerial work and then to high-level political leadership and a long career of public service, all the while closely attuned with and frequently traveling internationally to study and learn. In Mexico, engineers trained in the newly reformed National School of Engineers helped establish a Meteorological Observatory in 1877 and hosted a National Meteorological Congress in 1900 with the goal of merging cosmopolitan scientific knowledge with systematic weather observations in the service of the country's agricultural development. Meanwhile, in London, a relatively small number of highly mobile and influential mining and metallurgical engineers offered their services to a globally integrated mining industry as freelance (and often freewheeling) consultants, working at the interface of finance, technical advising, headhunting, and firm-level management. Together with prominent American consulting engineers, some of these British operators found their way to the gold boom in South Africa just before the turn of the century. Shaped by their experience with racially segregated labor regimes in globally dispersed mining districts, these consulting engineers working in the Witwatersrand district debated and designed a labor system that anticipated Apartheid institutions by several decades.

Back in Europe, we can watch the graduates of France's École des Mines of Saint-Étienne, one of the country's important engineering schools, move into employment by state agencies and corporate enterprises. Through the first half of the twentieth century and beyond, young graduates' career paths were not limited to technical work or even to the mining sector and included postings on every continent. On the other side of the world, the pages of the *Far Eastern Review*, published in Manila, provide a window onto the regional movement of technical talent as engineers and entrepreneurs worked to connect local projects with global networks, while Japan and China built domestic institutions to produce indigenous engineering expertise. At the same time, the Hunan province of China became by the 1920s a global leader in antimony production, a position founded on the interaction of engineers and institutions that linked entrepreneurs in Hunan with technicians in France and the United States, shaping the global supply of a critically strategic (but largely anonymous) metal. As a result, several Chinese engineers would become leading figures in US and global minerology and metallurgy. In the US Midwest, the newly established Armour Research Foundation, a technical research and training institute in Chicago, responded to the postwar direction of US foreign policy and over the following decades supported technical training initiatives for engineers and applied scientists in over a dozen countries around the world.

Though just a handful of the thousands of engineering stories that played out around the globe between the mid-nineteenth and the mid-twentieth centuries, these case studies begin to vividly illustrate the themes we identified in the Introduction to this volume. They demonstrate that engineers played diverse and significant roles in each history. Not merely highly trained and interchangeable technicians, engineers worked inside and outside of organizations to design, manage, plan, and sometimes claim ownership shares. They served as consultants, public intellectuals, and political figures. Most importantly, these chapters show that the locally focused problem-solving work of engineers was never conducted in isolation of global influences. The knowledge, expertise, and experience wielded by individual engineers in local workplaces arose out of organizations but depended on global interactions.

As we have noted, engineers and engineering have long been inseparable from histories of institutions and organizations: universities, associations, firms, nations, and empires. Histories of engineering are typically bounded by these kinds of organizations. Sometimes the logic of the historical frame is purely pragmatic, tied to the archival material readily available to the historian. In other cases, histories carry an implicit argument that these institutions define and delimit the profession as places where engineering knowledge, identities, cultures, and practices are forged and where engineering is practiced. Many make this argument explicit, but the studies in this volume tell a somewhat different story, one marked by connections outside organizational and national boundaries.

Around the world, engineers in the last decades of the nineteenth century and into the twentieth increasingly graduated from universities, belonged to national or subnational associations, were employed by private firms or government agencies, and regularly read one or more of the technical publications in their field. These institutional relationships became the hallmark of the professional engineer. Although some people continued to make their way to engineering work and identify themselves as "engineers" without these institutional affiliations, this older path became ever more uncommon. In the twentieth century, engineers typically built their professional reputations, achieved social status, and understood themselves as product and part of these professional bodies, and more importantly, they were the main actors in forming them. Engineers' professional lives were rooted in these institutions, which were situated in specific places or focused on delineated projects at local, national, or imperial scales. Their particular form varied among engineering fields and subfields and between countries, and no two engineering institutions or communities were exactly alike. But they all shared common traits and established dialogues and interactions across the broader, global landscape.

What was the place of the global in this history? In each of the studies in this volume, we see the global emerging as an internal characteristic of the different narratives. Global interactions and influences do not impinge on local action from the outside but materialize out of the work and relationships of engineers on the ground. The chapters in this volume vividly illustrate how modern engineering and the professional work of engineers emerged in the half-century or so between about 1870 and 1930 as a genuinely global phenomenon. This globality was not simply a larger scale of engineering practice, of local engineering initiatives expanding from local to national and global arenas. Nor was it simply a natural result of the universal nature of engineering knowledge. It was intrinsic to the nature of engineering, entwined in interactions and influences that crossed the boundaries of local, national, and imperial scales. In other words, we cannot understand the emergence of engineering without accounting for this global dimension. The *global* dimension of engineering was not a phenomenon engineered from below or imposed from above but rather from within the practice of engineering.

Conventional histories set within and bounded by institutions or nations provide one critical element of this history, but even in institutional and nationally bounded sources, we can find the global emerging from the mobility that characterized the profession. As we see in the preceding chapters, mobility constituted part and parcel of the knowledge that engineers drew on, the institutions that formed them, and the applied, problem-solving experiences and interactions that defined their work. Individual engineers moved; so too did the institutional forms of professional engineering: schools, associations, and technical publications. This mobility was not epiphenomenal, nor was it simply a consequence of engineers' training and work; it was in fact the very matrix out of which the profession emerged.

The mobility that made global influence and interaction possible took several forms. First, engineering knowledge flowed globally with and between individuals and organizations. Codified in print materials—textbooks, blueprints, patents, and technical journals, for example—scientific and technical knowledge carried formal claims of universal application to classrooms, offices, shop floors, work sites, and mining districts around the globe. But codified or explicit knowledge was typically insufficient alone. Problem-solving required adaptation to local circumstances and benefited from the prior experiences of others, contributing to the incessant mobility that has characterized the profession for over a century. Engineers served as vessels of embodied knowledge, carrying with them both book learning and the tacit know-how they accumulated through experience.[12] Second, institutional forms, constituted in university-based curricular programs, training centers, professional associations, and research institutes, were also globally mobile, "exported" through channels of empire and capital, and "imported" by locally based capacity-building initiatives around the world. Finally, individuals moved across borders to gain an education, join engineering associations, seek employment, and pursue career opportunities. The many engineers who did not move, who worked in situ, largely immobile, were themselves a product of mobile institutions and cross-border flows of technical knowledge. The mobility of knowledge, institutions, and individuals was mutually entangled and inseparable, despite our effort to discuss them separately here. All crossed organizational boundaries and extended networks of influence around the world. While engineers often focused their work on local problem-solving, the problem-solving environment transcended local and often national boundaries. The emergence of the engineering profession in the nineteenth and twentieth centuries, we argue, is inseparable from these global mobilities and interactions.

MOBILITY OF ENGINEERING KNOWLEDGE

Engineers' daily work to design, plan, build, troubleshoot, fix, manage, and direct drew from reservoirs of knowledge and expertise acquired through schooling and hands-on experience. Engineering knowledge, in other words, has both codified and tacit components, the former taught in curricular programs and conveyed in print materials, the latter accumulated through practice. In school curricula and in widely circulated textbooks and technical journals, young engineers-in-formation acquired the scientific and technical knowledge of their fields. They studied the principles of physics, mathematics, fluid mechanics, thermodynamics, materials science, and industrial chemistry, all laying claim to universal and scientific truth and relevance. Varying across the curriculum of engineering fields, this constituted a common body of knowledge, available for application in diverse industries, activities, and places. Yet engineers also learned that application and problem-solving in any particular setting presented new and unique challenges: adapting established best practice to a new environment, a different input mix, a distinct institutional or legal context, and—often of most importance—a different cost–benefit calculus necessary to maximize efficiency and minimize cost. Thus learning-by-doing—the acquisition of tacit knowledge—and learning from the experiences of others survived the surge in university-based education and has remained no less important than the acquisition of codified knowledge. Learning-by-doing was often embedded in curricula and continued in the workplace. The retention of some elements of an older tradition of artisanal, trial-and-error work and shop-floor learning recognizes the situation-specific nature of engineering. Of necessity, engineers' work combined universal knowledge with local learning. The mobility of knowledge embodied in individuals, institutions, and print materials has long served to disseminate this hard-won know-how.

The cross-border mobility of engineering knowledge is deeply embedded in each of the case studies presented in this volume. Educational institutions around the world borrowed and adapted the intellectual and scientific content of others' technical training programs and university curricula. School-based curricular programs in the United States, France, India, Mexico City, and China increasingly contained common scientific and technical elements. As new fields of scientific study and technical application were established in universities around the world, their founders and professors depended on access to codified knowledge. This was the case, for instance, of meteorological engineering in Mexico and agronomy in Chile as well as in technical training undertaken by railroad companies in China and industrial projects in India. In Mexico, recognition of the utility of meteorological science for agricultural planning and practices (as well as water management for power generation and flood control) originated in transatlantic flows of scientific and technical knowledge between individuals and scientific societies in the nineteenth century. The construction of irrigation systems and flood control in India, railway systems and machine works in China, cyanidation plants and racialized labor systems in South Africa, and technical research centers worldwide all drew on a shared awareness and circulation of codified engineering knowledge. Globally mobile, that knowledge was translated to local languages, taught in new school programs, acquired through travel and reading, and conveyed and applied directly in the workplace.

Engineering knowledge is rarely neutral or disinterested and always entangled with economic, social, and political interests across different spatial scales. Elisabeth Köll shows how linguistic control over engineering knowledge was a contentious area of dispute for English, French, and German engineers working in China, as well as for Chinese experts trained abroad.[13] The universal and codified knowledge learned in universities traveled widely, embodied in new institutions and in individuals who then served the interests of empire, capital, and nation. Moreover, this knowledge was not solely produced in the West. Historians have often been tempted to portray globalizing mobility as unidirectional, moving from European (or "Western," North Atlantic) centers of modern scientific and technical knowledge production to the rest of the world. Imperial projects and global networks of trade and capital provided powerful conduits for knowledge diffusion in the late nineteenth century. But West-to-the-Rest diffusion comprised only one factor in this history. Technological knowledge and its institutions were also sought, imported, adopted, and adapted by local interests around the world, as we have seen. At the same time, emerging engineering capacities in places like Mexico, India, and China, for example, drew from existing local practices and were shaped by local norms, local factor endowments, and local political economies, as well as by constant interactions with transnational influences.

As Mark Hendrickson demonstrates, the global knowledge frontier in antimony extraction and refining moved from France to China to the United States, with Chinese (specifically, from the Hunan province) metallurgists, entrepreneurs, and officials playing the lead role in what became a globalized twentieth-century story. Aparajith Ramnath's portrait of the secular patron of India's national "Engineer's Day" (September 15) illustrates how that country's engineering institutions and practices arose out of the interplay between British colonial institutions and early nationalist initiatives. Visvesvaraya put global influences to work building India's national capacity.[14] Doug Jones's chapter shows how British and American mining engineers debated and designed the production system for gold in the Witwatersrand basin of South Africa. Assuming a universal metric of cost reduction, they drew on the knowledge they gained in British and American classrooms and adapted it to local conditions, all the while drawing from their globetrotting experiences in other imperial settings, from California and Mexico to Central America and Venezuela. Engineering social labor in South Africa involved the global movement of both technical knowledge and social experiences across imperial and corporate boundaries to produce a highly institutionalized and segregated system based on claims of cost-efficiency, profit maximization, and scientific racism, just one more technical element in the global system of mineral extraction and processing.

The technical press across all engineering fields served a similar purpose, with more regularity and wider dissemination. Whether industry-specific, like the New York–based *Engineering and Mining Journal* (Tuffnell's chapter), or regionally based, like the Manila-based *Far Eastern Review* (Köll's chapter) or in new national publications like Mexico City's *Estudios de Meteorologia Comparada* (Wolfe's chapter), technically oriented publications occupied a critical, mediating space between the scientific and technological principles learned in classrooms and the more tacit, experiential expertise gained in the workplace. The pages of industry-oriented periodicals shared news of efforts to develop and innovate new techniques. Articles, notices, reports from the field, and letters in these journals conveyed information on problems encountered and lessons learned while implementing new technologies in new environments. In print, engineers shared knowledge they could codify into presentations and articles, reports, and letters. They also worked to communicate some of their hard-won and hard-to-codify experiential or tacit knowledge. The boundary between the two was often blurry. Many dozens of such publications appeared in the late nineteenth and early decades of the twentieth centuries in cities and industrial centers around the world. Some endured and enjoyed large regional and sometimes global circulation, while others lasted only months or a few years. Like engineering congresses and engineers' travel, technical publications constituted part of a globally connected network of knowledge exchange.[15]

In the mining sector, these publications were critical venues for the dissemination of both universal knowledge and local experience. Even well-established figures like John Hayes Hammond published anonymous pieces reporting on local work, disseminated worldwide. Mining engineers and metallurgists posed as both neutral arbiters of universal knowledge and as experts able to understand, decode, and explain local problems, as we see in the chapters by Tuffnell, Jones, and Hendrickson. Their role as knowledge intermediaries cemented their position as technical experts, whether situated in the City of London or testifying on labor and race issues in Johannesburg. It also often served their financial interests. This external and universal basis for engineering knowledge reinforced a diffusionist ideal that permeated to different actors in the technical world. Wolfe's and Ramnath's chapters show

how local technocrats adopted Eurocentric or orientalist views about science and technology that viewed the construction of national capacity as another step in the universal application of engineering knowledge. In the case of M.V. in India, planning an economy and building a nation were interchangeable terms for a larger industrialization endeavor, which involved reengineering the productive capacity in the colony. Likewise, the development of meteorology in Mexico, cost-saving mining techniques in South Africa, and the expansion of technical and engineering capacity in French colonies all involved applying claims of universal knowledge to solve local problems and legitimize local work. But trade and technical publications—and, as we will see below, the constant mobility of engineers themselves—also belied these claims. The continual publication of technical reports from engineers in the field highlights how local learning, tapping old and producing new local knowledge, remained central to the success of engineering projects in nearly any setting. As Tuffnell notes, "Global mobility was thus both real and idealized."[16]

Regardless of national origin, engineers' work in offices, factories, and field sites around the world increasingly shared a common foundation in scientific and technical knowledge, in the criteria and objectives of their work, and in the kinds of experiences they accumulated throughout their careers. This is not to say that a single engineering culture characterized all global workplaces. As others have demonstrated, important local, regional, and national priorities and practices shaped distinct engineering cultures and capacities in France and Britain (for example), between the southern and northern regions of the United States, or between India, China, and Japan.[17] However, as the chapters here illustrate, all local practice distilled systematic knowledge and experience through local conditions, exigencies, and experiences, and all attended to the ever-present imperatives of increasing efficiency and reducing costs. None worked in isolation from global patterns and influences. All adopted a common terminology and standardized routines, though specific standards and norms varied across sociopolitical boundaries.

MOBILITY OF INSTITUTIONAL FORMS

Global flows of technical knowledge were embedded in engineering institutions. From training institutes and university programs to professional associations and technical publications, institutional models "traveled," carried by the projects of capital and empire, adopted and adapted by nation builders around the world. Prior to the nineteenth century, highly skilled technical workers who directed large-scale projects—those whom we might call "engineers" in the pre-professional era—typically acquired expertise through apprenticeships and experience. In some places, training was institutionalized in informal workshops around a master practitioner. Technical training eventually found a place in universities. By the late eighteenth century and into the nineteenth, university-based forms of engineering and technical education and training began, slowly, to spread globally.

Initiatives for institutional diffusion often emerged within imperial systems. In 1792, the Spanish Crown established the first institutionalized center for technical training in the Western Hemisphere in Mexico City: the Real Seminario de Minería (the Royal Mining College). A century later, it would grow into by-then independent Mexico's National Engineering University and supply most of the technical personnel who would establish and lead Mexico's early meteorological services, as Wolfe notes in his chapter. In India, British initiatives created four engineering colleges to train Indian students in civil engineering, primarily for work in the colonial Public Works Department. The first of these was the College of Civil Engineering in Roorkee, established in 1847. In 1871, Britain's India Office established the Royal Indian Engineering College near London to train civil engineers for the Indian Public Works Department (and, by the 1880s, for work in Egypt and Uganda). It was closed in 1906 as the engineering colleges on the subcontinent and newer technical training centers and university programs increasingly satisfied demand from the Indian civil service and the private sector.[18]

In other places, initiative and investment came from locally based interests but also drew on international models for their design and orientation. In the United States, the military school at West Point offered the first formal courses in engineering in the 1810s. Colonel Sylvanus Thayer modeled its curriculum after the École Polytechnique in Paris, where he had traveled to study for two years following his education at Dartmouth College. Soon after, the first incarnation of the Renssellaer Polytechnic Institute modeled its early engineering program—still one of just a few in the early Republic—after France's École Centrale. From the Grand Tours of eighteenth- and nineteenth-century Europe to the institution-building efforts of the late nineteenth and early twentieth centuries, transnational efforts to study, borrow, and adapt institutional forms were the norm across North Atlantic countries and beyond.[19]

We see the continued practice of this kind of institutional borrowing in the preceding chapters. In South Africa, the School of Mines and Technology opened in Kimberley in 1896. In India, M. Visvesvaraya's first technical training came in the Poona College of Science, which was established, like the Roorkee school, to train engineers for public works employment in the Bombay Presidency. In China, formal technical training took place via both imperial and locally based efforts. The Jiangnan Arsenal opened in 1865 and under Chinese leadership quickly became an important center for the dissemination of technical knowledge from Europe, including extensive translation work and technical training, often employing European instructors. By the end of the century, European railroad construction companies developed their own in-house training and education programs for Chinese technicians, and the Chinese military school at Tianjin also offered technical training for engineers. Everywhere, new educational and training institutions borrowed inspiration, objectives, and curricular designs from abroad but sought to establish the foundations for local or national technical capacities.

In many countries, local interest in reducing the reliance of the state and private industry on foreign experts yielded a kind of import substitution process for engineering talent or the nationalization of technical capacity ("Mexicanization," "Indianization," or "Sinoization," for example). Mikael Wolfe notes the reforms undertaken by the Mexican national state between 1867 and 1890 to transform the former Royal School of Mines into the National School of Engineering (and by the 1920s into the National Autonomous University of Mexico, with its own Facultad de Ingeniería, soon joined by several new provincial programs). In 1877, graduates of the older engineering program were called upon to establish Mexico's Central Meteorological Observatory, which would advance data collection and meteorological study in Mexico and oversee a national system of regional observatories. The Mexican engineers who led this effort interacted with colleagues at the US Weather Bureau in Washington and the International Meteorological Organization in Vienna. Nearly a century later, that international organization would send a team of technical experts to a major hydrometeorological survey of the upper Nile River catchment basin during the construction of the Aswan High Dam.[20]

In both India and China, the expansion of locally promoted universities with engineering programs created a foundation for the education of Indian and Chinese engineers by the 1920s. Köll's chapter shows how, early on, the global connection between national interests and imperial ambitions generated intercolonial exchanges: in 1898, a Sino-foreign syndicate working in the Henan and Shaxi provinces recruited the chief engineer of the Public Works Department of Bengal, who employed ten Indian surveyors to assess the value of coal and iron ore deposits. In their previous work, Aparajith Ramnath and Elisabeth Köll have both described how British colonial schools (in India) and European railroad firms (in China) established institutions and practices of technical education and training before the turn of the century.[21] In their chapters here, we begin to see the results of these investments. In both places, nationally oriented investment in engineering capacity increased after 1911 as the regions gained a greater degree of political autonomy from European powers. In China, investment in engineering education came in waves coinciding with periods of relative political autonomy and stability, learning and adapting through the changing environment of technical opportunities in the late nineteenth and early twentieth centuries. Locally trained Indian and Chinese engineers increasingly filled public- and private-sector posts. The European instructors, experts, and institutions of the earlier years—typically focused on educating engineers for work in colonial public works departments or European firms—gradually gave way through the first half of the new century to locally supported institutions that sought to train Indian and Chinese engineers for the development of local infrastructure, work in local governments, and employment in new industries. At the same time, maturing technological capacity in Japan attracted attention throughout Asia. Japan often served as a new point of reference and model for emulation as well as, by the 1920s, a source of imperial expansion.

Hundreds of newly constituted professional associations and societies for civil, mechanical, and mining engineers appeared almost concurrently with the founding of university-based engineering programs around the world. In colonial settings, these were sometimes provincial units or offshoots of metropolitan societies. New, locally based associations drew on the experience of those who had studied abroad and been members of the older associations in Germany, Great Britain, or the United States. Many globally mobile engineers belonged to several at once, and some associations offered reciprocal memberships. Swedish emigrant engineers working in Chicago even set up their own professional society.[22] The publications of these associations frequently listed contacts and shared information with others, as Tuffnell notes. In India, the several London-based engineering associations provided a professional identity and status to the initially small number of Indian engineers, but after 1920, the India-based Institution of Engineers brought together all engineering fields to strengthen the Indian profession.[23] In Mexico, the Central Meteorological Observatory joined the international meteorological service only after invitation by the US Weather Bureau. In South Africa, expatriate British, US, and German mining engineers established the South African Association of Engineers (coexisting with the Association of Mine Managers, the Witwatersrand Chamber of Mines, and the Chemical and Metallurgical Society of South Africa, as both Jones and Tuffnell note). Graduates of France's engineering school in Saint-Étienne organized their own alumni association, which provided an institutional foundation for career-long interaction. Elisabeth Köll notes the rapid expansion of membership in Chinese engineering associations from very modest beginnings in 1912 to a robust professionalizing and nationalist agenda by 1931, when two older associations merged to form the Chinese Institute of Engineers, which reached over fourteen thousand members by the late 1940s.[24] Köll's chapter in this volume shows how a binational engineering group, the Association of Chinese and American Engineers, was deeply involved in the foundation and operation of the International Famine Relief Commission during North China's drought crisis in 1920–1921.

Professional congresses and meetings sponsored by national associations offered venues for travel, study, face-to-face interaction, and learning. Whether hosted in Mexico City (like the country's first Congreso de Meteorología Nacional in 1900), in Johannesburg, or in Boston, professional meetings often attracted international audiences and offered a forum for knowledge exchange and socialization. The annual meeting of the American Institute of Metals in Boston in 1917 gave K. C. Li, originally a math teacher at the Junan Technical High School who subsequently studied at London's Royal School of Mines, a forum to present his pathbreaking paper on antimony processing techniques. In 1905, the American Institute of Mining Engineers crossed national borders to hold its annual Congress in Mexico City, explicitly affirming the integration of North American mining and attracting engineers who worked on five continents. Its proceedings were published, disseminated, and subsequently read by engineers across the globe.

By the middle decades of the twentieth century, degree-granting programs were well established around the world, supported everywhere by local economic interests and constituting an integral part of nation-building projects in East Asia, Latin America, and across the Global South. In the years following World War II, new forms of institutionalized research institutes were established in the United States, Europe, and increasingly in other places around the world with both public and private funding. Aurora Gómez-Galvarriato's chapter shows the global reach of the Armour Research Foundation (ARF) in Chicago and its impact on the mobility of engineering knowledge and institutions. In the first "industrial exploration tour" of the ARF to South America, in 1942, the main objective was to counter the heavy influence of German engineers and technology in the Southern cone, which originated in Germany and Chile's military cooperation. The development mission, coordinated by Nelson Rockefeller, had as a main objective pushing US technical influence. Following the war, the institute became part of President Truman's Point Four international development policy and was partially funded by the International Bank for Reconstruction and Development (of the World Bank). In the second half of the century, the ARF organized information and study tours of technical facilities for delegations from around the world who sought to increase technical capacities back home. Like other US-based foundations working transnationally (Ford, Rockefeller, etc.), the ARF sent study teams to over a dozen countries around the world to develop reports and proposals on technical capacity building. It helped establish training centers and research centers in India, El Salvador, Burma (Myanmar), Colombia, Venezuela, and Mexico, where the Instituto Mexicano de Investigación Tecnológica (IMIT) played a major role in national technological innovation for over three decades.[25] The IMIT model proved influential in the construction of similar research institutes, supported by development funds, from El Salvador to Burma.

The history of ARF exemplifies institutional emulation and collaboration on a global scale. The ARF was not just focused on training engineers who could implement their programs abroad. They sought to form technicians and new institutions that would replicate a partnership model between economic planning and technical expertise and promote the creation of lateral, transnational, institutional networks. The key to the success of the ARF's efforts worldwide was the mobility of scientific and technical personnel: face-to-face extended interaction to gather, share, and assimilate technical expertise. In this sense, its work combined the mobility of scientific and engineering knowledge and institutional models, channeled through the circulation of individuals to achieve local capacity building.

MOBILITY OF INDIVIDUALS

Engineering knowledge circulated globally, embedded in formal institutions and carried, codified, in print materials. Mostly, however, it was embodied in the global mobility of thousands of individual engineers who followed cross-border, global pathways opened by increasing flows of trade, investment, and imperial projects in the late nineteenth and early twentieth centuries. These engineers (overwhelmingly men) were professionalized in institutions that arose out of local or national aspirations and initiatives, yet the institutions themselves were the product of cross-border movements as institutional models were "imported," adapted, and adopted in places like West Point and MIT as well as in schools in Asia, Latin America, and beyond. The majority of newly professionalized engineers were largely immobile. They studied and worked within national boundaries, were hired by firms or government agencies, or worked as independent consultants or contractors. But many thousands crossed the organizational and spatial borders of localities, nations, empires, and corporate entities to seek education, work, and opportunity. Highly mobile, they both followed and facilitated the global circulation of technical knowledge and institutional models.

Engineers' geographic mobility was in one sense nothing new. Individuals who combined skilled experience with technical knowledge sufficient to solve cutting-edge problems or undertake large-scale projects have always been relatively scarce. This was as true in the first millennium or late medieval Europe as in the twentieth century.[26] Many engineers have also long been highly mobile, moving to gain knowledge and experience, to associate (and learn from) others, and to parley their expertise into patronage, job offers, and career opportunities in diverse and often distant spaces. In the antebellum United States, skilled mechanics moved frequently across space and between firms, seeking new experience, new knowledge, and building expertise and professional relationships through their careers. They also sometimes traveled and studied in Europe or worked in the United States alongside European engineers. The result was a networked "indigenous American technical community, pursuing a learning trajectory to adapt

European technologies to the American setting."[27] In other words, the early US experience was not unlike that of Japan, India, China, and elsewhere not long after. The strategic mobility of many engineers served both to extend global networks but often originated in a combination of career-building and nation-based interests.

Long-distance mobility of engineers began in universities. In the eighteenth and nineteenth centuries, the first formalized university-based engineering programs in the Netherlands, Germany, and France attracted students from across Europe and, before the expansion of US engineering schools after 1870, from North America. The primary target of international travel for technical training had long been European schools. In mining and metallurgy, the university at Freiberg was the most popular, attracting many US and other international students through the nineteenth century and vividly illustrating the globalized nature of mining expertise.[28] The Kurfürstlich-Sächsische Bergakademie zu Freiberg (KSBF) alumni records indicate that in its first eighty years of existence (1766–1846), about 13 percent of the students came from places outside current-day Germany. By the second half of the century, the school had gained greater popularity as a destination for mining technicians. Between 1866 and 1929, more than half of the university's students came from outside Germany, and 20 percent came from outside Europe. Most of these came from the United States, with gradually descending percentages from Poland, Russia, Austria, Romania, Ukraine, and the United Kingdom, as well as virtually every country in Latin America and Asia, including Japan, India, Indonesia, China, and the Philippines.[29] At roughly the same time, over a third of prospective engineers from Norway, Sweden, Denmark, and Finland sought their formal education in Europe or the United States, hundreds of Latin Americans and many young men (and very occasionally women) from India and elsewhere traveled to study at MIT and other schools in the United States, and the engineering programs of Britain and France received dozens of students from their colonial dependencies.[30]

Freiberg was an exceptional case of integration in the world and provides a clear example of inter-imperial connections. It lay at the center of many of the expansion stories of mining from the mid-nineteenth-century gold rushes and into the new century. The KSBF had a unique international projection in the nineteenth century, hosting significantly higher percentages of international students than the most popular US schools, like MIT and the Colorado School of Mines (CSM), or the Royal School of Mines (RSM) in London. By the turn of the century, however, American institutions had nearly caught up to KSBF, and the RSM had also narrowed the gap.[31]

We have caught glimpses in this volume of young men from India, China, and Mexico whom their families or local sponsors sent to study in European universities and increasingly to the United States.[32] By the 1880s, promising young men in East Asia had more proximate choices, and many chose to study in new Japanese engineering schools, as Elisabeth Köll's chapter indicates. In 1908 alone, over 8,000 Chinese students studied in Tokyo, and around 300 obtained their degrees in the United States.[33] Engineering programs in both the United States and Japan were attracting international students in significant numbers by the 1880s, within a decade of their founding, while European universities continued to attract many others from within their own imperial orbits, as Marco Bertilorenzi notes for the École des Mines at Saint-Étienne. The international mobility of students studying engineering abroad continued into the twentieth century, from the two young Mexicans sent by their government to study meteorology in the United States and Europe in 1908, to the many thousands from India, China, and other countries who have continued to seek educational opportunities across international borders in the decades since.

Newly graduated engineers also crossed borders physically or figuratively to join professional associations. We catch glimpses of cross-border association membership in these chapters: young engineers in India joined Britain's Institution of Civil Engineers while those from Mexico joined the American Society for Civil Engineers or the American Institute of Mining Engineers (AIME) or the binational membership of China-based associations.[34] The AIME began incorporating international members almost immediately after its founding in 1871, reaching 7 percent by 1876, from thirteen different countries, including Mexican, Japanese, and Liberian nationals. In 1881, the association revised its membership rules in part to facilitate international membership, which nearly tripled by 1912. Of these, 25 percent came from Britain, 22 percent from Germany, 18 percent from France, and 15 percent from Belgium, with lower numbers from Austria, Mexico, Russia, Switzerland, Sweden, Liberia, Japan, and beyond.[35] Across all engineering fields, both national and regional associations provided venues for regular meetings where informational exchange occurred formally in the presentation and discussion of papers and informally in conversations and personal relationships. This informational exchange through professional associations and the symbolic construction of a shared identity was manifest in the formal publications of associations as well as in independent technical publications.[36] Written material in texts and technical journals provided a guide to others' experiences and best practices, but individual mobility and the varied experiences and interactions that mobility made possible provided opportunities to adapt and modify common practice, as well as develop tacit and experiential know-how on the shop floor or at the mine-face.[37]

American engineers proved particularly mobile, with mining and metallurgical engineers leading their peers in other fields. Doug Jones's chapter captures the ways in which mining engineers from the United States and Great Britain moved between global mining centers in western North America, South America, and southern Africa in search of employment and career opportunities. Along the way, they accumulated expertise in managing the technical aspects of mineral extraction and refining as well as in race-based strategies for incorporating labor management into the technical system of large-scale mining and refining operations. The majority were not as globally mobile as the elite consulting engineers featured in Steven Tuffnell's chapter. Although based in London (or New York or San Francisco), those men built careers and networks through work on multiple continents. Their human capital was both technical and reputational and depended on the network of globally dispersed relationships they built over time.

London constituted one central node for consulting engineers' extraordinary mobility, expansive networks, and outsized importance, but so too did locations in North America. In one sample of twenty-four prominent mining engineers in the first decade of the century, they averaged, individually, over a dozen career postings, moving back and forth across North America and frequently across international borders. Nearly half spent some time in Leadville or Cripple Creek Colorado, and many closed their careers in San Francisco, Denver, or New York. From their university training through each successive career move, conference attended, and extensive socializing on trains and ships between postings, they each built impressive continental and global networks of professional contacts. In extended interviews, each name-dropped on average thirty colleagues as business partners, role models, friends, and collaborators.[38] Across the ocean and on a larger scale, Marco Bertilorenzi's chapter presents his data set of about 2,600 engineering graduates from France's premier École des Mines de Saint-Étienne between the 1880s and the 1960s. Although based on a single institution, established firmly within a national context, and dedicated to the country's national and imperial aspirations, the careers of students from Saint-Étienne provide a vivid window into their professions' global mobility and the kinds of interactions and employments they achieved.

Mining was a particularly mobile profession, perhaps the most globalized in the period, although civil and mechanical engineers who designed, built, and managed large-scale infrastructure around the world in imperial, corporate, and developmentalist contexts might compete. According to the membership lists of the main professional associations in the United States, around 6 percent of members of the American Institute of Mechanical Engineers, 7 percent of members of the American Society for Civil Engineers, and 15 percent of the American Institute for Electrical Engineers worked abroad with a foreign address, compared to 21 percent of the AIME. These four associations enjoyed rapid and steady growth through the period, although in the first decades of the twentieth century, the newest disciplines, electrical and mechanical engineering, took the lead against the older fields of civil and mining engineering. By 1926, these US associations had a combined membership of over 53,000 engineers.[39]

Each chapter in this volume illustrates the professional mobility of engineers. Many worked directly or indirectly for agencies of imperial powers. English engineers traveled to India or other parts of the British Empire in the employment of colonial public works departments and other agencies. Meanwhile, young Indians gained technical training in England or India and traveled within imperial circuits for employment. Graduates of the École des Mines de Saint-Étienne and dozens of other European and US schools took positions in colonial or quasi-colonial public administration agencies, from Tunisia and Persia for the French engineers, to workplaces on every continent. But the employer of most were the private-sector firms around the world engaged in infrastructure development, extractive and agricultural industries, and, to a lesser extent in this era, manufacturing. While international work comprised a sizable minority of engineers from the industrialized nations, many developing countries saw a majority of national engineers work abroad. In Norway, this constituted a startling 75 percent; in Sweden, about 40 percent.[40] The multinational corporation itself depended on a global circulation of technicians and technical knowledge in the late nineteenth century, just as it would through the rest of the twentieth. Although most histories of multinational corporations describe their global spread as a post-1920 phenomenon, we see a strong presence in the half-century prior. Take, for example, the roughly two thousand mining and metallurgical firms listed in British and American corporate directories between 1880 and 1920. If we were to map the operational locations of these firms and the addresses of their technical employees, the lines connecting individuals to employing forms literally obliterate large sections of the world map, illustrating the density of global connections via corporate operations. The multi-locationality of corporate activities reflected the circulation of capital and commodities but also the circulation of technical knowledge and expertise.[41]

Mark Hendrickson's chapter on China's emergence as the leading center for the extraction and processing of antimony vividly illustrates the globalized influence, interactions, and mobilities on which this knowledge system was based. Engineers and engineering knowledge regarding antimony moved from the Hunan province in China to France to China to the United States and back to Hunan. A handful of young men from Hunan gained formal training and extended their education in Europe and the United States. They contributed to building technological capacity at home and built a network of professional relationships across Europe and the United States. They engaged all the major institutions of the profession on several continents: universities, associations, trade publications, private firms, and state regulators. In doing so, they learned, exchanged, applied, and contributed to a global body of knowledge in antimony and mineral processing.

We have argued that the emergence of modern, professional engineering cannot be understood apart from a global context, begging the question of what constitutes "the global." As we noted in the Introduction, the global is not a single phenomenon. The global dimension of engineering history was (and still is) the accretion of multiple forms of influences and interactions across temporal, spatial, and organizational scales. As the chapters in this book demonstrate, the emergence of modern, professional engineering in any one place and in any one country was shaped by interactions and influences that crossed institutional, organizational, and national boundaries.

Cross-border, transnational, and indeed globally situated patterns of mobility should not be seen as just a consequence of the emergence of modern professional engineering between 1870 and 1950. Engineers may have designed and built much of the infrastructure that facilitated the global expansion of empires and capital during this era. But engineering itself was a product of that globalization: of growing demand for technical expertise in the service of empire and capital, as well as the growing density of interactions and influences across organizational boundaries and spatial scales. This was the case at both institutional and individual levels. New institutions resulted from local initiatives and ambitions but adopted and adapted external models. Engineers drew on codified forms of scientific and technical knowledge that claimed universal application but engaged day-to-day in the locally focused work of problem-solving in particular settings. Engineers and engineering institutions worked in relationship with extensive cross-border networks for knowledge acquisition and sharing.

National boundaries mattered. Nationally based aspirations for industrial and military sovereignty drove investment in technical capacity in the nineteenth and twentieth centuries, from the United States to Japan, India, the Soviet Union, and nearly every other country around the world. National cultures of engineering emerged from locally rooted institutional and social environments. Engineers and the organizations that employed them frequently identified with national interests and sought to further national (or imperial) causes. Many thousands worked for government agencies. National communities and cultures of practice connected engineering expertise across universities, firms, and the state. The linkages between these institutions constituted "national innovation systems" and the foundation for nationally oriented technological "learning."[42]

But national boundaries—national capacities, cultures, and norms—did not fully contain the engineering profession, as the story that opened this chapter and those before illustrate. Nor was the emergence of professional engineering simply a function of state-building or imperialist aspirations, international trade, and outwardly expanding circles of capital investments from the industrializing countries around the North Atlantic. In contrast, cross-border and interorganizational mobility of individuals, institutional models, and engineering knowledge was integral to the emergence of professional engineering even in Europe and the United States, as it would be in every other part of the world. Descriptions of the "spread" or "diffusion" of technical expertise capture one part of what we can observe in these histories. Engineering knowledge, institutional models, and engineers themselves were carried through global channels formed by imperial and capital projects, typically originating in the countries around the North Atlantic. At the same time, local demand for technical expertise around the world sought access to technical knowledge, training, and expertise from those sources. But the spatial patterns of transnational interactions and influences took other forms and directions as well. "Exports" and "imports" of engineering talent were not simply unidirectional, "West to the rest." Borrowing, learning, and exchanging defined the development of engineering in Europe and the United States and continued to shape the development and practice of the profession around the world.

Furthermore, the emergence of professional engineering was not clearly and consistently sequential in nature. It was not the simple result of the diffusion of individuals, institutional models, and knowledge from the North Atlantic. Longstanding practices of engineering had local roots around the world. The first modern institutions to educate, train, and associate engineers were often established in places like Mexico or India at the same time (or sometimes before) as similar institutions in what later became the global frontier in Germany or the United States, for example. Technical training in Chile traces its origins to the early nineteenth century and in Japan well before that country emerged by the 1880s as a new exporter of engineering models, know-how, and technical expertise in Asia. Simultaneity, borrowing, adaptations, and iterative and interactive forms of nation-based initiatives were as common as center-to-periphery diffusion.

Engineers' mobility undermines traditional narratives set within competing organizations—whether corporations, nations, or empires—all of which were sovereign planning agencies. This movement suggests an underlying network (or networks) that shaped the expansion of professional engineering in the twentieth century. If we geo-locate the workplaces of tens of thousands of individual engineers working globally in the decades before and after 1900, along with the organizations that trained and employed them, and strip away all national and continental borders, these locations fully delineate the world's major landmasses. Engineers and their institutions constituted, in other words, a "great blueprint" of newly professionalized technical expertise. Just a few generations after their emergence as a modern profession, engineers worked nearly everywhere.[43] Virtually nonexistent before 1870, these professionalized, university-trained engineers and other technical experts became "the pivotal actors of modernity," devout believers in the universal "authority of their expertise," and essential to the technical, organizational, political, and cultural development of global capitalism in the twentieth century.[44]

NOTES

CHAPTER 1

1. For historically contextualized definitions of "engineering" and "technology," see Eric Schatzberg, *Technology: Critical History of a Concept* (Chicago: University of Chicago Press, 2018), 97–117; Ruth Oldenziel, *Making Technology Masculine: Men, Women and Modern Machines in America, 1870–1945* (Amsterdam: Amsterdam University Press, 1999), 14–16.

2. On the professionalization of engineering in the United States, see David F. Noble, *America by Design: Science, Technology, and the Rise of Corporate Capitalism* (New York: Oxford University Press, USA, 1979), as well as the now-classic works on engineering institutions, including Daniel Hovey Calhoun, *The American Civil Engineer: Origins and Conflict* (Cambridge: Technology Press, Massachusetts Institute of Technology, 1960); Monte A. Calvert, *The Mechanical Engineer in America, 1830–1910: Professional Cultures in Conflict* (Baltimore: Johns Hopkins Press, 1967); Edwin T. Layton, *The Revolt of the Engineers: Social Responsibility and the American Engineering Profession* (Baltimore: Johns Hopkins University Press, 1986); Bruce Sinclair, *A Centennial History of the American Society of Mechanical Engineers 1880–1980* (Toronto: University of Toronto Press, 2019).

3. See the role of technicians in several prominent efforts to assess twentieth-century capitalism: Thorstein Veblen, *The Engineers and the Price System*, vol. 31 (New Brunswick, NJ: Transaction Publishers, 1921); Lewis Mumford, *Technics and Civilization* (Chicago: University of Chicago Press, 2010); John Kenneth Galbraith, *The New Industrial State* (Princeton, NJ: Princeton University Press, 2007); Nicos Poulantzas, *Classes in Contemporary Capitalism* (London: Verso Books, 2018).

4. For recent collections with an international focus, see the special issues of *History of Technology* 23 (2007) and *IEEE Technology and Society Magazine* 25 (2006). For cases in Latin America, see Mílada Bazant, "La Enseñanza y La Práctica de La Ingeniería Durante El Porfiriato," *Historia Mexicana* 33, no. 3 (1984): 254–297; Adolfo Ibáñez

Santa María, "Los Ingenieros, El Estado y La Política En Chile: Del Ministerio de Fomento a La Corporación de Fomento: 1927–1939," *Historia* 18 (1983): 45–102; Raquel Anna Sapunaru, "Uma Breve História Da Engenharia e Seu Ensino No Brasil e No Mundo: Foco Minas Gerais," *Revista de Engenharia Da Universidade Católica de Petrópolis* 10, no. 1 (2016): 39–52; P. C. da Silva Telles, "A History of Engineering Education in Brazil," *IEEE Communications Magazine* 30, no. 11 (November 1992): 66–71; David Pretel and Lino Camprubí, *Technology and Globalisation: Networks of Experts in World History* (London: Springer, 2018); Manuel E. Contreras, *El surgimiento de la ingeniería en Bolivia: Formación y ejercicio de la profesión, 1900–1964* (La Paz, Bolivia: Plural editores, 2022). For Europe: Per-Olof Gronberg, *The Peregrine Profession: Transnational Mobility of Nordic Engineers and Architects, 1880–1930* (Leiden: Brill, 2019); Ben Marsden and Crosbie Smith, *Engineering Empires* (London: Palgrave Macmillan UK, 2005); André Grelon, "Les universités et la formation des ingénieurs en France (1870 à 1914)," *Formation Emploi* 27, no. 1 (1989): 65–88; Göran Ahlström, "Higher Technical Education and the Engineering Profession in France and Germany during the 19th Century: A Study on Technological Change and Industrial Performance," *Economy and History* 21, no. 2 (July 1978): 51–88; Peter Lundgreen, "Education for the Science-Based Industrial State? The Case for Nineteenth-Century Germany," *History of Education* 13, no. 1 (1984): 59–67; Henrik Björck, "A Distinguished Scientific Field? Pursuing Resources and Building Institutions for Engineering Research in Sweden, 1890–1945," *History and Technology* 32, no. 4 (October 2016): 315–348; Giovanni B. Stracca, "Historical Evolution of Engineering Education in Italy," *European Journal of Engineering Education* 5, no. 1 (July 1980): 215–224; Dmitri Gouzevitch and Irina Gouzévitch, "Les corps d'ingénieurs comme forme d'organisation professionnelle en Russie: Genèse, évolution, spécificité (xviiie-xixe siècles)," *Cahiers du monde russe* 41, no. 41/4 (October 2000): 569–614. For Asia: Aparajith Ramnath, *The Birth of an Indian Profession: Engineers, Industry, and the State, 1900–47* (London: Oxford University Press, 2017); Elisabeth Köll, "The Making of the Civil Engineer in China: Knowledge Transfer, Institution Building, and the Rise of a Profession," in *Knowledge Acts in Modern China: Ideas, Institutions, and Identities*, ed. Robert J. Culp, U. Eddy, and Wen-Hsin Yeh (Berkeley: University of California Press, 2016), 148–173; William C. Kirby, "Engineering China: Birth of the Developmental State, 1928–1937," in *Becoming Chinese: Passages to Modernity and Beyond*, vol. 23, ed. Wen-Hsin Yeh (Berkeley: University of California Press, 2020), 137–160; Kevin McCormick, *Engineers in Japan and Britain: Education, Training and Employment*, Nissan Institute/Routledge Japanese Studies Series (London: Routledge, 2002). For African examples: Michel Vandewiele, "History of Western Education in French-Speaking Africa: The Example of Senegal," *Psychological Reports* 53, no. 2 (October 1983): 507–515; W. T. S. Gould, "Technical Education and Migration in Tiriki, Western Kenya, 1902–1987," *African Affairs* 88, no. 351 (1989): 253–271; Osman Lotfy El-Sayed, Juan Lucena, and Gary Downey, "Engineering and Engineering Education in Egypt," *IEEE Technology and Society Magazine* 25, no. 2 (2006): 18–25; Clement Henry Moore, *Images of Development: Egyptian Engineers in Search of Industry* (Cambridge, MA: MIT Press, 1980).

5. There is a growing literature on the role of engineering and technical actors in the formation of standards and globalization. See Gilles Palsky, "The Debate on the Standardization of Statistical Maps and Diagrams (1857–1901). Elements of the History of Graphical Semiotics," *Cybergeo: European Journal of Geography* 85 (March 16, 1999): http://www.cybergeo.eu/index148.html; Nils Brunsson and Bengt Jacobsson, *A World of Standards* (New York: Oxford University Press, 2000); Christopher Spencer and Paul Temple, "Standards, Learning, and Growth in Britain, 1901–2009," *The Economic History Review* 69, no. 2 (2016): 627–652; JoAnne Yates and Craig N. Murphy, *Engineering Rules: Global Standard Setting since 1880* (Baltimore: JHU Press, 2019); Steven W. Usselman, *Regulating Railroad Innovation: Business, Technology, and Politics in America, 1840–1920* (New York: Cambridge University Press, 2002); also Eric Nystrom, *Seeing Underground: Maps, Models, and Mining Engineering in America* (Reno: University of Nevada Press, 2014).

6. Gary Lee Downey and Juan C. Lucena, "Engineering Studies," in *Handbook of Science and Technology Studies*, ed. Sheila Jasanoff, Gerald E. Markle, James C. Peterson, and Trevor Pinch (Thousand Oaks, CA: Sage, 2011), 167–188; John Rae, "Engineers Are People," in *The Engineer in America: A Historical Analogy from Technology and Culture*, ed. Terry S. Reynolds (Chicago: University of Chicago Press, 1991), 27, wrote that "we know a good deal more about engineering works than we do about engineers and much more about who promoted engineering works than about who actually designed and built them." See also Chris Smith and Peter Meiksins, "The Role of Professional Engineers in the Diffusion of 'Best Practice' Concepts: A Comparative Approach," *Economic and Industrial Democracy*, SAGE Social Science Collections, 16, no. 3 (1995): 399–427; Gary L. Downey, Arthur Donovan, and Timothy J. Elliott, "The Invisible Engineer: How Engineering Ceased to Be a Problem in Science and Technology Studies," in *Knowledge and Society: Studies in the Sociology of Science Past and Present*, ed. Lowell Hargens, Robert Alun Jones, and Andrew Pickering (Greenwich, CT: JAI Press, Inc., 1989); see also Alice Lam, "Engineers, Management and Work Organization: A Comparative Analysis of Engineers' Work Roles in British and Japanese Electronics Firms," *Journal of Management Studies* 33, no. 2 (1996): 185; Peter Meiksins and Chris Smith, "Introduction: Engineers and Comparative Research," in *Engineering Labour: Technical Workers in Comparative Perspective*, ed. Peter Meiksins and Chris Smith (London: Verso, 1996), 2, on the "limited scholarly attention" given to engineers.

7. On scientists as opposed to engineers in the study of technical expertise, see, for example, Sheila Jasanoff, "Ordering Knowledge, Ordering Society," in *States of Knowledge: The Co-Production of Science and Social Order*, ed. Sheila Jasanoff (New York: Routledge, 2004), 13–45; Harry Collins and Robert Evans, *Rethinking Expertise* (Chicago: University of Chicago Press, 2007).

8. While modern engineering is often linked to the rise of industrial capitalism, there is substantial work on engineers in socialist and fascist settings; see Timo Luks, "The Factory as Environment: Social Engineering and the Ecology of Industrial Workplaces

in Inter-War Germany," *European Review of History: Revue Europeenne d'histoire* 20, no. 2 (2013): 271–285; John C. Guse, "Nazi Ideology and Engineers at War: Fritz Todt's 'Speaker System,'" *Journal of Contemporary History* 48, no. 1 (2013): 150–174; Thomas R. Wellock, "The Children of Chernobyl: Engineers and the Campaign for Safety in Soviet-Designed Reactors in Central and Eastern Europe," *History and Technology* 29, no. 1 (2013): 3–32; Elena Kochetkova, "Seeing the Forest and the Trees: Western Forestry Systems and Soviet Engineers, 1955–1964," *Technology and Culture* 57, no. 3 (2016): 586–611. On the similarities of engineered landscapes and projects across capitalist and socialist societies, see Kate Brown, "Gridded Lives: Why Kazakhstan and Montana Are Nearly the Same Place," *The American Historical Review* 106, no. 1 (2001): 17–48; Kathryn L. Brown, *Plutopia: Nuclear Families, Atomic Cities, and the Great Soviet and American Plutonium Disasters* (New York: Oxford University Press, 2013).

9. Antoni Estevadeordal, Brian Frantz, and Alan M. Taylor, "The Rise and Fall of World Trade, 1870–1939," *The Quarterly Journal of Economics* 118, no. 2 (2003): 359–407; Kevin H. O'Rourke and Jeffrey G. Williamson, *Globalization and History: The Evolution of a Nineteenth-Century Atlantic Economy* (Cambridge, MA: MIT Press, 1999).

10. Despite a recent increase of scholarly interest in global "circuits of expertise," a global history of engineering has yet to be written. However, see Emily S. Rosenberg, *A World Connecting, 1870–1945* (Cambridge, MA: Belknap Press of Harvard University Press, 2012). Emily S. Rosenberg, "Circuits of Expertise," in *A World Connecting, 1870–1945* (Cambridge, MA: Belknap Press of Harvard University Press, 2012), 919–959; also Gary Lee Downey and Juan C. Lucena, "National Identities in Multinational Worlds: Engineers and 'Engineering Cultures,'" *International Journal of Continuing Engineering Education and Life-Long Learning* 15, no. 3/4/5/6 (2005): 252; Aparajith Ramnath et al., "The Historically Global Forming of Engineering Education," in *International Handbook of Engineering Education Research*, ed. Aditya Johri (New York: Routledge, 2023), 9–29.

11. Several scholars have recently highlighted the particular mobility of the engineering profession, especially in transnational or global contexts. See Gronberg, *The Peregrine Profession*; Ana Simões, Ana Carneiro, and Maria Paula Diogo, *Travels of Learning: A Geography of Science in Europe*, Boston Studies in the Philosophy of Science (Dordrecht: Kluwer Academic Publishers, 2003).

12. J. Rae and R. Volti, *The Engineer in History (Revised Edition)* (New York: Peter Lang, 2001); Richard Shelton Kirby, *Engineering in History* (New York: McGraw-Hill, 1990).

13. Peter Meiksins, Chris Smith, and Boel Berner, *Engineering Labour: Technical Workers in Comparative Perspective* (London: Verso, 1996).

14. On early patent history, see Pamela O. Long, *Openness, Secrecy, Authorship: Technical Arts and the Culture of Knowledge from Antiquity to the Renaissance* (Baltimore: JHU Press, 2001). On technology transfer in the early US, see David J. Jeremy, ed., *International Technology Transfer: Europe, Japan, and the U.S.A. 1700–1914* (London: Edward Elgar, 1991); also David J. Jeremy, "British Textile Technology Transmission

to the United States: The Philadelphia Region Experience, 1770–1820," *Business History Review* 47, no. 1 (1973): 24–52. On Scandinavia, see Kristine Bruland, *Technology Transfer and Scandinavian Industrialisation* (New York: St. Martin's Press, 1991).

15. Engineers, we note, were just one of a number of new types of professionalized technical experts that emerged in the nineteenth century and arguably the most numerous. Others included, for instance, public health and sanitation experts, new categories of medical professionals, economic geologists, and more.

16. O'Rourke and Williamson, *Globalization and History*.

17. Women remained a tiny minority of engineers until well into the second half of the century; see, for example, Nina E. Lerman, *From "Useful Knowledge" to "Habits of Industry": Gender, Race, and Class in Nineteenth-Century Technical Education* (Philadelphia: University of Pennsylvania, 1993); Ruth Oldenziel, "Decoding the Silence: Women Engineers and Male Culture in the U.S., 1878–1951," *History and Technology* 14, nos. 1–2 (November 1997): 65–95; Amy Sue Bix, "Feminism Where Men Predominate: The History of Women's Science and Engineering Education at MIT," *Women's Studies Quarterly* 28, nos. 1/2 (2000): 24–46; Amy Sue Bix, *Girls Coming to Tech! A History of American Engineering Education for Women* (Boston: MIT Press, 2014). Excellent comparative works on the histories of female engineering are Annie Canel and Ruth Oldenziel, *Crossing Boundaries, Building Bridges* (Florence: Taylor and Francis, 2005), and Ramnath et al., "The Historically Global Forming of Engineering Education."

18. Indeed, late medieval universities in Europe, the Islamic world, and East Asia may have been more "international" in their student body than many today. Terri Kim, "Shifting Patterns of Transnational Academic Mobility: A Comparative and Historical Approach," *Comparative Education* 45, no. 3 (2009): 387–403; also Gronberg, *The Peregrine Profession*.

19. On US students in Germany, see Warren Alexander Dym, "Freiberg and the Frontier: Louis Janin, German Engineering, and 'Civilisation' in the American West," *Annals of Science* 68, no. 3 (July 2011): 295–323. On students from India at MIT, see Ross Knox Bassett, *The Technological Indian* (Cambridge, MA: Harvard University Press, 2016); on the influence of MIT in Mexico, see Domingo García Garza, "Aportaciones Para El Análisis de La Cultura Empresarial En La Universidad Mexicana: El Caso Del Tec de Monterrey," *Revista Mexicana de Investigación Educativa* 18, no. 56 (2013): 191–221; also Gabriela Recio Cavazos, *Eugenio Garza Sada. Life and Legacy of a Mexican Businessman* (Monterrey: Editorial Digital del Tecnológico de Monterrey, n.d.).

20. Ramnath et al., "The Historically Global Forming of Engineering Education," 9.

21. For one example that highlights the multifaceted role of engineers, see Victor Seow, *Carbon Technocracy: Energy Regimes in Modern East Asia*, Studies of the Weatherhead East Asian Institute, Columbia University (Chicago: University of Chicago Press, 2022).

22. Mira Wilkins, *The Emergence of Multinational Enterprise: American Business Abroad from the Colonial Era to 1914* (Cambridge, MA: Harvard University Press, 1970).

23. Israel G. Solares and Edward Beatty, "Engineers & Corporate Management, ca 1870–1930: The Invisible Hand Redux," *Enterprise & Society* 25, no. 2 (2024): 1–26, 486–511; Megan Black, *The Global Interior: Mineral Frontiers and American Power* (Cambridge, MA: Harvard University Press, 2018).

24. Stephen Tuffnell, "Engineering Inter-Imperialism: American Miners and the Transformation of Global Mining, 1871–1910," *Journal of Global History* 10, no. 1 (March 2015): 53–76.

25. Edward Beatty, *Technology and the Search for Progress in Modern Mexico* (Berkeley: University of California Press, 2015), chap. 6. On the translation of technical print materials, see Ruselle Meade, "Translating Technology in Japan's Meiji Enlightenment, 1870–1879," *East Asian Science, Technology and Society: An International Journal* 9, no. 3 (September, 2015): 253–274. See also the exchange of technical knowledge among other kinds of experts in Kenya's coffee sector: Paul Ocobock, "Planters Progress" (manuscript, 2021).

26. On conceptualizing "flow" or "circulation" as opposed to "diffusion," see Stuart Alexander Rockefeller, "'Flow,'" *Current Anthropology* 52, no. 4 (August 2011): 557–578. One classic statement on diffusion is George Basalla, "The Spread of Western Science," *Science* 156, no. 3775 (1967): 611–622; in contrast, see Emily S. Rosenberg, "Circuits of Expertise," in *A World Connecting, 1870–1945* (Cambridge, MA: Belknap Press of Harvard University Press, 2012), 919–959.

27. See the large literature on the construction of those scales, including Peter J. Taylor, "The State as Container: Territoriality in the Modern World-System," *Progress in Human Geography* 18, no. 2 (1994): 151–162; David Delaney and Helga Leitner, "The Political Construction of Scale," *Political Geography* 16, no. 2 (February 1997): 93–97; Sallie A. Marston, "The Social Construction of Scale," *Progress in Human Geolography* 24, no. 2 (2000), 219–242; Adam Moore, "Rethinking Scale as a Geographical Category: From Analysis to Practice," *Progress in Human Geography* 32, no. 2 (April 2008): 203–225.

28. Engineering, like most industrial-scale economic activity, is an act of creative destruction; Peter Haff, "Humans and Technology in the Anthropocene: Six Rules," *The Anthropocene Review* 1, no. 2 (August 2014): 126–136; Joseph A. Schumpeter, *Capitalism, Socialism and Democracy* (New York: Routledge, 2013), chap. VII.

29. On the development of university curricula, see Jonathan Harwood, "Engineering Education between Science and Practice: Rethinking the Historiography," *History and Technology* 22, no. 1 (March 2006): 53–79. For the US case, see Karl D. Stephan, "All This and Engineering Too: History of Accreditation Requirements for Nontechnical Curriculum Content in U.S. Engineering Education 1933–2000," in *Proceedings International Symposium on Technology and Society* (Stamford, CT: IEEE, 2001), 155–161; James E. Brittain and Robert C. McMath, "Engineers and the New South Creed: The Formation and Early Development of Georgia Tech," *Technology and Culture* 18, no. 2 (April 1977): 175; Bruce Seely, "Research, Engineering, and Science in American

Engineering Colleges: 1900–1960," *Technology and Culture* 34, no. 2 (April 1993): 344. For international coverage from the 1920s, see also Society for the Promotion of Engineering Education (US), *Report of the Investigation of Engineering Education, 1923–1929: With Report on Summer Schools for Engineering Teachers, 1927–1933* (Pittsburgh, PA: Office of the Secretary of the Society, 1934).

30. Adi Ophir and Steven Shapin, "The Place of Knowledge a Methodological Survey," *Science in Context* 4, no. 1 (1991): 3–22.

31. See, for example, Israel G. Solares and Edward Beatty, "Engineers & Corporate Management, ca 1870–1930: The Invisible Hand Redux," *Enterprise & Society* 25, no. 2 (2024): 486–511, and the chapters in Peter Meiksins, Chris Smith, and Boel Berner, *Engineering Labour: Technical Workers in Comparative Perspective* (London: Verso, 1996); for the case of MIT, see John W. Servos, "The Industrial Relations of Science: Chemical Engineering at MIT, 1900–1939," *Isis* 71, no. 4 (1980): 531–549; for India, see Ramnath, *The Birth of an Indian Profession*; John Black, "The Military Influence on Engineering Education in Britain and India, 1848–1906," *The Indian Economic & Social History Review* 46, no. 2 (April 2009): 211–239. On Mexico, see García Garza, "Aportaciones Para El Análisis de La Cultura Empresarial En La Universidad Mexicana: El Caso Del Tec de Monterrey."

32. US Census counts of engineers reported in David M. Blank and George J. Stigler, "The Supply of Engineers," in *The Demand and Supply of Scientific Personnel*, ed. David M. Blank and George J. Stigler (New York: NBER, 1957), Annex C, 5 and table B-1; US population from US Bureau of the Census, *Historical Statistics of the United States, Colonial Times to 1970* (Washington, DC: GPO, 1975).

33. Engineering program enrollment from David M. Blank and George J. Stigler, *The Demand and Supply of Scientific Personnel* (New York: NBER, 1957), 157. In 1920, the Federated American Engineering Societies celebrated its first conference, comprising twenty-eight member organizations with an estimated total membership of 55,000 engineers; Herbert Hoover was appointed as the first president of the organization. Federated American Engineering Societies, *Organization and Purpose* (Washington, DC: Federated American Engineering Societies, 1922), 3.

34. Robert R. Locke, *The End of the Practical Man: Higher Education and the Institutionalization of Entrepreneurial Performance in Germany, France, and Great Britain—1880–1940* (Greenwich, CT: JAI Press, 1984), 41.

35. Guse, "Nazi Ideology and Engineers at War: Fritz Todt's 'Speaker System,'" 154.

36. European associations include the French Société des ingénieurs civils (1848), the British Society of Engineers (1854), the German Verein Deutscher Ingenieure (1856), and the Spanish Instituto de la Ingeniería de España (1905). On European cases, see Smith and Meiksins, *Engineering Labour: Technical Workers in Comparative Perspective* (London: Verso, 1996). On Germany, see Karl-Heinz Ludwig, Lothar Burchardt, and Verein Deutscher Ingenieure, *Technik, Ingenieure und Gesellschaft Geschichte d.*

Vereins Dt. Ingenieure 1856–1981, (Dusseldorf: VDI-Verlag, 1981). On France, see André Grelon, "Les Écoles d'ingénieurs et La Recherche Industrielle. Un Aperçu Historique," *Culture Technique*, no. 18 (1988): 232–238. On Britain, see R. A. Buchanan, "The Rise of Scientific Engineering in Britain," *The British Journal for the History of Science* 18, no. 2 (July 1985): 218–233. On Russia, see Dmitri Gouzevitch and Irina Gouzévitch, "Les corps d'ingénieurs comme forme d'organisation professionnelle en Russie: Genèse, évolution, spécificité (xviiie-xixe siècles)," *Cahiers du monde russe* 41, no. 41/4 (October 1, 2000): 569–614.

37. For the British Empire, see Denver A. Webb, "Lords of All They Surveyed? The Royal Engineers, Surveying, Mapping and Development in South Africa's Eastern Cape," *African Historical Review* 45, no. 1 (2013): 22–45. For an older account of technology and imperial influence, see Daniel R. Headrick, *The Tentacles of Progress: Technology Transfer in the Age of Imperialism, 1850–1940* (New York: Oxford University Press, 1988). From a large literature on the technology of formal and informal imperialism, with some references to the work of engineers, see, for example, Megan Black, *The Global Interior: Mineral Frontiers and American Power* (Cambridge, MA: Harvard University Press, 2018); Rebecca Tinio McKenna, *American Imperial Pastoral* (Chicago: University of Chicago Press, 2017); Thomas F. O'Brien, *The Revolutionary Mission: American Enterprise in Latin America, 1900–1945*, vol. 81 (New York: Cambridge University Press, 1999); Timothy Mitchell, *Rule of Experts: Egypt, Techno-Politics, Modernity* (Berkeley: University of California Press, 2002); Casper Andersen, *British Engineers and Africa, 1875–1914* (London: Routledge, 2015).

38. See Jan Todd, *Colonial Technology: Science and the Transfer of Innovation to Australia*, vol. 19 (London: Cambridge University Press, 1995). Compare, for example, Headrick, *The Tentacles of Progress*, with Rosenberg, "Circuits of Expertise." Conceptualizing "the West" and "the Rest" is, of course, problematic; lines of demarcation are blurry, overlapping, and dependent on a direction of influence that was historically contested and complicated, and it remains contested by historians today.

39. On technology transfer, see Beatty, *Technology and the Search for Progress in Modern Mexico*, chaps. 1, 8; Edward Beatty, "Approaches to Technology Transfer in History and the Case of Nineteenth-Century Mexico," *Comparative Technology Transfer and Society* 1, no. 2 (2003): 167–197.

40. On colonial Indian Schools, see Ramnath et al., "The Historically Global Forming of Engineering Education," 10; on new engineering schools in early Meiji Japan, see McCormick, *Engineers in Japan and Britain*.

41. For example, Beatty, *Technology and the Search for Progress in Modern Mexico*, chap. 3; Tessa Morris-Suzuki, *The Technological Transformation of Japan: From the Seventeenth to the Twenty-First Century* (New York: Cambridge University Press, 1994), 71–104; Bassett, *The Technological Indian*, 2016; Teresa Cribelli, *Industrial Forests and Mechanical Marvels: Modernization in Nineteenth-Century Brazil* (New York: Cambridge University Press, 2016).

42. See citations in note 4.

43. On engineers in state building: Ramnath, *The Birth of an Indian Profession*; Denis Bocquet, "Engineers and the Nation in Italy (1750–1922): Local Traditions and Different Conceptions of Unity and Modernity," *History and Technology* 23, no. 3 (September 2007): 227–240; Kees Gispen, *New Profession, Old Order: Engineers and German Society, 1815–1914* (New York: Cambridge University Press, 1989), as well as citations in note 4.

44. On students from India at MIT, see Bassett, *The Technological Indian*, 2016; on the influence of MIT in Mexico, see García Garza, "Aportaciones Para El Análisis de La Cultura Empresarial En La Universidad Mexicana." On the use of the MIT model in Iran, see Stuart W. Leslie and Robert Kargon, "Exporting MIT: Science, Technology, and Nation-Building in India and Iran," *Osiris* 21, no. 1 (January 2006): 110–130; Ramnath et al., "The Historically Global Forming of Engineering Education," 11.

45. William C. Kirby, "Engineering China: Birth of the Developmental State, 1928–1937," in *Becoming Chinese*, vol. 23, ed. Wen-Hsin Yeh (Berkeley: University of California Press, 2020).

46. On the creation of technological capacities, see the compilation of case studies in Martin Fransman, "Technological Capability in the Third World: An Overview and Introduction to Some of the Issues Raised in This Book," in *Technological Capability in the Third World*, ed. M. Fransman and K. King (London: Palgrave Macmillan, 1984), 3–30; also William F. Maloney and Felipe Valencia Caicedo, "Engineering Growth: Innovative Capacity and Development in the Americas," (February 20, 1917), *CESifo Working Paper Series*, No. 6339.

47. Derived from the case studies in this volume and the literature cited in note 4.

48. On this kind of technological nationalism, see the case of Argentina and the ambitions of military and car production after the Great Depression: Facundo Picabea and Alberto Lalouf, "'General, Si Usted Me Permite, Yo Le Voy a Fabricar Automóviles En El País': Un Nuevo Abordaje Sobre La Producción Automotriz En La Argentina (1946–1952)," *Apuntes de Investigación Del CECyP*, no. 21 (2012); Ruth Stanley, "Transferencia de tecnología a través de la migración científica: Ingenieros alemanes en la industria militar de Argentina y Brasil (1947–1963)," *Revista iberoamericana de ciencia tecnología y Sociedad* 1, no. 2 (2004): 27. On the role of nationalism in the history of Brazilian industrial research, see Simon Schwartzman and Maria Helena M. Castro, "Nacionalismo, Iniciativa Privada e o Papel Da Pesquisa Tecnológica No Desenvolvimento Industrial: Os Primórdios de Um Debate," *Dados–Revista de Ciências Sociais (Rio de Janeiro)* 28 (1985): 1. This nationalist movement permeated even the nature of traditional engineering in the Commonwealth; see Roy Macleod, "Colonial Engineers and the 'Cult of Practicality': Themes and Dimensions in the History of Australian Engineering," *History and Technology* 12, no. 2 (January 1995): 147–162.

49. For a comparison of engineering capacities in Latin America, see William F. Maloney, Osmel Manzano, and Andrew Warner, "Missed Opportunities: Innovation

and Resource-Based Growth in Latin America [with Comments]," *Economia* 3, no. 1 (2002): 111–167.

50. On the cultural connotation of "technology" broadly, see Nina E. Lerman, "Categories of Difference, Categories of Power: Bringing Gender and Race to the History of Technology," *Technology and Culture* 51, no. 4 (2010): 893–918; Carroll Pursell, "Seeing the Invisible: New Perceptions in the History of Technology," *Icon* 1 (1995): 9–15; Oldenziel, *Making Technology Masculine: Men, Women and Modern Machines in America, 1870–1945*, chap. 1.

51. See Amy Sue Bix's comments in Ramnath et al., "The Historically Global Forming of Engineering Education," 12–15.

52. See citations in note 17.

53. See the case of Bolivia and the early dominance of foreign engineers in the formation of the engineering schools in the twentieth century; Contreras, *El surgimiento de la ingeniería en Bolivia*.

54. For one example, see the comments of Christina Winberg on South Africa in Ramnath et al., "The Historically Global Forming of Engineering Education," 16–19.

55. For example, Beatty, *Technology and the Search for Progress in Modern Mexico*, chap. 8; Jonathan Curry-Machado, *Cuban Sugar Industry: Transnational Networks and Engineering Migrants in Mid-Nineteenth Century Cuba* (New York: Springer, 2011).

56. On engineering and the middle class, see John B. Rae, "Engineers Are People," *Technology and Culture* 16, no. 3 (1975): 404–418, as well as many of the citations in note 4.

57. Geoffrey Blainey, "Herbert Hoover's Forgotten Years," *Business Archives and History* 3, no. 1 (1963): 53–70; Veblen, *The Engineers and the Price System*; Oldenziel, *Making Technology Masculine*, chap. 1.

58. There is a voluminous historical literature on scientific management, including the classic accounts in Chandler, *Visible Hand*, and Noble, *America by Design*. On the global nature of Taylorism, Fordism, and technical dominance on the shop floor, see Stefan J. Link, *Forging Global Fordism: Nazi Germany, Soviet Russia, and the Contest over the Industrial Order*, America in the World (Princeton, NJ: Princeton University Press, 2020). On Henri Fayol, a French mining engineer developing management theory along some similar lines, see Jean-Louis Peaucelle and Cameron Guthrie, *Henri Fayol, the Manager* (London: Routledge, 2015).

59. On engineered racialized labor regimes, see James W. Martin, *Banana Cowboys: The United Fruit Company and the Culture of Corporate Colonialism* (Albuquerque: University of New Mexico Press, 2018); Philip A. Howard, *Black Labor, White Sugar: Caribbean Braceros and Their Struggle for Power in the Cuban Sugar Industry* (Baton Rouge, LA: LSU Press, 2015); Julie Greene, *The Canal Builders: Making America's Empire at the Panama Canal* (New York: Penguin, 2009). On Japanese racialized labor management in

hydroeletrical engineering, see Aaron Stephen Moore, "'The Yalu River Era of Developing Asia': Japanese Expertise, Colonial Power, and the Construction of Sup'ung Dam," *The Journal of Asian Studies* 72, no. 1 (February 2013): 115–139.

60. On social engineering as social reform, see Robbie Aitken, *Exclusion and Inclusion: Gradations of Whiteness and Socio-Economic Engineering in German Southwest Africa, 1884–1914*, vol. 6 (Oxford: Peter Lang, 2007); Uğur Ümit Üngör, "Geographies of Nationalism and Violence: Rethinking Young Turk 'Social Engineering,'" *European Journal of Turkish Studies*, no. 7 (September 23, 2008): 46–71; Aaron Stephen Moore, *Constructing East Asia: Technology, Ideology, and Empire in Japan's Wartime Era, 1931–1945* (Stanford, CA: Stanford University Press, 2020).

61. On the US technocracy movement, see William E. Akin, *Technocracy and the American Dream: The Technocrat Movement, 1900–1941* (Berkeley: University of California Press, 1977); on Europe, see Charles S. Maier, "Between Taylorism and Technocracy: European Ideologies and the Vision of Industrial Productivity in the 1920s," *Journal of Contemporary History* 5, no. 2 (April 1970): 27–61. On engineers in government generally, see Lino Camprubí, *Engineers and the Making of the Francoist Regime* (Cambridge, MA: MIT Press, 2014); Joel Andreas, *Rise of the Red Engineers: The Cultural Revolution and the Origins of China's New Class* (Stanford, CA: Stanford University Press, 2009); Loren Graham, *The Ghost of the Executed Engineer: Technology and the Fall of the Soviet Union* (Cambridge, MA: Harvard University Press, 1993); Susanne A. Wengle, "Engineers versus Managers: Experts, Market-Making and State-Building in Putin's Russia," *Economy and Society* 41, no. 3 (August 2012): 435–467; Janis Mimura, "Planning for Empire," in *Planning for Empire* (Ithaca, NY: Cornell University Press, 2011); Miguel Angel Centeno, *Democracy within Reason: Technocratic Revolution in Mexico* (Philadelphia: Penn State Press, 2010); Eve E. Buckley, *Technocrats and the Politics of Drought and Development in Twentieth-Century Brazil* (Chapel Hill, NC: UNC Press Books, 2017). On planning and developmentalism in nationalist China, see Kirby, "Engineering China," 139–143; on state planning in newly independent India, see Nikhil Menon, *Planning Democracy: Modern India's Quest for Development* (New York: Cambridge University Press, 2022).

62. Herbert Hoover as president of the US (1929–1933), Stanley Baldwin as prime minister of Britain (1923–1924, 1924–1929, 1935–1937), Pascual Ortíz Rubio as president of Mexico (1930–32), M. Visvesvaraya as Diwan (prime minister) of Mysore Kingdom in India (1912–1918), Antonio Mario da Silva as prime minister of Portugal (four times in the 1920s), and Pedro Nel Ospina as president of Colombia (1922–1926) are all examples of technical experts who acquired political power at roughly the same time, most of them in mining and metallurgy. One might add William Lyon Mackenzie King, prime minister of Canada (1921–1926, 1926–1930, 1935–1948), who built a reputation in industrial relations. The prominent place of engineers in state bureaucracies, planning offices, and political leadership in China and the Soviet Union (for instance) is also well documented.

CHAPTER 2

1. W. Valentine Ball, *The Law Affecting Engineers* (London: Constable and Company, 1909), 1.

2. Paul Lucier, *Scientists and Swindlers: Consulting on Coal and Oil in America, 1820–1890* (Baltimore: Johns Hopkins University Press, 2008), 33; Carroll Pursell, "Science and Industry," in *Nineteenth-Century American Science: A Reappraisal*, ed. George H. Daniels (Evanston, IL: Northwestern University Press, 1972), 231–249.

3. Lucier, *Scientists and Swindlers*, 32.

4. Pursell, "Science and Industry," 246.

5. Lucier, *Scientists and Swindlers*, 5.

6. John Hays Hammond, "Professional Ethics," *Transactions of the American Institute of Mine Engineers*, 38 (New York: American Institute of Mine Engineers, 1908), 627.

7. Much of this literature has focused on copper mining; see Logan Hovis and Jeremy Mouat, "Miners, Engineers, and the Transformation of Work in the Western Mining Industry, 1880–1930," *Technology and Culture* 37, no. 3 (1996): 429–456; Sarah E. M. Grossman, *Mining the Borderlands: Industry, Capital, and the Emergence of Engineers in the Southwest Territories, 1855–1910* (Reno: University of Nevada Press, 2018), chap. 5; Timothy LeCain, *Mass Destruction: The Men and Giant Mines That Wired America* (New Brunswick, NJ: Rutgers University Press, 2009); and Duncan Money, *White Mineworkers on Zambia's Copperbelt, 1926–1974* (Leiden: Brill, 2022), 28–42. For gold, see Benjamin Mountford and Stephen Tuffnell, eds., *A Global History of Gold Rushes* (Oakland: University of California Press, 2018), with essays by Andrew Isenberg, Erik Eklund, Cassandra Mark-Thiesen, and Bathsheba Demuth. The shift to nonselective mining also took place in the Malayan tin industry; see Corey Ross, *Ecology and Power in the Age of Empire: Europe and the Transformation of the Tropical World* (Oxford: Oxford University Press, 2017), 136–163; in Mexican silver mining, see Daviken Studnicki-Gizbert, "Exhausting the Sierra Madre: Mining Ecologies in Mexico over the Longue Durée," in *Mining North America: An Environmental History since 1522*, eds., Robert John McNeill and George Vrtis (Oakland: University of California Press, 2017), 19–46; in South Africa's open-pit diamond industry, see William H. Worger, *South Africa's City of Diamonds: Mine Workers and Monopoly Capitalism in Kimberley, 1867–1895* (London: University Press, 1987); and in taconite mining, see Nancy Langstrom, *Sustaining Lake Superior: An Extraordinary Lake in a Changing World* (New Haven, CT: Yale University Press, 2017), 79–106.

8. On the historical development of management consulting, tasks increasingly taken on by mining and engineering consultants, see Christopher D. McKenna, *The World's Newest Profession: Management Consulting in the Twentieth Century* (Cambridge, UK: Cambridge University Press, 2006), 12–13. On the questions of expertise and authority at the heart of consulting, see Graeme Gooday, *The Morals of Measurement: Accuracy,*

Irony, and Trust in Late Victorian Electrical Practice (Cambridge, UK: Cambridge University Press, 2004).

9. Questions of professional values were discussed across all branches of engineering; see Dale H. Porter and Gloria C. Clifton, "Patronage, Professional Values, and Victorian Public Works: Engineering and Contracting the Thames Embankment," *Victorian Studies* 31, no. 3 (Spring, 1988): 319–449; Stathis Arapostathis, "Morality, Locality and 'Standardization' in the Work of British Consulting Electrical Engineers, 1880–1914," *History of Technology* 28 (2008): 53–74.

10. "Admission to Mines," *Engineering and Mining Journal*, September 22, 1904, 459 (hereafter: *EMJ*).

11. Gary Lee Downey and Juan C. Lucena, "Engineering Studies," in *Handbook of Science and Technology Studies*, ed. Sheila Jasanoff, Gerald E. Markle, James C. Peterson, and Trevor Pincher (Thousand Oaks, CA: Sage, 2011), 167.

12. "Protecting Mine-Investors," *Mining and Scientific Press*, October 9, 1909, 479 (hereafter: *MSP*).

13. James H. Curle, *Gold Mines of the World: Containing Concise and Practical Advice for Investors Gathered from a Personal Inspection of the Mines of the Transvaal, India, West Australia, Queensland, New Zealand, British Columbia, and Rhodesia* (London: Waterlow and Sons, 1899), 7.

14. "Consulting Engineers and Their Work," *Engineering*, March 27, 1903, 428.

15. Chalmers and Hatch, *Gold Mines of the Rand*, 206.

16. On these engineers' approach to the quantification of racial capacity, see chapter 4 of this volume.

17. Alan Patrick Cartwright, *Gold Paved the Way: The Story of the Gold Fields Group of Companies* (London: Macmillan, 1967), 79, 94; Harry H. Webb and Pope Yeatman, "Milling Gold Ores on the Witwatersrand," *Engineering Magazine*, June 1898, 407 (hereafter *EM*); Witwatersrand Chamber of Mines, *The Mining Industry: Evidence and Report of the Industrial Commission of Enquiry* (Cambridge, MA: Cambridge University Press), 245; John Alexander Chalmers and Frederick Henry Hatch, *The Gold Mines of the Rand: Being a Description of the Mining Industry of Witwatersrand, South African Republic* (London: Macmillan, 1895), 206. To put S&J's mill in perspective, North Dakota's Homestake Mine had a 900-stamp mill, the Alaska Treadwell, 540.

18. John Yates, *Present Day Metallurgical Engineering on the Rand* (London: Mining Journal, 1898), 39; Manuel Eissler, *The Cyanide Process for the Extraction of Gold and Its Practical Application on the Witwatersrand Gold Fields in South Africa* (London: Crosby Lockwood and Son, 1895), 61.

19. Cartwright, *Gold Paved the Way*, 100.

20. William Charles Scully, *The Ridge of the White Waters* (London: Stanley Paul & Co., 1912), 70, 73, & 126. Figures from Yates, *Present Day*, plates.

21. Maryna Fraser, ed., *Johannesburg Pioneer Journals: 1888–1909* (Cape Town: Van Riebeeck Society, 1985), 41.

22. Hennen Jennings, *Mining as a Profession. Including First Stages of Metallurgy* (San Francisco: Mining and Scientific Press, 1914), 11.

23. Herbert Hoover, *Principles of Mining: Valuation, Organization, and Administration* (London: McGraw-Hill Book Company, 1909), 186. See also A. G. Charleton, "The General Principles of Successful Mine Management," *EM*, November 1900, 236; John Hays Hammond, *The Engineer* (New York: Scribner's, 1922), 123–128.

24. Wernher, Beit & Co., for instance, founded a subsidiary of Chicago's Fraser and Chalmers in London and a works at Erith, which was managed by the former chief mechanical engineer of DeBeers, Louis Irving Seymour. S. Neumann and Co., meanwhile, took control of several South African coal mines after the Boer War, including the Witbank Colliery.

25. Hugh Frederick Marriott, *Money and Mines: The Administration, Organisation and Economics of Precious and Non-Ferrous Metal Mines* (New York: MacMillan, 1925), 78.

26. Marriott, *Money and Mines*, 81.

27. Jennings, *Mining as a Profession*, 10. See also Hammond, *Engineer*, 16; Hoover, *Principles*, 185, 190.

28. Marriott, *Money and Mines*, 80.

29. Ruth Oldenziel, *Making Technology Masculine: Men, Women and Modern Machines in America, 1870–1945* (Amsterdam: Amsterdam University Press, 1999), 70–78. This is not to make the case that all elite engineers were Western white men, either—far from it; see essays by Köll, Hendrickson, and Ramnath. See also John Broich, "Was It Really the 'White Man's Burden'? The Non-British Engineers Who Engineered the British Empire," *Britain and the World* 9, no. 2 (2016): 197–212.

30. Hammond, *Engineer*, 122.

31. Charles B. Going, "The Engineer and the Policy of Imperialism," *EM*, January 1899, 527–532; Casper L. Redfield, "Industrial Ascendancy of the Anglo-American Peoples," *EM*, February 1901, 847–852. These claims were not devoid of nationalist content, however; see J. F. Holloway, "The Growing Importance of the Engineer," *EM*, April 1894, 96–101. For a study of masculine socialization into mine engineering at the Colorado School of Mines, see LeCain, *Mass Destruction*, 55–60.

32. Oldenziel, *Making Technology Masculine*, 91–119.

33. Marriott, *Money and Mines*, 85–86.

34. Edwin T. Layton, *The Revolt of the Engineers: Social Responsibility and the American Engineering Profession* (Baltimore: Johns Hopkins University Press, 1986), 48, 21.

35. A. J. Wilson, *The Professionals: The Institution of Mining & Metallurgy, 1892–1992* (London: IMM, 1992), 52.

36. Stephen Tuffnell, "Engineering Gold Rushes: Engineers and the Mechanics of Global Connectivity," in *Global History of Gold Rushes*, ed. Benjamin Mountford and Stephen Tuffnell (Berkeley: University of California Press, 2018), 229–251.

37. William Beinart and Saul Dubow, *Scientific Imagination in South Africa: 1700 to the Present* (Cambridge, UK: Cambridge University Press, 2021), 130.

38. Jeffrey Bartos, "The Accumulated Knowledge of a Thousand Generations, 1885–1920," *Technology and Culture* 62, no. 4 (2021): 1172–1198.

39. See the discussion of "trustworthiness" in Ben Marsden and Crosbie Smith, *Engineering Empires: A Cultural History of Technology in Nineteenth-Century Britain* (Basingstoke: Palgrave Macmillan, 2005), 7–8; Gooday, *Morals of Measurement*.

40. Quoted in David Kynaston, *The City of London: Volume II. The Golden Years, 1890–1914* (London: Pimlico, 1996), 8.

41. Ian Phimister, "Corners and Company-Mongering: Nigerian Tin and the City of London, 1909–12," *Journal of Imperial and Commonwealth History* 28, no. 2 (2000): 23.

42. Phimister, "Corners and Company-Mongering," 23.

43. On French investment on the LSE, see Robert Kubicek, "Finance Capital and South African Goldmining, 1886–1914," *Journal of Imperial and Commonwealth History* 3, no. 3 (1975): 386–395.

44. Jean-Jacques van Helten, "Mining, Share Manias and Speculation: British Investment in Overseas Mining, 180–1913," in *Capitalism in a Mature Economy: Financial Institutions, Capital Exports and British Industry, 1870–1939*, ed. Youssef Cassis and van Helten (Aldershot: Elgar, 1990), 160–161.

45. See Casper Andersen, *British Engineers and Africa, 1875–1914* (London: Pickering & Chatto, 2011), 57–87; Oskar Broberg and Klas Rönnbäck, *Capital and Colonialism: The Return on British Investments in Africa 1869–1969* (Cham, Switzerland: Palgrave Macmillan, 2019), 29.

46. Richard Hartley, "Bewick Moreing in Western Australian Gold Mining 1897–1904: Management Policies & Goldfield Responses," *Labour History* 65 (1993): 1.

47. Ernest A. Wiltsee to John Hays Hammond, June 28, 1896, Folder 2, Box 2, John Hays Hammond Papers, Sterling Memorial Library, Yale University, New Haven, CT.

48. For a history of the Consolidated and its tense relationship with the NYSE, see Ranald Michie, *The London and New York Stock Exchanges, 1850–1914* (London: Allen & Unwin, 1987), 204.

49. "American World of Copper" is the model proposed by Robrecht Declerq, Duncan Money, and Hans Otto Frøland, eds., *Born with a Copper Spoon: A Global History of Copper, 1830–1980* (Vancouver: University of British Columbia Press, 2022).

50. Anonymous, *The New York Mining Directory* (New York: Hollister & Goddard, 1880).

51. John Hays Hammond, *The Autobiography of John Hays Hammond,* 2 vols. (New York: Farrar & Reinhart, 1935), II.479. The Mills building was constructed by Darius Ogden Mills, a mining speculator and banker from San Francisco who made a fortune from Comstock silver mining and logging in the Tahoe forests, with whom Hammond later formed the Milham Exploration Company to prospect for oil.

52. Joseph D. Grant, *Redwoods and Reminiscences* (San Francisco: Save the Redwoods League and Menninger Foundation, 1973), 13; Maureen A. Jung, "Capitalism Comes to the Diggings: From Gold-Rush Adventure to Corporate Enterprise," in *A Golden State: Mining and Economic Development in Gold Rush California,* ed. James J. Rawles and Richard Orsi (Berkeley: University of California Press, 1999), 69. Gray Brechin, *Imperial San Francisco: Urban Power, Earthly Ruin* (Berkeley: California University Press, 2006), 45.

53. Clark C. Spence, "British Investment and the American Mining Frontier, 1860–1914," *New Mexico Historical Review* 36, no. 2 (1961): 121. In turn, it was Californian engineers and technical innovations that transformed the British Empire's mineral industries; Jessica B. Teisch, *Engineering Nature: Water, Development, and the Global Spread of American Environmental Expertise* (Chapel Hill: University of North Carolina Press, 2011).

54. "English Money for American Mines," *MSP,* September 21, 1895, 185.

55. Lionel Frost, "'Metallic Nerves': San Francisco and Its Hinterland during and after the Gold Rush," *Australian Economic History Review* 50, no. 2 (2010): 129–147.

56. Mariusz Lukasiewicz, "From Diamonds to Gold: The Making of the Johannesburg Stock Exchange, 1880–1890," *Journal of Southern African Studies* 43, no. 4 (2017): 715–732.

57. W. R. Lawson, "Kaffir Finance," *The National Review,* May 1896, 412; van Helten, "Mining, Share Manias and Speculation," 170.

58. Charles van Onselen, *Studies in the Social and Economic History of the Witwatersrand, 1886–1914. 1. New Babylon* (New York: Longman, 1982), 1.

59. Charles Harvey and John Press, "The City and International Mining, 1870–1914," *Business History* 32, no. 3 (1990): 108. The 1898 *Stock Exchange Yearbook* listed more than 1,000 mining and exploration firms; Darwin, *Empire Project,* 119.

60. Its activities are carefully detailed in Robert Vicat Turrell and Jean-Jacques van Helten, "The Rothschilds, the Exploration Company and Mining Finance," *Business History* 28, no. 2 (1986): 181–205.

61. John Hays Hammond to Charles D. Lane, September 24, 1897, Volume 3, Box 4, Hammond Papers, Yale University. Bradley was a colleague of Hammond's at the Bunker Hill mine, Idaho, and succeeded him as manager in 1893.

62. Turrell and van Helten, "The Rothshchilds," 191.

63. A. Moil, *Mining and Mining Investments* (London: Methuen & Co., 1904), 109.

64. Turrell and van Helten, "The Rothshchilds," 186.

65. Quoted in John Darwin, *The Empire Project: The Rise and Fall of the British World System, 1830–1970* (Cambridge, UK: Cambridge University Press, 2009), 119.

66. Hammond to Thomas H. Leggett, May 17, 1895, Reel 1, John Hays Hammond Papers, Weston Library, University of Oxford, Oxford, UK. Leggett was subsequently appointed consulting engineer to S. Neuman and Company. See also Hammond to Pope Yeatman, May 27, 1895, Reel 1, Hammond Papers, Weston Library.

67. Hammond to Richard A. Parker, May 29, 1895, Reel 1, Hammond Papers, Weston Library.

68. Hammond, *Autobiography*, II.514.

69. Thomas Arthur Rickard, *Interviews with Mining Engineers* (San Francisco: Mining and Scientific Press, 1922), 414.

70. Lionel Phillips, *Transvaal Problems: Some Notes on Current Politics* (London: J. Murray, 1905), 75.

71. Hammond, *Autobiography*, I.298.

72. George H. Nash, *The Life of Herbert Hoover: The Engineer, 1874–1914* (New York: W. W. Norton, 1983), 476, 727, 6 & 7. John Hays Hammond wrote similarly to correct misinformation to restore public confidence, for example, "The Consolidated Gold Fields of South Africa," *The Economist*, January 8, 1898, 52.

73. Hammond, *Autobiography*, I.278.

74. Broberg and Rönnbäck, *Capital and Colonialism*, 288.

75. Broberg and Rönnbäck, *Capital and Colonialism*, 295–296.

76. Cartwright, *Gold Paved the Way*, 119.

77. Ian Phimister and Jeremy Mouat, "Mining, Engineers, and Risk Management: British Overseas Investment, 1894–1914," *South African Historical Journal* 49, no. 1 (2003): 2.

78. Quoted in Phimister and Mouat, "Mining, Engineers, and Risk Management," 18.

79. Van Helten, Mining, "Share Manias and Speculation," 174–175; Turrell and van Helten, "The Rothschilds, the Exploration Company," 181–205; Layton, *Revolt of the Engineers*, 3–55.

80. Turrell and van Helten, "The Rothschilds, the Exploration Company," 183.

81. Maryna Fraser and Alan Jeeves, eds., *All That Glittered: Selected Correspondence of Lionel Phillips, 1890–1924* (Cape Town: Africana Publishing Corporation, 1977), 8.

82. Edward J. Balleisen, *Fraud: An American History from Barnum to Madoff* (Princeton, NJ: Princeton University Press, 2017), 87.

83. Balleisen, *Fraud*, 76.

84. "To the Mining Student," *EMJ*, October 6, 1904, 539.

85. "Financial Journalism," *Mining Magazine*, February 1913, 123 (hereafter: *MM*); "Plaintive Shareholders," *MM*, January 1913, 30.

86. "Professional Responsibility," *EMJ*, April 28, 1904, 673.

87. See David C. Itzkowitz, "Fair Enterprise or Extravagant Speculation: Investment, Speculation, and Gambling in Victorian England," in *Victorian Investments: New Perspectives on Finance and Culture*, ed. Nancy Henry and Cannon Schmitt (Bloomington: Indiana University Press, 2009), 98–119. See Balleisen, *Fraud*, 76; Peter Knight, *Reading the Market: Genres of Financial Capitalism in Gilded Age America* (Baltimore: Johns Hopkins University Press, 2016); Richard White, "Information, Markets, and Corruption: Transcontinental Markets in the Gilded Age," *Journal of American History* 90, no. 1 (2003): 19–43.

88. See "Experts and Experts," *EMJ*, October 20, 1904, 618.

89. "Proper Procedure," *MM*, December 1909, 263.

90. Rickard, *Interviews*, 425.

91. "Professional Ethics Again," *MSP*, July 18, 1908, 68–69. Rickard also collected together a range of essays by himself and others on responsible mining finance in *The Economics of Mining* (London: Hill Publishing Company, 1907).

92. Hammond, "Professional Ethics," 626.

93. It is perhaps little wonder that Hammond later moved into examining questions of national political economy; see Mark Hendrickson, "'The Sesame That Opens the Door of Trade': John Hays Hammond and Foreign Direct Investment in Mining, 1880–1920," *Journal of the Gilded Age and Progressive Era* 16, no. 3 (2017): 325–346.

94. Hammond, "Professional Ethics," 626.

95. John Hays Hammond, "Commencement Address," *Quarterly of the Colorado School of Mines*, August 1909, 5.

96. Hammond, "Commencement Address," 6.

97. White, "Information, Markets, and Corruption," 20.

98. White, "Information, Markets, and Corruption," 20.

99. Phimister and Mouat, "Mining, Engineers, and Risk Management," 4.

100. "Counsels of Imperfection," *MM*, December 1909, 266.

101. "Counsels of Imperfection," *MM*, January 1910, 40. As he had written in *The Principles of Mining*, it was "the duty of every engineer to diminish the opportunity of the vulture so far as possible"; see "Counsels of Imperfection," *MM*, January 1910, 171. Debate subsequently continued; see "Engineer as Promoter," *MM*, February 1910, 115–116; "Professional Ethics," *MM*, March 1910, 170–171; "Counsels of Imperfection," *MM*, March 1910, 222.

102. Nash, *Life*, 390–92, 479. Later, also in the pages of the *Mining Magazine*, Hoover revealed he had a great deal of common ground with the Bookie when he dismissed investors as "idiots"; see "Economics in a Boom," *MM*, May 1912, 371. On the efforts of Hoover, Hammond, and Jennings to become public intellectuals, see Bartos, "Accumulated Knowledge," 1172–1198.

103. Herbert Hoover, *Memoirs of Herbert Hoover*, 3 vols. (London: Hollis and Carter, 1951–1953), I.100. On Hoover's financiering and speculating, see Jeremy Mouat and Ian Phimister, "The Engineering of Herbert Hoover," *Pacific Historical Review* 77, no. 4 (2008): 553–584.

104. Studnicki-Gizbert, "Exhausting the Sierra Madre," 19–20.

105. Hugh Ferguson and Mike Chrimes, *The Consulting Engineers: The British Consulting Engineers Who Created the World's Infrastructure* (London: ICE Publishing, 2020), 9.

CHAPTER 3

1. "How the Word 'Scientist' Came to Be," *NPR*, May 21, 2010, sec. Science, https://www.npr.org/2010/05/21/127037417/how-the-word-scientist-came-to-be.

2. See Edward Beatty, *Technology and the Search for Progress in Modern Mexico* (Berkeley: University of California Press, 2015), for a comprehensive account of this history. For Mexican positivism more generally, see the classic works of Charles Hale.

3. A small but growing literature on the history of meteorology exists for Mexico, but none to my knowledge examine the role of engineers in the development of meteorology. See Rodrigo Vega y Ortega et al., "Aportaciones a la historia de la meteorología a través de los estudios ambientales de las asociaciones científicas de la ciudad de México, 1857–1910," *Letras históricas*, no. 15 (2016): 99–121; Luis Alberto Arrioja Díaz Viruell, "Historia y Clima En México: Perspectivas y Horizontes Desde La Historiografía," *Estudis d'història Agrària*, no. 33 (2021): 13–31; Luis Alberto Arrioja and Armando Alberola, *Clima, Desastres y Convulsiones Sociales En España e Hispanoamérica, Siglos XVII-XX* (Alicante: Universidad de Alicante/El Colegio de Michoacán, 2016); Gustavo G. Garza Merodio, *Variabilidad Climática En México a Través de Fuentes Documentales (Siglos XVI al XIX)* (Mexico City: UNAM, Instituto de Geografía, 2023).

4. Although I will discuss the gendered dimensions of Mexican engineering-cum-meteorology more below, it is worth quoting Ruth Oldenziel for the case of the United States: "Technology is a neologism. Only in the 1930s did technology become a keyword of American grammar and an all-encompassing concept that could explain human life itself." Ruth Oldenziel, *Making Technology Masculine: Men, Women and Modern Machines in America, 1870–1945* (Amsterdam: Amsterdam University Press, 1999), 14. Oldenziel's description above can be generally applied to Mexico as well. For example, in contrast to *ingeniero*, for which a search in the Universidad Nacional Autónoma de México's (UNAM's) digital newspaper database yields nearly 94,000 references from 1876 to 1911, technology was seldom used as a term in Mexican

newspapers and journals until after World War II. The UNAM's digital database of hundreds of newspapers yields only 377 references for *tecnología* from 1876 to 1911 and approximately 550 from 1911 to 1946, but more than 36,000 from 1946 to 1981. Hemeroteca Nacional Digital de México, UNAM, accessed December 2022.

5. Edwin T. Layton Jr., "American Ideologies of Science and Engineering," *Technology and Culture* 17, no. 4 (1976): 695.

6. In 1909, on the eve of the Mexican Revolution, only one woman was enrolled in the National Engineering School, and she never graduated. Domínguez, *La ingenieria civil en México*, 43. In this sense, Mexico was similar to the United States at the time, where Oldenziel observes that technology "came to denote the useful application of scientific knowledge for the benefit of humankind, and engineers were designated as the sole bearers of that form of knowledge. Rather than a neutral term, technology is itself part of a narrative production or plot of modernism, in which men are the protagonists and women have been denied their part." Oldenziel, *Making Technology Masculine*, 14.

7. On engineering in Mexico after the Revolution of 1910–1920, see J. Justin Castro and James A. Garza eds., *Technocratic Visions: Engineers, Technology, and Society in Mexico* (Pittsburgh, PA: University of Pittsburgh Press, 2022).

8. "Ley Orgánica de la Instrucción Pública en el Distrito Federal," *Diario Oficial de la Federación*, December 2, 1867.

9. See Frank Safford, *The Ideal of the Practical: Colombia's Struggle to Form a Technical Elite* (Austin: University of Texas Press, 1976), on this topic. For Mexico, see Beatty, *Technology and the Search for Progress in Modern Mexico*, and Castro and Garza, *Technocratic Visions: Engineers, Technology, and Society in Mexico*; for Latin America more broadly, see Eden Medina, Ivan Costa Marques, and Christina Holmes eds., *Beyond Imported Magic. Essays on Science, Technology, and Society in Latin America* (Cambridge, MA: MIT Press, 2014), and David Pretel, ed., *History of Technology*, vol. 34 (New York: Bloomsbury Academic, 2019).

10. Anabel Velasco Reynaga, *Mariano Santiago de Jesús de La Bárcena Ramos: De Minerales, Fósiles y Plantas* (Guadalajara, México: Editorial Universidad de Guadalajara, 2020), 11–12. See also "Mariano Bárcena," *La Patria*, April 12, 1899.

11. Reynaga, *Mariano Santiago de Jesús de La Bárcena Ramos*, 8–10.

12. Observatorio Meteorológico Central [Mariano Bárcena], *Instrucciones que comunica el Observatorio Meteorológico Central a los observatorios foráneos* (Mexico City: Imprenta de Francisco Díaz de León, 1877), 4.

13. Rafael Aguilar Santillán, *Bibliografía Meteorológica Mexicana: Que comprende las publicaciones de meteorología, física del globo y climatología hechas hasta fines de 1889* (Mexico City: Imprenta del Gobierno, 1890), 5–7.

14. Aguilar Santillán, *Bibliografía Meteorológica Mexicana*, 5–7.

15. Vicente Riva Palacio, *Memoria presentada al Congreso de la Union por el Secretario de Estado y del Despacho de Fomento, Colonización, Industria y Comercio de la República Mexicana* (Mexico City: Imprenta de Francisco Díaz de León, 1877), 486.

16. Velasco Reynaga, *Mariano Bárcena*, 14–15. Bárcena served as both governor and director of the observatory simultaneously for a few months.

17. Mariano Bárcena and Miguel Pérez, *Estudios de Meteorologia Comparada, Tomo 1* (Mexico City: Secretaría de Fomento, 1885), IV.

18. Bárcena and Pérez, *Estudios de Meteorologia Comparada*, 7.

19. Bárcena and Pérez, *Estudios de Meteorologia Comparada*, VIII.

20. Ángel Anguiano, "A *El Tiempo*," *El Siglo Diez y Nueve*, December 29, 1892.

21. Mónica de la Guardia Durán, "El astrónomo Ángel Anguiano: Un experto aprendiz," *Revista digital universitaria* 16, no. 4 (April 2015): 2–14.

22. *Gaceta Médica de México*, May 15, 1894.

23. "National Health Board: Seventh Anniversary of Its Establishment in Its Present Form," *The Two Republics*, August 3, 1898.

24. William Gordon, "The Place of Climatology in Medicine," *Proceedings of the Royal Society of Medicine* 6 (1913): 103–124. By this time, germ theory was ascendent over miasma theory and was therefore gradually eclipsing the perceived importance of the environment in diagnosing and preventing disease (except to the extent that certain disease vectors, such as mosquitoes, thrived in distinct climates).

25. Agustín Aragón, "El territorio de México y sus habitantes," in *México: Su evolución social*, ed. Justo Sierra (Mexico City: J. Ballesca y compania, 1902).

26. Mariano Bárcena, "La meteorología y el campesino," *El Agricultor Mexicano*, January 1, 1896, 26–27, 86–93.

27. For an overview, see Esther Katz and Annamária Lammel, "Introducción," in *Aires y lluvias. Antropología del clima en México*, ed. Annamária Lammel, Marina Goloubinoff, and Esther Katz (Mexico City: CIESAS, 2008), 27–50.

28. Manuel Moreno y Anda and Antonio Gómez, *El Clima de la República en el Año 1896* (Mexico City: Oficina de la Secretaría de Fomento, 1900), 4–10.

29. Manuel E. Pastrana, *El Servicio Meteorológico de La República Mexicana* (Mexico City: Secretaría de Fomento, 1906), 3. The capitalization of Meteorology is from the original.

30. José María Romero, "La Publicación de Los Resultados de Las Observaciones Meteorológicas Con Relación al Pronóstico Del Tiempo," in Congreso Meteorológico Nacional and Sociedad Científica Antonio Alzate, *Actas, Resoluciones y Memorias Del Congreso Meteorológico Nacional Iniciado Por La Sociedad Científica 'Antonio Alzate'* (Mexico City: Secretaría de Fomento, 1901), 89–106. Quotations in the following pages are from this source unless otherwise noted.

31. Romero, "La Publicación de Los Resultados de Las Observaciones Meteorológicas Con Relación al Pronóstico Del Tiempo."

32. Romero, "La Publicación de Los Resultados de Las Observaciones Meteorológicas Con Relación al Pronóstico Del Tiempo."

33. Romero, "La Publicación de Los Resultados de Las Observaciones Meteorológicas Con Relación al Pronóstico Del Tiempo," 104.

34. The national literacy rate was 4 percent in 1895, rising to 20 percent in 1910, while in Mexico City, it rose from 38 percent in 1895 to 50 percent in 1910. Nora Pérez-Rayón Elizundia, *México 1900: Percepciones y valores en la gran prensa capitalina* (Mexico City: Universidad Autónoma Metropolitana, 2001), 35.

35. For how such "politico-environmental" press coverage of extreme weather events, as I term it, may have impacted the eruption of the Mexican Revolution, see Mikael Wolfe, "The Climate of Conflict: Politico-Environmental Press Coverage and the Eruption of the Mexican Revolution," *Hispanic American Historical Review* 99, no. 3 (August 2019): 467–499.

36. Rodrigo Vega y Ortega, "Las conferencias geográficas impartidas por las alumnas de la Escuela Normal para Profesoras de la Ciudad De México, 1894–1905," *Redes* 19, no. 36 (June 2013): 135.

37. Maria Luisa Crespo, "Estudios Atmosfericos," June 8, 1895. My liberal translation fails to fully capture the eloquence of the Spanish original: "La Meteorología, una de las ciencias que presta más utilidad al hombre, y que atrae con frecuencia su atención, se presenta sin obstáculos á aquel que quiere descubrir sus secretos é investigar sus causas; tiene por sí misma un verdadero adelanto, pues se encuentra al alcance de todas las inteligencias. Su dominio es inmenso, no cesa de investigar los inumerables fenómenos que á cada paso se nos presentan tanto en el seno de la atmósfera como en las profundidades de los océanos; precisa sus efectos y cada día hace notar con mayor interés la ponderosa influencia que ejerce en todas nuestras acciones, preservándonos al mismo tiempo de infinidad de peligros que han podido evitarse por medio de la observación y el estudio. La marina, la agricultura, la hygiene, y por consiguiente nuestra salud, se encuentran sujetas á sus leyes; dando al mismo tiempo ponderosas instrucciones para impedir la producción de epidemias, y en fin, coronando la obra, resuelve uno de los problemas que por mucho tiempo ha ejercitado la sagacidad de los sabios y de los primeros físicos: 'la prevision del tiempo.'"

38. Pastrana, *El Servicio Meteorológico de La República Mexicana*, 7.

39. Pastrana, *El Servicio Meteorológico de La República Mexicana*, 14–20.

40. Isabel González García, *Los progresos de la Meteorología en México de 1810 a 1910* (Mexico City: Tip. de la Vda. De F. Díaz de León, 1911); Vega, "Las conferencias," 138–139; Mariana Espinosa Aldama, "La propagación de la cultura científica a través de la Sociedad Astronómica de México (1910–1916)," MA thesis, UNAM, 2010, 15.

41. González García, *Los progresos*, 11–15.

CHAPTER 4

1. Better known simply as "the Rand," from which South Africa's currency gets its name, the Witwatersrand remains the largest known ore province on Earth and has produced roughly a third of the gold ever mined. Hartwig E. Frimmel, "The Witwatersrand Basin and Its Gold Deposits," in *The Archaean Geology of the Kaapvaal Craton, Southern Africa*, ed. Alfred Kröner and Axel Hofmann (Cham, Switzerland: Springer, 2019), 255–275. Key works on gold mining in this period include Frederick A. Johnstone, *Class, Race, and Gold: A Study of Class Relations and Racial Discrimination in South Africa* (London: Routledge, 1976); Robert V. Kubicek, *Economic Imperialism in Theory and Practice: The Case of South African Gold Mining Finance, 1886–1914* (Durham, NC: Duke University Press, 1979); J. S. Crush, Alan Jeeves, and David Yudelman, *South Africa's Labor Empire: A History of Black Migrancy to the Gold Mines* (Boulder, CO: Westview Press, 1991); Alan H. Jeeves, *Migrant Labour in South Africa's Mining Economy: The Struggle for the Gold Mines' Labour Supply, 1890–1920* (Kingston, ON: McGill-Queen's University Press, 1985); Charles van Onselen, *New Babylon, New Nineveh: Everyday Life on the Witwatersrand 1886–1914* (Johannesburg: Jonathan Ball, 2001), 1–46; Benjamin Mountford and Stephen Tuffnell, eds., *A Global History of Gold Rushes* (Oakland: University of California Press, 2018); Mingwei Huang, "The Chinese Century and the City of Gold: Rethinking Race and Capitalism," *Public Culture* 33, no. 2 (May 2021): 193–217; Mae Ngai, *The Chinese Question: The Gold Rushes and Global Politics* (New York: W. W. Norton, 2021), 187–269.

2. "Address by E. J. Laschinger, M. E., Vice-President," *Proceedings of the South African Association of Engineers* 13 (August 3, 1907): 2–5.

3. On "boundary objects," see Geoffrey C. Bowker and Susan Leigh Star, *Sorting Things Out: Classification and Its Consequences* (Cambridge, MA: MIT Press, 1999), 15–16, 285–317. On engineers' position between capital and labor, see Edwin T. Layton, *The Revolt of the Engineers: Social Responsibility and the American Engineering Profession* (Cleveland, OH: Press of Case Western Reserve University, 1971); David F. Noble, *America by Design: Science, Technology, and the Rise of Corporate Capitalism* (Oxford: Oxford University Press, 1977); Peter Whalley, *The Social Production of Technical Work: The Case of British Engineers* (Albany: State University of New York Press, 1986); Peter Way, *Common Labour: Workers and the Digging of North American Canals, 1780–1860* (Cambridge, UK: Cambridge University Press, 1993), 57–59; Carl Mitcham, *Thinking through Technology: The Path between Engineering and Philosophy* (Chicago: University of Chicago Press, 1994), 238–240; Jessica M. Smith, *Extracting Accountability: Engineers and Corporate Social Responsibility* (Cambridge, MA: MIT Press, 2021).

4. Alpheus F. Williams, *American Engineers in South Africa* (Cape Colony: Kimberley, 1902); Stephen Tuffnell, "Engineering Inter-Imperialism: American Miners and the Transformation of Global Mining, 1871–1910," *Journal of Global History* 10, no. 1 (March 2015): 59.

5. Geoffrey Wheatcroft, *The Randlords* (London: Weidenfeld & Nicolson, 1985), 121.

6. Rhodes to E. R. Tymms, November 11, 1900, Oxford, Bodleian Libraries, Rhodes papers, MSS Afr. lett. s. 227, 213.

7. Edward J. Way to George Heys, January 11, 1892, MS.727/L5, Brenthurst Library, Johannesburg, South Africa. Emphasis in original.

8. "Farewell Dinner to Mr. J. Hennen Jennings," *Proceedings of the South African Association of Engineers* 8 (May 16, 1903): 161–168.

9. Lionel Phillips to Wernher, Beit, January 3, 1890, Maryna Fraser and Alan Jeeves, eds., *All That Glittered: Selected Correspondence of Lionel Phillips, 1890–1924* (Cape Town: Oxford University Press, 1977), 34.

10. Lionel Phillips to Wernher, Beit, January 30, 1890, *All That Glittered*, 36.

11. Phillips to Eckstein, December 17, 1892, *All That Glittered*, 66. Emphasis in original.

12. Lionel Phillips, *Some Reminiscences* (London: Hutchinson, 1924), 114.

13. John Hays Hammond, "South Africa and Its Future," *North American Review* 164, no. 483 (February 1897): 233–248; Charles van Onselen, *The Cowboy Capitalist: John Hays Hammond, the American West, and the Jameson Raid in South Africa* (Johannesburg: Jonathan Ball, 2017).

14. Witwatersrand Chamber of Mines, *The Mining Industry: Evidence and Report of the Industrial Commission of Enquiry* (Johannesburg: Witwatersrand Chamber of Mines, 1897), 445–458.

15. Shula Marks and Stanley Trapido, "Lord Milner and the South African State," *History Workshop Journal* 8 (Autumn 1979): 50–81; Shula Marks and Stanley Trapido, "Lord Milner and the South African State Reconsidered," in *Imperialism, the State and the Third World*, ed. Michael Twaddle (London: British Academic Press, 1992), 80–94; Keith Breckenridge, *Biometric State: The Global Politics of Identification and Surveillance in South Africa, 1850 to the Present* (New York: Cambridge University Press, 2014). Breckenridge calls attention to how the demands of mine engineers for pass laws "became the distinguishing characteristic of the South African state." *Biometric State*, 73.

16. Exceptions include Tuffnell, "Engineering Inter-Imperialism," 65–67; Morley Nkosi, *Black Workers, White Supervisors: The Origins of the Labor Structure in South Africa* (Trenton, NJ: Africa World Press, 2017), 137–140. Historians have increasingly taken note of the presence of US mining engineers in South Africa, but few have considered what role engineering as a discipline played or understood engineers in a global context. Elaine Katz has shown their important technical contributions, as well as the role of the first Transvaal state mining engineer, a German, in originating the "colour bar" in her "The Contributions of American Mining Engineers and Technologies to the Witwatersrand Gold Mining Industry, 1890–1910," *South African Journal of Economic History* 20, no. 2 (2005): 48–82, and "Revisiting the Origins of the Industrial Colour Bar in the Witwatersrand Gold Mining Industry, 1891–1899," *Journal of Southern African*

Studies 25, no. 1 (March 1999): 73–97, respectively. Others have focused on a different camp of mining engineers affiliated with Consolidated Gold Fields, led by Hammond, emphasizing their role in the failed Jameson Raid, whereas this chapter emphasizes the role of engineers employed at Corner House, unofficially led by Jennings. See Jessica Teisch, "'Home Is Not So Very Far Away': Californian Engineers in South Africa, 1868–1915," *Australian Economic History Review* 45, no. 2 (July 2005): 139–160; Charles van Onselen, *The Cowboy Capitalist: John Hays Hammond, the American West, and the Jameson Raid in South Africa* (Johannesburg: Jonathan Ball, 2017). John Higginson focuses on the role of US mining engineers William Honnold and Herbert Hoover in "Privileging the Machines: American Engineers, Indentured Chinese and White Workers in South Africa's Deep-Level Gold Mines, 1902–1907," *International Review of Social History* 52, no. 1 (April 2007): 1–34.

17. "The Transvaal Mines Commission," *Financial Times*, June 8, 1897.

18. Hammond to Jennings, June 16, 1897, John Hays Hammond, Sr. Papers (MS 259), Box 4, Folder 2, Private Letter Book 3, Manuscripts and Archives, Beinecke Rare Book and Manuscript Library, Yale University.

19. John Yates, *Present Day Metallurgical Engineering on the Rand* (London: The Mining Journal, 1898), 109–136.

20. *Industrial Commission*, 44.

21. *Industrial Commission*, 91.

22. *Industrial Commission*, 92. Hammond boasted of a similar method devised by a fresh University of California graduate he recruited to the Rand solve the shaft sinking problem with fresh eyes, which "soon established world records." Hammond, *Autobiography of John Hays Hammond*, vol. 1 (New York: Farrar & Rinehart, 1935), 300.

23. *Industrial Commission*, 218–219.

24. Hamilton Smith, "The Witwatersrand Gold-Fields," *The Times*, January 17, 1893; Union B. White, "America's Greatest Gold-Mining Engineer a Native of Louisville," *Courier-Journal* (Louisville, KY), August 30, 1908.

25. Mark Hendrickson, "Advance Agent of Expanding Empires: George F. Becker and Mineral Exploration in South Africa and the Philippines," *History and Technology* 35, no. 3 (2019): 237–265.

26. *Industrial Commission*, 215.

27. *Industrial Commission*, 219.

28. *Industrial Commission*, 228.

29. *Industrial Commission*, 220.

30. Phillips to J. P. FitzPatrick, June 12, 1897, *All That Glittered*, 105.

31. For more on colonial Mozambique, see Patrick Harries, *Work, Culture, and Identity: Migrant Laborers in Mozambique and South Africa, c. 1860–1930* (Portsmouth,

NH: Heinemann, 1994), 109–140; Eric Allina, *Slavery by Any Other Name: African Life under Company Rule in Colonial Mozambique* (Charlottesville: University of Virginia Press, 2012); Charles van Onselen, *The Night Trains: Moving Mozambican Miners to and from the Witwatersrand, 1902–1955* (Johannesburg: Jonathan Ball, 2019).

32. *Mining Industry*, 449.

33. *Chamber of Mines of the South African Republic: Ninth Annual Report for the Year Ending 31st December 1897* (Johannesburg: Argus Printing, 1898), 75–86.

34. T. A. Rickard, *Interviews with Mining Engineers* (San Francisco: Mining and Scientific Press, 1922), 248.

35. Stanley Trapido, "Imperialism, Settler Identities, and Colonial Capitalism: The Hundred-Year Origins of the 1899 South African War," in *The Cambridge History of South Africa, Volume 2: 1885–1994*, ed. Robert Ross, Anne Kelk Mager, and Bill Nasson (Cambridge, UK: Cambridge University Press, 2011), 84–88, 96–97.

36. "Mining Stocks," *Engineering and Mining Journal*, March 20, 1897, 295–296.

37. *Chamber of Mines of the South African Republic*, 5.

38. Yates, *Metallurgical Engineering on the Rand*, 11–12.

39. Marks and Trapido, "Lord Milner and the South African State," 72–73.

40. Sidney Jennings to H. Eckstein, October 31, 1900, Volume 233, Archives of H. Eckstein & Co., Barlow Rand Archives, Sandton, South Africa (hereafter cited as BRA).

41. H. O'Kelly Webber, *The Grip of Gold: A Life Story of a Dominion* (London: Hutchinson, 1936), 130–131.

42. Lionel Phillips, *Transvaal Problems: Some Notes on Current Politics* (London: John Murray, 1905), 107.

43. "Recruiting of Barbadian Labour for Transvaal Mines in South Africa," CO 28/262/9, The National Archives, Kew, January 15, 1904.

44. Rachel K. Bright, *Chinese Labour in South Africa, 1902–10: Race, Violence, and Global Spectacle* (Houndmills, UK: Palgrave Macmillan, 2013), 30.

45. Phillips, *Some Reminiscences*, 114.

46. For an excellent treatment of this history in the US South, see Moon-Ho Jung, *Coolies and Cane: Race, Labor, and Sugar in the Age of Emancipation* (Baltimore: Johns Hopkins University Press, 2006).

47. Herbert Hoover to Theodore Hoover, 1899, Box 2, Theodore J. Hoover Manuscripts, Herbert Hoover Presidential Library.

48. Bright, *Chinese Labour in South Africa*, 32; Ngai, *Chinese Question*, 188.

49. Hennen Jennings does not appear to have given his own report, but he did reply that he would attend the meeting at the Chamber of Mines. Jennings to Chamber of Mines Secretary, October 4, 1902, Elaine Katz Papers, A3424 D3.2, Historical Papers

Research Archive, William Cullen Library, University of the Witwatersrand, Johannesburg, South Africa (hereafter cited as Katz Papers).

50. Leggett to Chamber of Mines Secretary, August 29, 1902, Katz Papers. Umqombothi has a much lower alcohol content than the high-proof spirits produced by government concessions pre-1899.

51. This was true of Hennen Jennings, and it is unlikely any of the other Americans were better informed than he was. See "Notes on Mr. Hennen Jennings' Letter, Dated the 1st. September 1902, Dealing with the Labour Problem," September 1902, Volume 250, BRA.

52. T. H. Leggett to Chamber of Mines Secretary, August 29, 1902, Katz Papers.

53. Fred Hellmann, *Determination of the Present Value of a Mine on the Rand* (London: Institution of Mining and Metallurgy, 1898).

54. Fred Hellman to Chamber of Mines Secretary, August 29, 1902, Katz Papers.

55. On white racism in Australian gold mining, see David Dorward, "'N***** Driver Brothers': Australian Colonial Racism in the Early Gold Coast Mining Industry," *Ghana Studies* 5, no. 1 (2002): 197–214.

56. Denny to Chamber of Mines Secretary, October 9, 1902, Katz Papers.

57. All Creswell quotes in this paragraph are from F. H. P. Creswell, *The Chinese Labour Question from Within* (London: P. S. King, 1905); David Yudelman, *The Emergence of Modern South Africa: State, Capital, and the Incorporation of Organized Labor on the South African Gold Fields, 1902–1939* (Cape Town: David Philip, 1984), 64, 218–232; Simone Lisa McCallum, "Radical Racism in the Transvaal: F. H. P. Creswell and the White Labour Movement, 1902–1912," master's thesis, Queen's University, 1994; Ngai, *Chinese Question*, 251–252, 264–265.

58. On white labourism, see Jonathan Hyslop, "The Imperial Working Class Makes Itself 'White': White Labourism in Britain, Australia, and South Africa before the First World War," *Journal of Historical Sociology* 12, no. 4 (1999): 398–421; William Kenefick, "Confronting White Labourism: Socialism, Syndicalism, and the Role of the Scottish Radical Left in South Africa before 1914," *International Review of Social History* 55, no. 1 (April 2010): 29–62; Lucien van der Walt, "The First Globalisation and Transnational Labour Activism in Southern Africa: White Labourism, the IWW, and the ICU, 1904–1934," *African Studies* 66, nos. 2–3 (August–December 2007): 223–251; Dagmar Engelken, "A White Man's Country: The Chinese Labour Controversy in the Transvaal," in *Wages of Whiteness and Racist Symbolic Capital*, ed. Wulf D. Hund, Jeremy Krikler, and David Roediger (Berlin: LIT Verlag, 2010), 161–193.

59. "Statement on Supply of Labour," 1902, Katz Papers.

60. "Report of Committee on White and Coloured Labour," December 2, 1902, Katz Papers.

61. G. R. Bozzoli, *Forging Ahead: South Africa's Pioneering Engineers* (Johannesburg: Witwatersrand University Press, 1997), 75–78.

62. Further Correspondence Relating to The Affairs of the Transvaal and Orange River Colony, Cd. 1895, 331–343.

63. Rickard, *Interviews with Mining Engineers*, 247.

64. London Secretary of the Chamber of Mines, *Notes on the Labour Position in the Transvaal* (London, 1904), 6–7.

65. Jennings to Wernher, Beit & Co., October 27, 1902, Volume 250, File 139, Number 53, BRA.

66. Jennings to Joseph Chamberlain, January 21, 1903, File 139, Number 83, BRA.

67. As in the rest of the British Empire, emancipation took place in 1834. For more on this in context of the Cape Colony, see Wayne Dooling, *Slavery, Emancipation and Colonial Rule in South Africa* (Athens: Ohio University Press, 2008).

68. Further Correspondence Relating to The Affairs of the Transvaal and Orange River Colony, Cd. 1895, 76–88.

69. Reports of the Transvaal Labour Commission: Minutes of Proceedings and Evidence, Cd. 1897, 61.

70. Peter Richardson, *Chinese Mine Labour in the Transvaal* (London: Macmillan, 1982); Mae M. Ngai, "Trouble on the Rand: The Chinese Question in South Africa and the Apogee of White Settlerism," *International Labor and Working-Class History* 91 (Spring 2017): 59–78.

71. John Hays Hammond, *Autobiography of John Hays Hammond*, vol. 2 (New York: Farrar & Rinehart, 1935), 511.

72. Rickard, *Interviews with Mining Engineers*, 226.

73. Extract from Messrs. Wernher, Beit & Co's Letter, September 9, 1903, Volume 208, BRA.

74. Browne to Wernher, Beit, November 9, 1905, Volume 208, BRA.

75. Ross E. Browne, "Working Costs of the Mines of the Witwatersrand," *Proceedings of the South African Association of Engineers* 12 (1907): 297.

76. Browne, "Working Costs," 330–331.

77. Browne, "Working Costs," 341.

78. Phillips to J. Wernher, May 20, 1906, *All That Glittered*, 161.

79. Phillips to Wernher, Beit & Co., June 18, 1906, *All That Glittered*, 164.

80. On the strike, see Van Onselen, *New Babylon, New Nineveh*, 309–367; Elaine N. Katz, "The Underground Route to Mining: Afrikaners and the Witwatersrand Gold Mining Industry from 1902 to the 1907 Miners' Strike," *Journal of African History* 36, no. 3 (November 1995): 467–489.

81. Honnold to Leggett, May 21, 1906, Box 78, William L. Honnold Papers, H.Mss. 0381, Special Collections at the Claremont Colleges.

82. Yudelman, *Emergence of Modern South Africa*, 142.

83. South African Association of Engineers, *Proceedings of the South African Association of Engineers*, vol. 13 (Johannesburg: SAAE, 1907), 11–28.

84. All quotations in this paragraph are from Transvaal Chamber of Mines Annual Report for the Year 1908, 19 (Johannesburg, 1909), 523–525.

85. *Mining Industry Commission Minutes of Evidence* (Pretoria: Government Printing and Stationery Office, 1908), 202.

86. Yudelman, *Emergence of Modern South Africa*, 65–66.

87. "Notes Re the Working of Government Claims in the Modderfontein District," 1909, 233, BRA.

88. Phillips to Eckstein, September 25, 1913, *All That Glittered*, 270–271.

89. "Mr. Ross E. Browne Pays a Compliment," *South African Mining Journal*, September 1912, 32.

90. "President's Valedictory Address," *Proceedings of the South African Association of Engineers* 14 (July 1909): 241, 247–248.

91. Anonymous, "Transvaal Mine Labor," *Engineering and Mining Journal* 88, no. 22 (November 27, 1909): 1068.

92. Jeremy Krikler, *White Rising: The 1922 Insurrection and Racial Killing in South Africa* (Manchester: Manchester University Press, 2005), 291–292.

93. Stanley Trapido, "South Africa in a Comparative Study of Industrialisation," *Journal of Development Studies* 17, no. 3 (April 1971): 310.

94. Walter G. Vincenti, *What Engineers Know and How They Know It: Analytical Studies from Aeronautical History* (Baltimore: Johns Hopkins University Press, 1990), 200–220; Dominique Vincq, *Everyday Engineering: An Ethnography of Design and Innovation* (Cambridge, MA: MIT Press, 2003); Antoine Picon, "Engineers and Engineering History: Problems and Perspectives," *History and Technology* 20, no. 4 (December 2004), 429–431. On South Africa as an exemplar of racial capitalism, see especially Martin Legassick and David Hemson, *Foreign Investment and the Reproduction of Racial Capitalism in South Africa* (London: Anti-apartheid Movement, September 1976); Neville Alexander, "Nation and Ethnicity," *Work in Progress* 28 (August 1983): 6–13; Stuart Hall, "Race, Articulation and Societies Structured in Dominance," in *Sociological Theories: Race and Colonialism* (Paris: UNESCO, 1980): 305–345; Harold Wolpe, "Capitalism and Cheap Labour-Power in South Africa: From Segregation to Apartheid," *Economy and Society* 1, no. 4 (November 1972): 425–456; Hugo ka Canham, *Riotous Deathscapes* (Durham, NC: Duke University Press, 2023).

CHAPTER 5

1. Benjamin A. Elman, *On Their Own Terms: Science in China, 1550–1900* (Cambridge, MA: Harvard University Press, 2005); William C. Kirby, "Engineering China: Birth of the Developmental State, 1928–1937," in *Becoming Chinese: Passages to Modernity and Beyond*, ed. Wen-hsin Yeh (Berkeley: University of California Press, 2000), 137–160.

2. Elisabeth Köll, "The Making of the Civil Engineer in China: Railroad Companies, Knowledge Transfer, and Professional Management," in *Knowledge Acts in Modern China: Ideas, Institutions, and Identities*, ed. Robert J. Culp, Eddy U., and Wen-hsin Yeh (Berkeley: Institute of East Asian Studies, University of California, Berkeley, 2016), 148–173.

3. On the issue of semi-colonialism and imperialism in late imperial and Republican China, see Anne Reinhardt, *Navigating Semi-Colonialism: Shipping, Sovereignty, and Nation-Building in China, 1860–1937* (Cambridge, MA: Harvard University Asia Center, 2018); Stephen R. Halsey, *Quest for Power: European Imperialism and the Making of Chinese Statecraft* (Cambridge, MA: Harvard University Press, 2015).

4. For examples, see R. A. Buchanan, *The Engineers: A History of the Engineering Profession in Britain, 1750–1914* (London: Kingsley, 1989); Terry S. Reynolds, ed., *The Engineer in America: A Historical Anthology from Technology and Culture* (Chicago: University of Chicago Press, 1991); Kevin McCormick, *Engineers in Japan and Britain: Education, Training, and Employment* (London: Routledge, 2000); Kees Gispen, *New Profession, Old Order: Engineers and German Society, 1815–1914* (Cambridge, UK: Cambridge University Press, 1989).

5. Weili Ye, *Seeking Modernity in China's Name: Chinese Students in the United States, 1900–1927* (Stanford, CA: Stanford University Press, 2001), 8.

6. Ye, *Seeking Modernity in China's Name*, 51; Thomas Edward La Fargue, *China's First Hundred* (Pullman: State College of Washington, 1942); Anonymous, "Peking-Kalgan-Tatung-Fu Railway," *Far Eastern Review* 10, no. 3 (August 1913): 82.

7. Benjamin A. Elman, *On Their Own Terms*, 360.

8. Hsien-ch'un Wang, *Western Technology and China's Industrial Development: Steamship Building in Nineteenth-Century China, 1828–1895* (New York: Palgrave Macmillan, 2022), 135–174.

9. Wang, *Western Technology and China's Industrial Development.*

10. See also Hsien-ch'un Wang, "Discovering Steam Power in China, 1840s–1860s," *Technology and Culture* 51, no. 1 (2010): 31–54.

11. Shiuon Chu, "The Longer Abolition of the Chinese Imperial Examination System (1900s–1910s)," *International Journal of Asian Studies* 20, no. 2 (2023): 721–737.

12. Ye, *Seeking Modernity*, 9.

13. Madeine Y. Hsu, "Chinese and American Collaborations through Educational Exchange during the Era of Exclusion, 1872–1955," *Pacific Historical Review* 83, no. 2 (2014): 314–332, 320.

14. Ye, *Seeking Modernity*, 51.

15. Hsu, "Chinese and American Collaborations," 321.

16. J. G. H. Glass, *Report on the Concessions of the Pekin Syndicate, Limited, in the Provinces of Shansi and Honan, China, with Estimates of Cost of Railways and Other Works Necessary for Their Development* (London: 1899), 3–4.

17. Glass, *Report on the Concessions of the Pekin Syndicate*, 3–5; on railroad engineers working in India, see Ian J. Kerr, *Engines of Change: The Railroads that Made India* (Westport, CT: Praeger, 2007).

18. E. T. Williams (transl.), *Recent Chinese Legislation relating to Commercial, Railway and Mining Enterprises with Regulations for Registration of Trade Marks, and for the Registration of Companies* (Shanghai: Shanghai Mercury Limited, 1904), 95.

19. Elisabeth Köll, *Railroads and the Transformation of China* (Cambridge, MA: Harvard University Press, 2019), 19–52; Lin Cheng, *The Chinese Railways: A Historical Survey* (Shanghai: China United Press, 1935); Jin Jiafeng, *Zhongguo jiaotong zhi fazhan ji qi quxiang* (The Development and Trends in China's Transportation System) (Nanjing: Zhengzhong shuju, 1937).

20. Elisabeth Kaske, ed., *China um 1900: Aufzeichnungen eines Krupp-Direktors* (Köln: Böhlau, 2005).

21. Köll, *Railroads and the Transformation of China*, 63–66.

22. Naofumi Nakamura, "Trading Locomotives between the US and Japan: A Case of Okura & Co. around the Turn of the 19th and 20th Centuries," in Institute of Social Science Discussion Paper Series, F-198, July 2022, quote p. 21.

23. Steven J. Ericson, "Taming the Iron Horse: Western Locomotive Makers and Technology Transfer in Japan, 1870–1914," in *Public Spheres, Private Lives in Modern Japan, 1600–1950*, ed. Gail Lee Bernstein, Andrew Gordon, and Kate Wildman Nakai (Cambridge, MA: Harvard University Asia Center, 2005), 185–217; Steven J. Ericson, *Sound of the Whistle: Railroads and the State in Meiji Japan* (Cambridge, MA: Harvard University Asia Center, Harvard University, 1996).

24. Ericson, "Taming the Iron Horse," 185–217, 185.

25. Naofumi Nakamura, "Railway Engineers' Groups in Early Meiji Japan," *Japanese Research in Business History* 28 (2011): 11–33.

26. Nakamura, "Railway Engineers' Groups in Early Meiji Japan," 13.

27. Graeme J. N. Gooday and Morris F. Low, "Technology Transfer and Cultural Exchange: Western Scientists and Engineers Encounter Late Tokugawa and Meiji Japan," *Osiris* 13 (1998): 99–128, 127.

28. Gooday and Low, "Technology Transfer and Cultural Exchange."

29. Köll, *Railroads and the Transformation of China*, 76–77; Ghassan Moazzin, *Foreign Banks and Global Finance in Modern China: Banking on the Chinese Frontier, 1870–1919* (Cambridge, UK: Cambridge University Press, 2022); Lin, *The Chinese Railways.*

30. Chih Meng, "The American Returned Students of China," *Pacific Affairs* 4, no. 1 (January 1931): 1–16; *Who's Who of American Returned Students* (*You Mei tongxue lu*) (Peking: Tsing Hua College, 1917).

31. H. G. W. Woodhead, ed., *The China Yearbook, 1912–1930* (Tianjin: North-China Daily News and Herald, 1931), 530.

32. "China's Railways—Tasks of the Foreign Advisor," *North-China Herald*, April 20, 1929.

33. Masanori Wada, "Seeking Employment during Japan's Early Industrialisation: New Engineering Graduates and their Struggles before 1900," *Journal of the Royal Asiatic Society*, published online March 11, 2024, 1–15, 9.

34. Tiedaobu zongwusi wenshuke (Central Office of the Ministry of Railways), *Tiedaobu zhiyuanlu* (Employee Directory of the Ministry of Railways) (Beijing: Ministry of Railways, 1929), general office department (*zongwusi*), 9A–20B and operations department (*gongwusi*), 26A–32B.

35. Köll, *Railroads and the Transformation*, 174.

36. *The China Press*, October 12, 1911, advertisement p. 12.

37. *The China Press*, October 12, 1911, advertisement p. 12.

38. J. A. Eley, "The Education and Training of Chinese Engineers," *Eastern Engineering—Supplement to The China Express and Telegraph*, February 22, 1923, 53. Published address by J. A. Eley to the Engineering Society of China.

39. Eley, "The Education and Training of Chinese Engineers."

40. Aaron Stephen Moore, *Constructing East Asia: Technology, Ideology, and Empire in Japan's Wartime Era, 1931–1945* (Stanford, CA: Stanford University Press, 2013); Janis A. Mimura, *Planning for Empire: Reform Bureaucrats and the Japanese Wartime State* (Cornell, NY: Cornell University Press, 2017).

41. Victor Seow, *Carbon Technocracy: Energy Regimes in Modern East Asia* (Chicago: Chicago University Press, 2021), 163–165.

42. Koji Hirata, "Made in Manchuria: The Transnational Origins of Socialist Industrialization in Maoist China," *The American Historical Review* 126, no. 3 (2021): 1072–1101.

43. Seow, *Carbon Technocracy*, 43–44, 80–87, quote 94.

44. *The Far Eastern Review: Commerce, Engineering, Finance* (hereafter *FER*), vols. 1–38 (Manila, Shanghai, and Yokohama: 1904–1941).

45. Between 1904 and 1941, about 500 *FER* articles directly addressed topics related to engineering and engineers.

46. Leslie Eaton Clark, *George Bronson Rea, Propagandist: The Life and Times of a Mercenary Journalist* (Lexington, KY: Rowman & Littlefield, 2017).

47. Anonymous, "A Review of the Railway Situation in China," *FER* 5, no. 4 (September 1908): 101–131.

48. See Rea's foreword to the first issue in *FER* 1, no. 1 (June 1904): 4.

49. In the 1920s, Rea worked as a consultant for the Chinese government and, during the 1930s, supported the puppet regime in Manchukuo as an official lobbyist until his death in 1936. See George Bronson Rea, *The Case for Manchukuo* (New York: D. Appleton-Century Company, 1935). In return for favorable reporting on Japanese investment and trade in China, the Yokohama Specie Bank bought advertising space, including the full-page back cover for every *FER* issue.

50. Leslie Eaton Clark, *George Bronson Rea*, 190.

51. Anonymous, "Approved Water System for Manila, P. I.," *FER* 1, no. 1 (June 1904): 5–7; Anonymous, "Municipal Improvements, Manila," *FER* 2, no. 7 (November 1905): 172; Rebecca Tinio McKenna, *American Imperial Pastoral: The Architecture of US Colonialism in the Philippines* (Chicago: University of Chicago Press, 2017), Henry Cabot Lodge quoted on p. 4.

52. "The American Engineer," *FER* 9, no. 10 (March 1913): 435; "The Young Chinese Engineer," *FER* 9, no. 10 (March 1913): 443.

53. "The University of Hong Kong," *FER* 10, no. 4 (September 1913): 137, 159.

54. See, for example, "A Monument to the Japanese Engineer," *FER* 21, nos. 6–7 (June–July 1925): 243–246.

55. George Bronson Rea, *The Independence of Manchoukuo* (Washington, DC: The George Washington Law School, 1933), quote 13. Paper read by Rea as Counsellor of the Ministry of Foreign Affairs of the Government of Manchukuo before members of the legal Fraternity of Gamma Eta Gamma.

56. Anonymous, "Soviet Oil Industry for 1932–33," *FER* 28, no. 3 (March 1932): 126–136; "Civil Aviation in the U.S.S.R.," *FER* 28, no. 6 (June 1932): 290–291; "Soviet Is Operating the World's Largest Hydro-Electric Plant," *FER* 28, no. 9 (September 1932): 409–413.

57. Jeremy Goh and Koh Keng We, "Industrialisation and Chinese Big Business in Colonial Singapore, Malaya, and China: The Transnational Enterprises of Lim Peng Siang (1904–41)," *Journal of Southeast Asian Studies* 54, no. 1 (April 2023): 89–115.

58. "Chinese Engineering Development in Shanghai," *FER* 19, no. 3 (March 1923): 165–171.

59. "Hangchow-Kiangshan Railway," *FER* 28, no. 4 (April 1932): 177, 181.

60. H. H. Ling, "The Canton-Hankow Railway," *FER* 33, no. 3 (March 1937): 123–125.

61. Mao Yisheng, *Qiantang jiang jianqiao huiyi* (Remembering the Building of Qiantang River Bridge) (Beijing: Wenshi ziliao chubanshe, 1982), 2–3.

62. London Public Records Office (hereafter London PRO), FO 228–2803: "Extract of Ministry of Railways' Letter no. 12201," December 1, 1934, archivist's pagination 48–49.

63. London PRO, FO 228–2803: "Locomotive Manufacturers' Association: Chinese Students," (December) 1934, archivist's pagination 68–70.

64. London PRO, FO 228–2803: "Locomotive Manufacturers' Association."

65. London PRO, FO 228–2803: "A Brief for the Meeting with the Locomotive Manufacturers' Association, 29th October 1936," archivist's pagination 25–35, 27.

66. London PRO, FO 228–2803: "A Brief for the Meeting with the Locomotive Manufacturers' Association," 29.

67. London PRO, FO 228–2803: "A Brief for the Meeting with the Locomotive Manufacturers' Association," 30.

68. "Chiaotung University—Where China's Young Engineers Are Being Trained," *FER* 29, no. 10 (October 1933): 466–468, 479.

69. C. A. Middleton Smith, "Engineers and Doctors in China," *FER* 29, no. 10 (October 1933): 540–544, photograph 540.

70. Smith, "Engineers and Doctors in China," 541.

71. Sir Cecil Clementi, "Engineering in HK and China," *Hong Kong University Engineering Journal* 1, no. 1 (April 1929), quoted in Vivian Kong, *Multiracial Britishness: Global Networks in Hong Kong, 1910–1945* (Cambridge, UK: Cambridge University Press, 2024), 104.

72. C. A. Middleton Smith, "Engineering Development in China," *FER* 33, no. 2 (February 1937), 48–50, 63.

73. Ziyuan weiyuanhu (comp.), *Zhongguo gongcheng renminglu* (Directory of Chinese Engineers), (Changsha(?): 1941), foreword, no pagination.

74. "Engineering in China," *The North-China Herald*, December 21, 1912, 791–792, quote 791.

75. "Announcement by the British Engineers Association," *FER* 10, no. 4 (July 1913): 79.

76. Elman, *On Their Own Terms*, 434. According to Elman, CES had 254 members in 1923, 304 in 1924, and more than 700 in 1927.

77. "The Shanghai Convention," *Journal of the Association of Chinese & American Engineers* 3, no. 4 (April–May 1922): 1–3, quote 1.

78. *Journal of the Association of Chinese & American Engineers* 3, no. 4 (April–May 1922): no page number.

79. O. J. Todd, *Two Decades in China: Comprising Technical Papers, Magazine Articles, Newspaper Stories and Official Reports Connected with Work under His Own Observation* (Peking: Association of Chinese and American Engineers, 1938), 396.

80. O. J. Todd, *Two Decades in China*, 203–204, 214; see also "Road Making for Victims," *North-China Herald*, October 2, 1920, 22.

81. O. J. Todd, "China's Call to the Engineer: The International Famine Relief Commission and Its Program," *Journal of the Association of Chinese & American Engineers* 4, no. 10 (December 1923): 22–24; Thorben Pelzer, *Engineering Trouble: US–Chinese Experiences of Professional Discontent, 1905–1945* (Leiden: Brill, 2023), 145–173.

82. Viola Smith and Anselm Chuh, *Motor Roads in China* (Washington, DC: US Government Printing Office, 1931), 7–9; O. J. Todd, "The Yellow River Dike Breaks of 1935," *FER* 32, no. 10 (October 1936): 454–460.

83. Robert J. Culp, Eddy U., and Wen-hsin Yeh, ed., *Knowledge Acts in Modern China: Ideas, Institutions, and Identities* (Berkeley: Institute of East Asian Studies, University of California, Berkeley, 2016).

84. *Journal of the Association of Chinese & American Engineers* 3, no. 4 (April–May 1922): title page.

85. Chinese Ministry of Information, *China Handbook, 1937–1945* (New York: The MacMillan Company, 1947), section "National Holidays, Anniversaries and Memorial Days," no page number.

86. Gispen, *New Profession, Old Order*, 45–46.

87. Ziyuan weiyuanhui, *Zhongguo gongcheng renminglu*, appendix, no pagination.

88. Ziyuan weiyuanhui, *Zhongguo gongcheng renminglu*.

89. "The Engineer as Peacemaker," *FER* 25, no. 10 (October 1929): 433–435, 447.

90. "The Engineer as Peacemaker"; the table on p. 447 only lists the number of foreign delegates, omitting Japan as the host country.

91. Köll, *Railroads and the Transformation of China*, 185–190.

92. David Strand, "Calling the Chinese People to Order: Sun Yat-sen's Rhetoric of Development," in *Reconstructing Twentieth-Century China: State Control, Civil Society, and National Identity*, ed. Kjeld Eric Brodsgaard and David Strand (Oxford: Clarendon Press, 1998), 33–68.

93. William C. Kirby, "Engineers and the State in Modern China," in *Prospects for the Professions in China*, ed. William P. Alford, William C. Kirby, and Kenneth Winston (New York: Routledge, 2010), 283–312, 298, and "Technocratic Organization and Technological Development in China: The Nationalist Experience and Legacy, 1928–1953," in *Science and Technology in Post-Mao China*, ed. Denis Fred Simon and Merle Goldman (Cambridge, MA: Council on East Asian Studies, Harvard University, 1989), 23–44, 41.

CHAPTER 6

1. In this article, Chinese words and names are transcribed in *pinyin*. Chinese names are presented according to the customary order where one's family name comes

before the given name. Exceptions are forms of names in English commonly known to the English-language press, such as K. C. Li (Li Guoqin), C. Y. Wang (Wang Chongyou), and Wah Chang (Huachang). Thanks to Shumeng Han for translating source material. And thanks to Shumeng and Micah Muscolino for assistance in the sections of this chapter that examine China.

2. Walter Renton Ingalls, *The Mineral Industry: Its Statistics, Technology, and Trade During 1906* (New York: McGraw-Hill Book Company, Inc., 1914), 38, and G. A. Roush, *The Mineral Industry: Its Statistics, Technology, and Trade During 1918* (New York: McGraw-Hill Book Company, Inc., 1919), 32–33.

3. Roush, *The Mineral Industry: Its Statistics, Technology, and Trade During 1918*, 33.

4. G. A. Roush, *The Mineral Industry: Its Statistics, Technology, and Trade During 1913* (New York: McGraw-Hill Book Company, Inc., 1914), 28–29, and Roush, *The Mineral Industry: Its Statistics, Technology, and Trade During 1918*, 35.

5. Jeff Nesbit, "China's Continuing Monopoly over Rare Earth Minerals," *U.S. News and World Report*, April 2, 2013.

6. There are many, many examples of excellent work on mining and minerals to choose from, but I draw particular inspiration from Mimi Sheller, *Aluminum Dreams: The Making of Light Modernity* (Cambridge, MA: The MIT Press, 2014); Tyler Priest, *Global Gambits: Big Steel and the U.S. Quest for Manganese* (New York: Praeger, 2003); Robin S. Gendron, Mats Ingulstad, and Espen Storli, eds., *Aluminum Ore: The Political Economy of the Global Bauxite Industry* (Vancouver: UBC Press, 2013); Thomas G. Andrews, *Killing for Coal America's Deadliest Labor War* (Cambridge, MA: Harvard University Press, 2008); Kent A. Curtis, *Gambling on Ore: The Nature of Mining in the United States, 1860–1910* (Boulder: University Press of Colorado, 2013); Paul Lucier, *Scientists and Swindlers: Consulting on Coal and Oil in America, 1920–1890* (Baltimore: The Johns Hopkins University Press, 2008); Megan Black, *The Global Interior: Mineral Frontiers and American Power* (Cambridge, MA: Harvard University Press, 2018); Ronald H. Limbaugh, *Tungsten in Peace and War, 1918–1946* (Reno: University of Nevada Press, 2010); Julie Michelle Klinger, *Rare Earth Frontiers: From Terrestrial Subsoils to Lunar Landscapes* (Ithaca, NY: Cornell University Press, 2017); Victor Seow, *Carbon Technocracy: Energy Regimes in Modern East Asia* (Chicago: University of Chicago Press, 2021); Shellen Xio Wu, *Empires of Coal: Fueling China's Entry into the Modern World Order, 1860–1920* (Stanford, CA: Stanford University Press, 2015); Clark Spence, *Mining Engineers and the American West: The Lace-Boot Brigade, 1849–1933* (Moscow: University of Idaho Press, 1993); Thomas Miller Klubock, *Contested Communities: Class, Gender, and Politics in Chile's El Teniente Copper Mine, 1904–1951* (Durham, NC: Duke University Press, 1998); Timothy J. LeCain, *Mass Destruction: The Men and Giant Mines That Wired America and Scarred the Planet* (New Brunswick, NJ: Rutgers University Press, 2009); and Eric C. Nystrom, *Seeing Underground: Maps, Models, and Mining Engineering in America* (Reno: University of Nevada Press, 2015).

7. Joseph Gambogi, "Rare Earths," in National Minerals Information Center, *2017 Minerals Yearbook* (Washington, DC: USGS, 2020), 60.2.

8. Kateryna Klochko, "Antimony," in National Minerals Information Center, *2016 Mineral Yearbook* (Washington, DC: USGS, 2019), 6.2

9. European Commission, *Study on the EU's List of Critical Raw Materials (2020) Final Report*, 3.

10. Although they don't deal with antimony in detail, a number of scholars have examined aspects of the history of the mineral dependence and scarcity issues. See, for example, Tyler Priest, *Global Gambits: Big Steel and the U.S. Quest for Manganese* (New York: Praeger, 2003); Alfred E. Eckes, *The United States and the Global Struggle for Minerals* (Austin: University of Texas Press, 1979); and Ronald Limbaugh, *Tungsten in Peace and War, 1918–1946* (Reno: University of Nevada Press, 2010). For a more contemporary analysis that brings to bear the tools of mineral economics, see Committee on Critical Mineral Impacts on the U.S. Economy, *Minerals, Critical Minerals, and the U.S. Economy* (Washington, DC: National Academies Press, 2008).

11. Edson S. Bastin, "Antimony," in US Bureau of Mines, *Mineral Resources of the United States: 1917* (Washington, DC: GPO, 1921), 653.

12. W. C. Butterman and J. F. Carlin, *Mineral Commodity Profiles: Antimony* (Washington, DC: USGS, 2004), 26.

13. R. R. Seal II, K. J. Schulz, and J. H. DeYoung Jr., with contributions from David M. Sutphin, Lawrence J. Drew, James F. Carlin Jr., and Byron R. Berger, "Antimony," in *Critical Mineral Resources of the United States—Economic and Environmental Geology and Prospects for Future Supply*, ed. K. J. Schulz, J. H. DeYoung Jr., R. R. Seal II, and D. C. Bradley, US Geological Survey Professional Paper 1802 (Washington, DC: 2017), C13, https://doi.org/10.3133/pp1802C.

14. USGS, *Mineral Commodity Summaries 2020* (Washington, DC: USGS, 2020), 5.

15. Cornelis Klein and Barbara Dutrow, *The 23rd Edition of the Manual of Mineral Science (after James D. Dana)* (New York: John Wiley & Sones, Inc.), 46.

16. Mineral Resources of the United States, *1882* (Washington, DC: GPO, 1883), 439.

17. Corby G. Anderson, "Antimony Production and Commodities," *SME Mineral Processing and Extractive Metallurgy Handbook* (Englewood, CO: SME, 2019), 432; Corby Anderson, "The Metallurgy of Antimony," *Geochemistry*, July 2012, 3–8; and Elmira Moosavi-Khoonsari, Sina Mostaghel, Andreas Siegmund, and Jean-Pierre Cloutier, "A Review on Pyrometallurgical Extraction of Antimony from Primary Resources: Current Practices and Evolving Processes," *Process*, August 2022, 1–35.

18. Shellen Xiao Wu, *Empires of Coal: Fueling China's Entry into the Modern World Order, 1860–1920* (Stanford, CA: Stanford University Press, 2015), 131.

19. Wu, *Empires of Coal*, 133–134; En-Han Lee, "China's Response to Foreign Investment in Her Mining Industry (1902–1911)," *The Journal of Asian Studies* 28, no. 1 (1968): 55–76; and Seow, *Carbon Technocracy*, 39.

20. Wu, *Empires of Coal*, 146–147.

21. Agnus W. McDonald Jr., *The Urban Origins of Rural Revolution: Elites and the Masses in Hunan Province, China, 1911–1927* (Berkeley: University of California Press, 1978), 70. On the successful effort to resist foreign investment in Hunan's mining industry—even in the face of the terms laid out in the McKay Treaty between China and Great Britain in 1902 and the 1903 Sino-American Commercial Treaty—see En-Han Lee, "China's Response to Foreign Investment in Her Mining Industry (1902–1911)," 60–61.

22. McDonald, *The Urban Origins of Rural Revolution*, 69–70.

23. McDonald, *The Urban Origins of Rural Revolution*, 70.

24. McDonald, *The Urban Origins of Rural Revolution*, 79.

25. Peisu Liang, "Hunan huachang liankuang gongsi jingguo lüeshu" [A brief overview of the history of Hunan Wah Chang Smelting Company], *Shiye zazhi* [Industry magazine], no. 178 (February 1933): 5–12; Zhu Youzhi and Guo Qin, *Hunan jinxiandai shiye renwu zhuanlüe* [Brief biographies of eminent people in Hunan's modern industry] (Changsha: Zhongnan daxue chubanshe, 2011), 82–98. Translated by Shumeng Han.

26. Liang Peisu, "Hunan huachang liankuang gongsi jingguo lüeshu," 5–12. Translated by Shumeng Han, 5–12. On the 1906 price spike, see *Mineral Resources*, 1908, 710.

27. Zhu Youzhi and Guo Qin, *Hunan jinxiandai shiye renwu zhuanlüe*, 82–98. Translated by Shumeng Han.

28. Youzhi and Qin, *Hunan jinxiandai shiye renwu zhuanlüe*.

29. Gavin Wright, "The Origins of American Industrial Success, 1879–1940," *American Economic Review* 80 (September 1990): 651–668.

30. J. W. Furness, "Antimony," in US Bureau of Mines, *Mineral Resources of the United States, 1925—Part I: Metals* (Washington, DC: GPO, 1928), 15.

31. Weili Ye, *Seeking Modernity in China's Name: Chinese Students in the United States, 1900–1927* (Stanford, CA: Stanford University Press, 2001), 8–10.

32. Karl Gerth, *China Made: Consumer Culture and the Creation of the Nation* (Cambridge, MA: Harvard University Press, 2003), 33.

33. "C. Y. Wang," accessed July 24, 2024, https://prabook.com/web/c.y.wang/1345835; "C. Y. Wang (王寵佑)—Boxer Indemnity Scholars," accessed July 24, 2024, https://boxerindemnityscholars.wordpress.com/2018/05/27/c-y-wang-%E7%8E%8B%E5%AF%B5%E4%BD%91/. China Association for Science and Technology, ed., *Brief Biographies of China's Scientific and Technical Experts*, Book of Engineering Technology, Collection of Metallurgy, vol. 1 (Beijing: China Science & Technology Press, 1995), 621.

34. In 1888, Kemp, along with thirteen other American geologists, founded the Geological Society of America and would serve in leadership positions in virtually every geological and mining organization in the late nineteenth and early twentieth centuries. To name just a few of the many positions Kemp held: president of the

American Institute of Mining and Metallurgical Engineers (1912), president of the Mining and Metallurgical Society of America (1912), president of the Geological Society of America (1921), and president of the Society of Economic Geologists. He was also a member of the National Academy of Sciences and a member of various geological societies in London, Stockholm, Belgium, Oslo, and Canada. He also took work as a consulting engineer with the Anaconda Copper Company, Calumet and Hecla Consolidated, Spanish American Iron Company, and New Jersey Zinc Company, among others. On Kemp, see Waldemar Lindgren, "James Furman Kemp; to His Memory," *Economic Geology* 22 (January 1927): 89–90, and Frank Dawson Adams, "Biographical Memoir of James Furman Kemp, 1859–1926," in National Academy of Sciences of the United States of America, *Biographical Memoirs Volume XVI—First Memoir* (Washington, DC: National Academy of Sciences, 1933), 8.

35. C. Y. Wang, *Antimony: Its History, Chemistry, Minerology, Geology, Metallurgy, Uses, Preparations, and Valuation: with Complete Bibliographies*, 2nd ed. (London: Charles Griffen & Company, Limited, 1919), vi; C. Y. Wang, "The Herrenschmidt's Process of Antimony Smelting," *The Mining Journal*, July 25, 1908, 108–109. For evidence of his early expertise in the United States, see C. Y. Wang, "The Antimony Industry of China," *EMJ*, October 26, 1912, 777. For his work on tungsten, see C. Y. Wang, "Wolfram Mining in China," *EMJ*, January 3, 1920, 1.

36. Thomas Thronton Read, *The Development of Mineral Industry Education in the United States*, with Foreword by Herbert Hoover (New York: The American Institute of Mining and Metallurgical Engineers, 1941), vii; C. Y. Wang, *Antimony: Its History, Chemistry, Minerology, Geology, Metallurgy, Uses, Preparations, and Valuation: With Complete Bibliographies*, 2nd ed. (London: Charles Griffen & Company, Limited, 1919), vi.

37. C. Y. Wang, "The Herrenschmidt's Process of Antimony Smelting," *The Mining Journal* 84 (July 25, 1908): 108; "Chung Yu Wang Obituary," accessed July 29, 2021, https://www.nmrs.org.uk/resources/obituaries-of-members/obituaries-of-members-names-beginning-with-w/chung-yu-wang/.

38. China Association for Science and Technology, ed., *Brief Biographies of China's Scientific and Technical Experts*, Book of Engineering Technology, Collection of metallurgy, vol. 1 (Beijing: China Science & Technology Press, 1995), 621.

39. Wang, "The Herrenschmidt's Process of Antimony Smelting"; "Chung Yu Wang Obituary."

40. K. C. Li and C. Y. Wang, *Tungsten* (New York: Reinhold Publishing Corporation, 1943).

41. H. T. Liang [Liang Huanting], "Opportunities in China," *The Far Eastern Review*, October 1915, 194.

42. "Antimony in China," *New York Times*, June 11, 1915, 16.

43. Li Lien-Fung, *A Daughter Remembers* (Singapore: Marshall Cavendish Editions, 2011), 42.

44. "K. C. Li," in *Who's Who in China*, 5th ed. (Shanghai: The China Weekly Review, 1932), 140–141.

45. "Opportunities in China," *The Far-Eastern Review*, October 1915, 194. For more on Huanyi's promotion of Chinese antimony in the United States, see "Antimony in China," *NYT*, June 11, 1915, 16.

46. Nelson T. Johnson, "Antimony Production in Hunan Province," *The Far Eastern Review*, December 1918, 506; William F. Collins, *Mineral Enterprise in China* (Tientsin: Tientsin Press, 1922), 97.

47. McDonald, *The Urban Origins of Rural Revolution*, 72.

48. Li, *A Daughter Remembers*, 89.

49. Zhao Tian-cong, *The Metallurgy of Antimony* (Changsha: Central South University of Technology Press, 1988), 201. For more recent reviews of antimony processing, see Corby Anderson, "The Metallurgy of Antimony," *Geochemistry*, July 2012, 3–8; Corby Anderson, "Antimony Production Commodities," in *Mineral Processing & Extractive Metallurgy Handbook*, ed. Society for Mining Metallurgy and Exploration (Englewood, CO: SMME, 2019), and Elmira Moosavi-Khoonsari, Sina Mostaghel, Andreas Siegmund, and Jean-Pierre Cloutier, "A Review on Pyrometallurgical Extraction of Antimony from Primary Resources: Current Practices and Evolving Processes," *Process*, August 2022, 1–35.

50. My description is drawn from Li's 1917 *EMJ* article, which was reprinted and quoted widely. See K. C. Li, "Metallurgy of Antimony" *EMJ*, October 20, 1917, 683–689; K. C. Li, , "Antimony, Its Metallurgy and Uses," *The Metal Record and Electroplater*, September 1917, 197–198; and K. C. Li, "Antimony, Its Metallurgy and Uses," *Journal of the American Institute of Metals* 11 (June 1917): 92–112. For a more extensive review of antimony processing and refining methods, see Wang's volume, Chung Yu Wang, *Antimony* (London: Charles Griffin & Company, 1919), particularly 69–144.

51. Li, "Metallurgy of Antimony," 687.

52. *Mining and Engineering World*, July 31, 1917, 183.

53. "Mining in the Hunan Province," *The Far Eastern Review*, February 1920, 79–80.

54. William F. Collins, *Mineral Enterprise in China* (Tientsin: Tientsin Press, 1922), 99–101.

55. Collins, *Mineral Enterprise in China*, 100–101.

56. Liang Peisu, "Hunan huachang liankuang gongsi jingguo lüeshu," 5–12. (translation by Shumeng Han).

57. Nelson T. Johnson, "Antimony Production in Hunan Province," *The Far Eastern Review*, December 1918, 506.

58. J. W. Furness, "Antimony," in US Bureau of Mines, *Mineral Resources of the United States, 1927* (Washington, DC: GPO, 1930), 42–43; "Changsha's Antimony Trade," *The Far Eastern Review*, August 1916, 119.

59. William C. Kirby, *Germany and Republican China* (Stanford, CA: Stanford University Press, 1984), 106.

60. Li, *A Daughter Remembers*, 88–89. For more on Zhidong and Shi-kai's role in the history of mining in China, see Wu, *Empires of Coal*.

61. H. G. Ferguson and D. A. Hall, "Antimony," in *Political and Commercial Geology and the World's Mineral Resources*, ed. Josiah Edward Spurr (New York: McGraw-Hill Book Company, Inc., 1920), 182–184. On the inability of US producers to compete with lower-priced antimony, see also J. E. Spurr, "Weekly Report to the Director the Week Ending October 26, 1918," File "Committee on Mineral Imports and Exports, 1918," Josiah Edward Spurr Papers, American Heritage Center, University of Wyoming, Collection Number 2479, Box 2.

62. McDonald, *The Urban Origins of Rural Revolution*, 73; Zhu Youzhi and Guo Qin, *Hunan jinxiandai shiye renwu zhuanlüe*, 82–98, 5–12. On the Cookson assertion, see Liang Peisu, "Hunan huachang liankuang gongsi jingguo lüeshu," 5–12 (translations by Shumeng Han).

63. Peisu, "Hunan huachang liankuang gongsi jingguo lüeshu," 121.

64. "Other Metals," *EMJ*, March 16, 1918, 533.

65. My larger book-length project on this issue will discuss the continued rise of the New York WCC and the decline of the WCC in China.

66. On Wah Chang Mining and Smelting Company claims, see also "China Begins to Advertise," *New York Tribune*, August 22, 1916, 10.

67. "The Antimony Industry at Changsha," *Millard's Review of the Far East*, April 5, 1919, 226.

68. McDonald, *The Urban Origins of Rural Revolution*, 74.

69. McDonald, *The Urban Origins of Rural Revolution*, 74.

70. Frank C. Schrader, "Antimony," in US Bureau of Mines, *Mineral Resources of the United States, 1920* (Washington, DC: GPO, 1922), 82.

71. "Should the United States Foster the Domestic Production of Antimony," *EMJ*, June 15, 1918, 1088.

72. "Occurrence and Utilization of Antimony Ores," *EMJ*, June 9, 1917, 1017.

73. Eckes, *The United States and the Global Struggle for Minerals*, 49.

74. Baruch to Yeatman, November 1, 1917, and Baruch to Yeatman, November 6, 1917, Yeatman Papers, American Heritage Center, University of Wyoming, Folder 2, Part 2.

75. Cuff, *War Industries Board*, 117.

76. On the development of El Teniente, see Klubock, *Contested Communities*. On Hammond's work in South Africa, see Mark Hendrickson, "'The Sesame That Opens the Door of Trade': John Hays Hammond and Foreign Direct Investment in Mining, 1880–1920," *Journal of Gilded Age and Progressive Era* 16 (July 2017): 325–346.

77. Minerals and Metals for War Purposes, *Statement of Hon. Pope Yeatman, Chief of Nonferrous Metals Section, War Industries Board*, May 11, 1918, 197.

78. Minerals and Metals for War Purposes, *Statement of Hon. Pope Yeatman*, 213–215.

79. See Courtenay De Kalb, "International Control of Mines," *Mining and Scientific Press*, January 18, 1919, 75–76; Staley, *Raw Materials in Peace and War*; Wilkins, *The Maturing of Multinational Enterprises*, 52–53; Rosenberg, *Spreading the American Dream*, 126–127; McGrath, *Charles Kenneth Leith*, 95–110; Tyler Priest, *Global Gambits: Big Steel and the U.S. Quest for Manganese* (New York: Praeger, 2003); and Alfred E. Eckes, *The United States and the Global Struggle for Minerals* (Austin: University of Texas Press, 1979).

80. William C. Redfield, *Dependent America: A Study of the Economic Basis of Our International Relations* (Boston: Houghton Mifflin Company, 1926), 158–160, 212–213.

81. Mining and Metallurgical Society of America, *International Control of Minerals* (Baltimore: The Lord Baltimore Press, 1925), 37.

82. "Men You Should Know," *EMJ*, July 12, 1924, 71.

83. *The Colorado School of Mines Magazine*, November 1930, 38.

84. C. K. Leith and H. Foster Bain, *Ores and Industry in the Far East* (New York: Council on Foreign Relations, 1927). See also "Ores and Industry in the Far East," *The Far Eastern Review*, November 1927, 493.

85. "Men You Should Know," *EMJ*, January 9, 1921, 73; "In Remembrance | Association of American State Geologists," accessed July 24, 2024, https://www.stategeologists.org/remembrance.

86. McGrath, *Charles Kenneth Leith*, 194.

87. Waldemar T. Schaller, "Memorial of Frank Charles Schrader," *American Mineralogist* 30, nos. 3–4 (1945): 148–152.

88. *International Control on Minerals*, 24.

89. *International Control of Minerals*, 36.

90. *International Control of Minerals*, 23–24.

91. *International Control of Minerals*, 37.

92. *International Control of Minerals*, 23–25.

93. *International Control of Minerals*, 37.

94. Emily S. Rosenberg, *Transnational Currents in a Shrinking World* (New Haven: Harvard University Press, 2014), 2.

95. Wright, "The Origins of American Industrial Success."

CHAPTER 7

1. My sincere acknowledgment to Ted Beatty and Israel Solares for having invited me to contribute to this volume and for their comments on previous versions of the

chapter, which has also benefited from comments from Dominique Barjot, Simone Fari, Elisa Grandi, Pierre-Cyrill Hautcoeur, Jean-Philippe Passaqui, and Francesca Sanna. All mistakes and inaccuracies are my own.

2. Anne-Françoise Garçon, *Entre l'État et l'usine: l'École des Mines de Saint-Étienne au XIXe siècle* (Rennes: Presses universitaires de Rennes, 2004); Anne-Françoise Garçon, "Existe-t-il un esprit Mines?" in *Les ingénieurs des Mines: cultures, pouvoirs pratiques*, ed. Anne-Françoise Garçon and Bruno Belhoste (Paris: Comité pour l'histoire économique et financière de la France, 2012).

3. André Grelon, "The Training and Career Structures of Engineers in France, 1880–1939," in *Education, Technology and Industrial Performance in Europe, 1850–1939*, ed. Robert Fox and Anna Guagnini (Cambridge, UK: Cambridge University Press, 1993), 42–64; André Thépot, *Les ingénieurs des Mines du XIXe siècle, histoire d'un corps technique d'État, Tome I: 1810–1914* (Paris: Éditions Eska, 1998); Garçon, "Existe-t-il un "Esprit Mine"; Isabelle Laboulais, *La maison des mines Genèse d'un corps d'ingénieurs civils (1794–1814)* (Rennes: Presses Universitaires de Rennes, 2013).

4. Garçon, *Entre l'État et l'usine*; Marco Bertilorenzi, Anne-Françoise Garçon, and Jean-Philippe Passaqui, "Introduction," in *Entre technique et gestion. Une histoire des Ingénieurs civils des mines*, ed. Marco Bertilorenzi, Anne-Françoise Garçon, and Jean-Philippe Passaqui (Paris: Presses des mines, 2016).

5. Garçon, *Entre l'Etat et l'Usine*; Dominique, "Public Utilities and Private Initiative: The French Concession Model in Historical Perspective," *Business History*, 53, no. 5 (2011): 782–800.

6. Garçon, *Entre l'Etat et l'Usine*; Jean Philippe Passaqui, *Les voyages forment l'ingénieur. Les houillères du Centre-Midi de la France (1851–1873)* (Paris: Garnier, 2015).

7. Françoiase Birck, *L'École des Mines de Nancy (ensmn) 1919–2012. Entre université, grand corps d'État et industrie* (Nancy: Presses universitaires de Nancy/Éd. universitaires de Lorraine, 2014).

8. Garçon, *Entre l'État et l'usine*.

9. Garçon, *Entre l'État et l'usine*.

10. Garçon, *Entre l'État et l'usine*; Marco Bertilorenzi, "L'École des Mines de Saint-Étienne et son conseil de perfectionnement," in *Entre technique et gestion. Une histoire des Ingénieurs civils des mines*, ed. Marco Bertilorenzi, Anne-Françoise Garçon, and Jean-Philippe Passaqui (Paris: Presses des mines, 2016); Hervé Joly, *Diriger une grande entreprise au XXe siècle: l'élite industrielle française* (Tours: Presses universitaires François-Rabelais, 2013).

11. Garçon, *Entre l'Etat et l'Usine*; Bertilorenzi et al., "Introduction."

12. Israel Solares and Edward Beatty, "Engineers & Corporate Management, ca 1870–1930: The Invisible Hand Redux," *Enterprise & Society* 25, no. 2 (2024): 486–511.

13. Bertilorenzi et al., "Introduction."

14. Jean-Luc Peaucelle and Cameron Guthrie, *Henri Fayol, The Manager* (London: Pickering and Chatto, 2017); Marco Bertilorenzi, Nadine Dubruc, and Jean-Philippe

Passaqui, "Introduction générale," in *Henri Fayol. Les multiples facettes d'un manager*, ed. Marco Bertilorenzi, Nadine Dubruc, and Jean-Philippe Passaqui (Paris: Presses des mines, 2019); Marco Bertilorenzi and Anne-Françoise Garçon, "De La Pratique à Une Science? La 'nouvelle Doctrine' d'Henri Fayol Dans Son Contexte Conceptuel," *Entreprises et Histoire* 2 (2016): 13–35.

15. Henri Fayol, *L'Administration industrielle et générale* (Saint-Étienne: Bulletin de la Société de l'Industrie minerale, 1916); Marco Bertilorenzi and Jean-Philippe Passaqui, "Entreprise et société. Aux sources de l'œuvre d'Henri Fayol," in *Le traité et la notice. Relire Fayol avec Fayol*, ed. Henri Fayol (Paris: Presses des Mines 2020); Armand Hatchuel and Blanche Segrestin, "Relire Fayol à l'heure de l'entreprise responsable," in *Le traité et la notice. Relire Fayol avec Fayol*, ed. Henri Fayol (Paris: Presses des Mines 2020).

16. Jacques Marseille, *Empire colonial et capitalisme français: Histoire d'un divorce* (Paris: l'Harmattan, 1994); Rondo Cameron, *France and the Economic Development of Europe, 1800–1914* (Princeton, NJ: Princeton University Press, 1978); Gerard Chastagnaret, *L'Espagne, puissance minière dans l'Europe du XIXe siècle* (Madrid: Casa Vélasquez, 2000); Hubert Bonin, *L'empire colonial français: de l'histoire aux héritages (XIXe-XXIe siècles)* (Paris: Armand Colin, 2018); Robert Fitzgerald, *The Rise of the Global Company: Multinationals and the Making of the Modern World* (Cambridge, UK: Cambridge University Press, 2015).

17. Mimi Sheller and John Urry, "The New Mobilities Paradigm," *Environment and Planning A* 38, no. 2 (2006): 207–226, https://doi.org/10.1068/a37268.

18. Francesca Sanna, "Organiser La Méditerranée Minière: Gestion d'entreprise, Contrôle Des Ressources et Rationalisation Du Travail à La Société Minière et Métallurgique Peñarroya (1881–1950)," PhD diss., Université de Paris, 2020; Álvaro Silva, "Organizational Innovation in Nineteenth-Century Railway Investment: Peripheral Countries in a Global Economy," *Business History Review* 88, no. 4 (2014): 709–736.

19. Yamina Bettahar, "Les ingénieurs civils et le processus d'industrialisation de l'Algérie, 1850–1960," in *Entre technique et gestion. Une histoire des Ingénieurs civils des mines*, ed. Marco Bertilorenzi, Anne-Françoise Garçon, and Jean-Philippe Passaqui (Paris: Presses des mines, 2016).

20. Yann Bencivengo, *Nickel. La naissance de l'industrie calédonienne* (Tours: Presses universitaires François-Rabelais, 2014); Yann Bencivengo, "Les ingénieurs civils des mines dans l'industrie du nickel en Nouvelle-Calédonie," in *Entre technique et gestion. Une histoire des Ingénieurs civils des mines*, ed. Marco Bertilorenzi, Anne-Françoise Garçon, and Jean-Philippe Passaqui (Paris: Presses des mines, 2016).

21. Miguel Lopez-Morell and Miguel Pérez de Perceval Verde, "French Civil Engineers in the Spanish Mining Industry," in *Entre technique et gestion. Une histoire des Ingénieurs civils des mines*, ed. Marco Bertilorenzi, Anne-Françoise Garçon, and Jean-Philippe Passaqui (Paris: Presses des mines, 2016); Ana Cardoso da Matos, "Les élèves portugais de l'École des Mines de Paris," in *Entre technique et gestion. Une histoire des Ingénieurs civils*

des mines, ed. Marco Bertilorenzi, Anne-Françoise Garçon, and Jean-Philippe Passaqui (Paris: Presses des mines, 2016); Sanna, *Organiser la Méditerranée minière.*

22. Garçon, *Entre l'Etat et l'Usine.*

23. Garçon, *Entre l'Etat et l'Usine*; Bertilorenzi, "L'École des Mines de Saint-Étienne"; Hervé Jacquemin and Remi Révillon, *L'Ecole des Mines de Saint-Etienne: 200 ans d'histoire* (Saint-Étienne: Villes et Pays d'art et d'histoire, 2016).

24. The outcome can be found here: https://storymaps.arcgis.com/stories/cacb82e4c-dae425a8a05123d9cf2e59e. Many thanks to Dr. Marco Orlandi, at Unipd's Mobility Lab, for his support in adapting the data set to ArcGis.

25. Per-Olof Grönberg, *The Peregrine Profession: Transnational Mobility of Nordic Engineers and Architects* (Leiden: Brill, 2019).

26. Bertilorenzi, "L'Ecole des Mines de Saint-Etienne."

27. Other studies described the transformations of ICMs as a path from mining toward generalist engineers: for instance, Robert Germinet, *La République des ingénieurs* (Paris: Jacob Duvenet, 2003), but this idea is not supported by this data set. Perhaps an extension of the data set to the promotions trained at EMSE after the 1960s might support this thesis.

28. Marseille, *Empire colonial et capitalisme français.*

29. Bertilorenzi, "L'École des Mines de Saint-Etienne."

30. Marseille, *Empire colonial et capitalisme français.*

31. Solares and Beatty, "Engineers & Corporate Management."

32. Garçon, *Entre l'Etat et l'Usine.*

CHAPTER 8

1. "Engineer, *n.*" (Definition 2a), OED Online, June 2022, Oxford University Press, www.oed.com/view/Entry/62225, accessed June 29, 2022. This chapter draws on various parts of my book on Visvesvaraya. Aparajith Ramnath, *Engineering a Nation: The Life and Career of M. Visvesvaraya (1861–1962)* (Gurugram: Viking, forthcoming). For helpful feedback, I thank the editors of the present volume, the other participants in the "Blueprint for Modernity" project, the peer reviewers for this volume, and my co-panelists and other attendees at the Annual Meeting of the Society for the History of Technology (2021), where I presented a version of this paper.

2. See John B. Rae, "Engineering as Preparation for Management: A Study of M.I.T. Alumni," *Business History Review* 29, no. 1 (1955): 64–74; Aparajith Ramnath, *The Birth of an Indian Profession: Engineers, Industry, and the State, 1900–47* (Delhi: Oxford University Press, 2017), esp. chap. 3; Meritt Roe Smith, [Review of *America by Design: Science, Technology, and the Rise of Corporate Capitalism,* by D. F. Noble], *American Historical Review* 83, no. 3 (1978): 817–818.

3. "P. C. Mahalanobis," *Encyclopedia Britannica* online, n.d., https://www.britannica.com/biography/P-C-Mahalanobis, accessed November 18, 2021. For a recent work highlighting Mahalanobis's contribution to planning, see Nikhil Menon, *Planning Democracy: How a Professor, an Institute, and an Idea Shaped India*, Kindle Edition (n.p.: Viking, 2022).

4. See various editorials/articles in *Science and Culture* associated with physicist M. N. Saha. For example, "Planning and Prospect," *Science and Culture* 9, 11 (May 1944): 457–462, here 457; "Planning Again," *Science and Culture* 15, no. 8 (February 1950): 292–295, here 293.

5. See the discussion in Edward Beatty and Israel G. Solares's Introduction and Conclusion to the present volume.

6. In a lecture published in 2001, historian of science Dhruv Raina points out that "little attention has been paid to the importance of [Visvesvaraya's time as an engineer] in shaping his views and his vision." While he offers a conjecture that Visvesvaraya's "exposure to the industrializing and technical culture prevalent in Bombay Presidency and his travels abroad" would have been crucial in molding his thinking, the details of how this influence worked remain to be explored. Dhruv Raina, *Visvesvaraya as Engineer-Sociologist and the Evolution of His Techno-Economic Vision*, NIAS Lecture, L1–2001 (Bangalore: National Institute of Advanced Studies, 2001), 10.

7. In other words, how exactly did it come about that, as historian Benjamin Zachariah puts it, "as a result of his career and earlier successes, he [Visvesvaraya] commanded a ready audience on matters related to economic life, national regeneration, industrialization and nation-building"? Benjamin Zachariah, *Developing India: An Intellectual and Social History, c.1930–50* (New Delhi: Oxford University Press, 2005), 244.

8. Again, scholars have speculated on the importance of Visvesvaraya's foreign trips. See, for example, the quote from Raina in an earlier footnote.

9. Cecil O. Smith Jr., "The Longest Run: Public Engineers and Planning in France," *American Historical Review* 95, no. 3 (1990): 657–692; Justin C. Castro, *Apostle of Progress: Modesto C. Rolland, Global Progressivism, and the Engineering of Revolutionary Mexico* (Lincoln: University of Nebraska Press, 2019).

10. H. Rangachar and P. Kodanda Rao, "A Brief Life Sketch," in *"M. V.": Dr. M. Visvesvaraya: Birth Centenary Commemoration Volume* (Bangalore: Visvesvaraya Centenary Celebration Committee, 1960), 293–334, here 293–294.

11. See section by S. N. Mukerji under "The Spread of Western Educational Practices to Asian Countries: India," *Encyclopedia Britannica* online, https://www.britannica.com/topic/education/The-spread-of-Western-educational-practices-to-Asian-countries#ref303195, accessed July 23, 2022; P. L. Rawat, *History of Indian Education* (Agra: Ram Prasad and Sons, 1973 [1956]), chaps. 12–14; B. Lewis Rice, *Mysore: A Gazetteer Compiled for Government: Vol. I. Mysore in General*, rev. ed. (Westminster: Archibald Constable and Company, 1897), 747–750. There were some prominent English-medium-educated Indians in earlier generations, like Dadabhai Naoroji

(1825–1917). But schools like Naoroji's were not (although they might receive government aid) part of a concerted government policy. See Dinyar Patel, *Naoroji: Pioneer of Indian Nationalism* (Cambridge, MA: Harvard University Press, 2020), 20–21.

12. He belonged to a Telugu-speaking community, the Mulakanadu Brahmins; his ancestors had long been settled in the state of Mysore. Pandri Nath, *Mokshagundam Visvesvaraya: Life and Work* (Bombay: Bharatiya Vidya Bhavan, 1987), 1. Visvesvaraya's private papers make it clear that he corresponded almost exclusively in English. See M. Visvesvaraya papers, several volumes, Institution of Engineers (India), Mysore Local Centre, Mysuru.

13. Rangachar and Kodanda Rao, "A Brief Life Sketch," 294.

14. Entry in list of past graduates, *University of Madras. The Calendar for 1903–1904*, vol. II (Madras: Higginbotham & Co., 1903), 393 (and 39–40 for details of Central College, Bangalore); Visvesvaraya's nomination papers for admission as Associate Member (1877) and Full Member (c. 1904) of the London-based Institution of Civil Engineers (ICE Archives, London); *The Bombay University Calendar for the Year 1884–85* (Bombay: Thacker & Co. 1884), 397. On Poona College of Science, see *Technical and Industrial Education in the Bombay Presidency: Final Report of the Committee Appointed by Government, 1921–1922* (Bombay: Government Central Press, 1923), 77; *The Bombay University Calendar for the Year 1884–85* (Bombay: Thacker & Co., 1884), 381–382.

15. Mirza Ismail, "Speech at the Wesleyan Mission High School," December 22, 1934, in *Speeches by Amin-ul-Mulk Sir Mirza M. Ismail, Kt., K.C.I.E., O.B.E., Dewan of Mysore*, vol. II (Bangalore: Superintendent, Government Press, 1936), 390–395, here 391.

16. Based on the syllabus for the year before Visvesvaraya took the BA exam. *The Madras University Calendar 1879–80* (Madras: Higginbotham & Co.), 38–44. The writers named here were among those prescribed for the batch of students immediately preceding Visvesvaraya's. See syllabus for 1880, listed in *The Madras University Calendar, 1877–78* (Madras: Higginbotham & Co., 1877), 110–111.

17. See H. Rangachar and P. Kodanda Rao, "A Brief Life Sketch," 295; "News-letters," *The Times of India*, June 16, 1874, 2; "The Bangalore Regatta," *The Times of India*, September 16, 1881, 3; Aparajith Ramnath, "Engineers' Day: The Story of the Irishman Who Moulded Visvesvaraya's Alma Mater," *The Wire*, September 15, 2019, https://thewire.in/the-sciences/engineers-day-theodore-cooke-mokshagundam-visvesvaraya-civil-engineering-college, accessed July 21, 2022.

18. *Speeches by Sir M. Visvesvaraya, K.C.I.E. Dewan of Mysore. 1910–11 to 1916–17* (Bangalore: Government Press, 1917).

19. Author's visit to Muddenahalli, 2018.

20. In a speech given in 1914, Visvesvaraya said, "The Hindu ideal of life is that this world is a preparation for the next and not a place to stay in and make ourselves comfortable . . . we must give up aimless activities and . . . spread education in all

grades, multiply occupations and increase production and wealth. All other activities should conform themselves to the economic ideal." "Address to the Mysore Economic Conference," June 3, 1914, in *Speeches by Sir M. Visvesvaraya, K.C.I.E. Dewan of Mysore 1910–11 to 1916–17* (Bangalore: Government Press, 1917), 142–153, here 149. An example of Visvesvaraya's emphasis on grassroots development is the "village improvement scheme" he instituted as dewan of Mysore. "Mysore Village Schemes," *The Times of India*, February 14, 1914, 10.

21. Rangachar and Rao, "A Brief Life Sketch," 296; *Bombay University Calendar for the Year 1884–85*, 396; M. Visvesvaraya, *Memoirs of My Working Life* (Bangalore: M. Visvesvaraya, 1951), chaps. I–IV.

22. Ramnath, *The Birth of an Indian Profession*, chap. 3.

23. See, for instance, M. Venkatesa Iyengar, "A Hard Task-Master," in *"M. V.": Dr. M. Visvesvaraya: Birth Centenary Commemoration Volume*, 257–264. On Visvesvaraya's conception of bureaucracy, see Narendar Pani's Introduction in *A Bureaucrat in Princely Mysore: Selections from the Private Diary of K.R. Srinivasa Iyengar 1904–1941*, ed. Narendar Pani, typescript report, Project No. 098 (Bangalore: Indian Institute of Management Bangalore, 1982).

24. Ramnath, *The Birth of an Indian Profession,* chap. 3.

25. Visvesvaraya, *Memoirs*, 9–11. For an example of his reports, see M. Visvesvaraya, "Memorandum on Irrigation Works in the Bombay Presidency, Excluding Sind," November 1901, in PWD Irrigation No. 101 of 1903, Part I, Maharashtra State Archives, Mumbai.

26. On the block system, see Ramnath, *Engineering a Nation*, chap. 5. As Raina notes, Visvesvaraya's larger approach to development also took the form of "social engineering." Raina, *Visvesvaraya as Engineer-Sociologist*, 27.

27. See, for instance, M. Visvesvaraya, "Memorandum on Irrigation Works in the Bombay Presidency, Excluding Sind," November 1901, para 54, in PWD Irrigation No. 101 of 1903, Part I, Maharashtra State Archives, Mumbai. On "productive" versus "protective" irrigation, see, for instance, Aditya Ramesh, "The Value of Tanks: Maintenance, Ecology and the Colonial Economy in Nineteenth-Century South India," *Water History* 10, no. 4 (December 2018): 267–289. See also Bret Wallach, "British Irrigation Works in India's Krishna Basin," *Journal of Historical Geography* 11, no. 2 (April 1, 1985): 155–173, on irrigation canals that were built with the aim of preventing famine.

28. See Raj Sekhar Basu, "The World of an Engineer: M. Visvesvaraya and Irrigation Engineering in Twentieth-Century India," in *Science and Society in India 1750–2000*, ed. Arun Bandopadhyay (Delhi: Manohar, 2010), 249–298; see, for instance, Jean Drèze, "Famine Prevention in India," WIDER Working Papers, World Institute for Development Economics Research (WP 45, May 1988), 21–27. For a detailed discussion, see Ramnath, *Engineering a Nation*, chap. 5.

29. He expressed these beliefs in an increasingly direct manner in later years. See, for instance, M. Visvesvaraya, *Reconstructing India* (London: P. S. King & Son, Ltd., 1920), 187; M. Visvesvaraya, *Planned Economy for India* (Bangalore: The Bangalore Press, 1934), 338–389.

30. Visvesvaraya, *Memoirs*, chap. IV; G. A. Mankar, *A Sketch of the Life and Works of the Late Mr. Justice M. G. Ranade*, vol. I (Bombay: Caxton Printing Works, 1902), especially chap. V.

31. Visvesvaraya, *Memoirs*, chap. IV. Ranade was born in 1842. Mankar, *A Sketch*, vol. I, 15.

32. G. A. Mankar, *A Sketch of the Life and Works of the Late Mr. Justice M. G. Ranade*, vol. II (Bombay: Caxton Printing Works, 1902), 16–22; Bipan Chandra, "Colonial India: British versus Indian Views of Development," *Review (Fernand Braudel Center)* 14, no. 1 (1991): 81–167; Barbara D. Metcalf and Thomas R. Metcalf, *A Concise History of Modern India*, 3rd ed. (Cambridge, UK: Cambridge University Press, 2012), chap. 5.

33. Manu Goswami, "From Swadeshi to Swaraj: Nation, Economy, Territory in Colonial South Asia, 1870 to 1907," *Comparative Studies in Society and History* 40, no. 4 (October 1998): 609–636, quoted text from p. 617. I thank Karthik Rao-Cavale for prompting me to look up the influence of List.

34. Goswami, "From Swadeshi to Swaraj," 616 (quoted text from here); Mankar, *A Sketch*, vol. II, chap. 1.

35. For instance, see Visvesvaraya, *Planned Economy for India*, 337–338.

36. Quoted (including the ellipsis) in Manorama Barnabas, "Study in the Philosophy of Social Change: The Ideas of Some Liberal Thinkers of Nineteenth Century Maharashtra and Their Relevance to Modernization," unpublished PhD thesis, University of Poona, 1974, 319. Barnabas writes that the article from which this quote comes "could well number among the last of Ranade's anonymous contributions to the Independent Section of the *Journal*." Barnabas, "Study in the Philosophy of Social Change," 316.

37. M. Visvesvaraya, *Planned Economy for India*, 53. Raghabendra Chattopadhyay also places Visvesvaraya in a line of thinkers including M. G. Ranade and Dadabhai Naoroji. Raghabendra Chattopadhyay, "The Idea of Planning in India, 1930–1951," PhD diss., Canberra, Australian National University, 1985, 16.

38. Visvesvaraya, *Memoirs*, chap. IV. On Tilak, see Metcalf and Metcalf, *A Concise History of Modern India*, chap. 5.

39. See Ross Bassett, *The Technological Indian* (Cambridge, MA: Harvard University Press, 2016), chap. 1.

40. Visvesvaraya, *Memoirs*, 31 and 121. On Indians drawing inspiration from Japan, see, for instance, Arpita Mathur, *India-Japan Relations: Drivers, Trends and Prospects* (Singapore: S. Rajaratnam School of International Studies, 2012), 4–6.

41. See, for instance, M. Visvesvaraya, *Reconstructing India*; M. Visvesvaraya, *Planned Economy for India*. On his second visit to Japan, see Visvesvaraya, *Memoirs*, 123.

42. Sydney Crawcour, "*Kogyo iken:* Maeda Masana and His View of Meiji Economic Development," *The Journal of Japanese Studies* 23, no. 1 (1997): 69–104. Quoted text from p. 70.

43. Visvesvaraya, *Memoirs*, 31–33, 42–43, and 121–122.

44. Aparajith Ramnath, "Reservoir Gods," *FiftyTwo.in*, September 17, 2021, https://fiftytwo.in/story/reservoir-gods. Vinod Vyasulu also mentions that this "insistence on investment by the state has a clear Keynesian flavor for a writer in the early 20th century." Vinod Vyasulu, "Nehru and the Visvesvaraya Legacy," *Economic and Political Weekly* 24, no. 30 (29 July 1989): 1700–1704, here 1701.

45. M. Shama Rao, *Modern Mysore: From the Coronation of Chamaraja Wodeyar X in 1868 to the present time* (Bangalore: Higginbothams, 1936), 247–248; Visvesvaraya, *Memoirs*, 56.

46. Bjorn Hettne, *Political Economy of Indirect Rule: Mysore 1881–1947* (New Delhi: Ambika Publications, 1978), chap. 8 (the quoted phrase is from p. 277).

47. "Speech at the Inauguration of the Mysore Economic Conference on 10th June 1911," in Krishnaraja Wadiyar, *Speeches by His Highness Maharaja Sri Krishnaraja Wadiyar Bahadur, G.C.S.I., G.B.E., Maharaja of Mysore: 1902–1933* (Mysore: Government Branch Press, 1934), 87–93; "Address to the Mysore Economic Conference," June 3, 1914, in *Speeches by Sir M. Visvesvaraya*, 142–153; Visvesvaraya, *Memoirs*, 44–45; "Mysore Economic Conference," *The Times of India*, October 2, 1911, 6; "Progressive Mysore. Economic Conference Opens," *The Times of India*, June 10, 1916, 10. See also Hettne, *Political Economy of Indirect Rule*, 259–262. Hettne (p. 259) suggests that the Mysore Economic Conference may have been modeled on Japan's Economic Investigation Commission (1910).

48. Ramnath, *Engineering a Nation*, chap. 8; "Progressive Mysore. Economic Conference Opens," *The Times of India*, June 10, 1916, 10; untitled report beginning "The Committee in Charge of the Industrial Section," *The Times of India*, July 31, 1911, 8; "Economic Conference. Mysore Dewan's Closing Speech," *The Times of India*, June 17, 1916, 9.

49. Vinod Vyasulu argues that "the Mysore Economic Conference . . . performed a function similar to the Planning Commission in the 1950s." Vyasulu, "Nehru and the Visvesvaraya Legacy," 1702.

50. *Report of the Indian Economic Enquiry Committee 1925*, vol. I (Calcutta: Government of India Central Publication Branch, 1925); *Report of the Committee Appointed by the Government of India to Enquire into the Bombay Back Bay Reclamation Scheme 1926* (Bombay: Government Central Press, 1927); "Visvesvaraya Inquiry. Preparation of Report within a Month," *The Times of India*, November 12, 1928, 12; *Technical and Industrial Education in the Bombay Presidency: Final Report of the Committee Appointed by*

Government, 1921–1922 (Bombay: Government Central Press, 1923); *City of Bombay: Municipal Retrenchment and Reform: Final Report by the Retrenchment Advisor* (Bombay: Indian Daily Mail Press, 1925); "Karachi's Civic Needs. Sir M. Visvesvaraya's Recommendations," *Bombay Chronicle*, April 20, 1926, 11.

51. Ramnath, *Engineering a Nation*, esp. chaps. 6 and 15.

52. M. Visvesvaraya, *Reconstructing India* (London: P. S. King & Son, Ltd., 1920); M. Visvesvaraya, *Planned Economy for India* (Bangalore: The Bangalore Press, 1934); M. Visvesvaraya, *Nation Building* (Bangalore: The Bangalore Press, 1937).

53. Visvesvaraya, *Reconstructing India*, vi. He also mentioned his travels abroad in the Preface to *Planned Economy for India*.

54. Visvesvaraya, *Memoirs*, 126; Visvesvaraya, *Reconstructing India*, vi.

55. Visvesvaraya, *Reconstructing India*, 2.

56. Visvesvaraya, *Reconstructing India*, 280.

57. Visvesvaraya, *Reconstructing India*, 133.

58. Visvesvaraya, *Reconstructing India*, "Contents," vii–viii.

59. Visvesvaraya, *Reconstructing India*, chap. XVII (quoted text from p. 321).

60. Or, as Raghabendra Chattopadhyay argues (writing from an implicitly Marxist perspective), Visvesvaraya's *Reconstructing India* "specified and rationalized all the demands of the Indian bourgeoisie." Chattopadhyay, "The Idea of Planning in India," 18.

61. See, for example, his article in *Capital*, dated April 25, 1929 (clipping in Visvesvaraya Papers, Institution of Engineers (India), Mysore Local Centre, Mysuru, vol. 13).

62. "Mysore University: A Suggestion," *Times of India*, March 20, 1912, 7.

63. Visvesvaraya, *Planned Economy*, 16.

64. *Report of the Indian Economic Enquiry Committee 1925*, vol. I (Calcutta: Government of India Central Publication Branch, 1925), i. On Kaul, see "Raja Hari Kishan Kaul Dead," *Times of India*, January 27, 1942, 5. Burnett-Hurst is described as "Professor of Commerce in the University of Allahabad" on the title page of A. R. Burnett Hurst, *Labour and Housing in Bombay: A Study in the Economic Conditions of the Wage-Earning Classes in Bombay* (London: P.S. King & Son, Ltd., 1925).

65. *Report of the Indian Economic Enquiry Committee 1925*, vol. I, chap. I; Chattopadhyay, "The Idea of Planning in India," 76 (n. 80). For a list of the members of the central legislature around this time, see *India Office List* 1924, 5–8. The *India Office List* was an annual compendium listing important officials in colonial India. While I have not been able to pinpoint when Visvesvaraya first got to know Purshotamdas Thakurdas, he knew Lallubhai Samaldas from at least 1916: see Ramnath, *Engineering a Nation*, chap. 12 and endnotes to chap. 19.

66. *Report of the Indian Economic Enquiry Committee 1925*, vol. II, part I (printed along with vol. I, cited above), 1–7.

67. *Report of the Indian Economic Enquiry Committee 1925*, vol. II.

68. *Report of the Indian Economic Enquiry Committee 1925*, vol. I, 64–68.

69. "Note of Dissent by Professor A. R. Burnett-Hurst," *Report of the Indian Economic Enquiry Committee*, vol. I, 91–111.

70. *Report of the Indian Economic Enquiry Committee*, vol. I, 64.

71. Chattopadhyay mentions that the Enquiry Committee was a kind of "precursor of the later efforts towards planning by the [colonial] Government." Chattopadhyay, "The Idea of Planning in India," 59.

72. On Saha, see Pratik Chakrabarti, *Western Science in Modern India: Metropolitan Methods, Colonial Practices* (Delhi: Permanent Black, 2004), 286; Deepak Kumar, "Reconstructing India: Disunity in the Science and Technology for Development Discourse, 1900–1947," *Osiris*, 2nd Series, 15 (2000): 241–257, here 248–250.

73. Lionel Robbins, *Economic Planning and International Order* (London: Macmillan and Co., Ltd., 1937), 3.

74. Peter Temin, "Soviet and Nazi Economic Planning in the 1930s," *The Economic History Review* 44, no. 4 (1991): 573–593.

75. Daniel Ritschel, *The Politics of Planning: The Debate on Economic Planning in Britain in the 1930s* (Oxford: Clarendon Press, 1997).

76. Marcia L. Balisciano, "Hope for America: American Notions of Economic Planning between Pluralism and Neoclassicism, 1930–1950," *History of Political Economy* 30, supplement (1998): 153–178.

77. Medha M. Kudaisya, *Tryst with Prosperity: Indian Business and the Bombay Plan of 1944* (Gurgaon: Penguin Random House India, 2018), chap. 4.

78. Zachariah, *Developing India*, 255.

79. Edwin Layton, "Veblen and the Engineers," *American Quarterly* 14, no. 1 (1962): 64–72. Layton argues that although such engineers were seen by the sociologist Thorstein Veblen as potential social revolutionaries, they were no social radicals. They were, however, interested in applying technology to social ends.

80. Balisciano, "Hope for America," 162.

81. Visvesvaraya, *Memoirs*, 126.

82. See William E. Leuchtenberg, *Herbert Hoover* (New York: Times Books, 2011 [2009]).

83. For instance, in a speech in 1912, Visvesvaraya declared, "As compared with Europe, our climate and traditions all predispose us to a life of inaction and ease." "Speech at Bangalore Central College Day Meeting," March 16, 1912, in *Speeches by Sir M. Visvesvaraya, K.C.I.E.: Dewan of Mysore: 1910–11 to 1916–17* (Bangalore: Government Press, 1917), 17–31, here 23.

84. Visvesvaraya, *Planned Economy for India*, v.

85. Chattopadhyay, "The Idea of Planning in India," 29–34 and 62–63.

86. Visvesvaraya, *Planned Economy for India*, 266–276 and 300. Quoted text from pp. 266, 268, 269, and 270.

87. Visvesvaraya, *Planned Economy for India*, 268–269.

88. Visvesvaraya, *Planned Economy for India*, 279–280.

89. Visvesvaraya, *Planned Economy for India*, 281–286.

90. Visvesvaraya, *Planned Economy for India*, 293–295.

91. Visvesvaraya, *Planned Economy for India*, 297–298.

92. Visvesvaraya, *Planned Economy for India*, 339–342.

93. Metcalf and Metcalf, *A Concise History of Modern India*, chap. 6.

94. Visvesvaraya, *Nation Building*, esp. chap. 1 and chap. 5.

95. Visvesvaraya, *Nation Building*, 65–66.

96. Visvesvaraya, *Nation Building*, 3–4.

97. Visvesvaraya, *Nation Building*, 40–42.

98. Chattopadhyay, "The Idea of Planning in India," chap. III. On the provincial Congress governments' resignation, see Metcalf and Metcalf, *A Concise History of Modern India*, chap. 7.

99. For example, M. Visvesvaraya, *Prosperity through Industry: Move towards Rapid Industrialization*, 3rd ed. (Bombay: The All-India Manufacturers' Organization, 1943); M. Visvesvaraya, *Reconstruction in Post-War India: A Plan of Development All Round* (Bombay: All-India Manufacturers' Organization, 1944). For Visvesvaraya's tenure as president of the AIMO, see "Past Presidents," website of the All India Manufacturers' Organisation, https://www.aimoindia.com/about/past-presidents/, accessed June 23, 2024.

100. Visvesvaraya, *Memoirs*, 130.

101. Kudaisya, *Tryst with Prosperity*, esp. chap. 5.

102. Visvesvaraya, *Planned Economy*, 8. Italics in original. Zachariah writes that Visvesvaraya "was convinced of the necessity to industrialize and less committed to questions of social organization in terms of capitalism or socialism." Zachariah, *Developing India*, 255.

103. See J. R. D. Tata's Foreword to Shakuntala Krishnamurthy, *Sir Mokshagundam Visvesvaraya* (Bangalore: n.p., 1992 [1980]), iii.

104. M. Visvesvaraya, *Reconstruction in Post-War India: A Plan of Development All Round* (Bombay: All-India Manufacturers' Organization, 1944).

105. Visvesvaraya, *Reconstruction in Post-War India*, 5 and 12–13.

106. Visvesvaraya, *Reconstruction in Post-War India*, i.

107. Visvesvaraya, *Reconstruction in Post-War India,* ii and 5.

108. Visvesvaraya, *Reconstruction in Post-War India,* 49.

109. Kudaisya, *Tryst with Prosperity,* 176–179.

110. Nehru's government awarded Visvesvaraya India's highest civilian honor, the Bharat Ratna, in 1955; Nehru attended Visvesvaraya's birth centenary celebrations in Bangalore in 1960. "Mr. Nehru Receives Nation's Highest Award: Five-Word Citation at Investiture," *The Times of India,* September 8, 1955, 1; "Glowing Tributes by Prime Minister: 'Life of Integrity in Cause of Nation,'" *The Times of India,* September 16, 1960, 1 and 7.

111. See Gyanesh Kudaisya, *A Republic in the Making: India in the 1950s* (New Delhi: Oxford University Press, 2017), 162–168; Vyasulu, "Nehru and the Visvesvaraya Legacy," 1701.

112. Vyasulu, "Nehru and the Visvesvaraya Legacy." See also Raina's discussion of Vyasulu's article: Raina, *Visvesvaraya as Engineer-Sociologist,* Table 2 (28).

CHAPTER 9

1. David R. Headrick, *The Tools of Empire* (New York: Oxford University Press, 1981).

2. David Pretel and Lino Camprubí, "Technological Encounters: Locating Experts in the History of Globalisation," in *Technology and Globalisation: Networks of Experts in World History,* ed. David Pretel and Lino Camprubí (London: Springer, 2018), 7.

3. Ben Marsden and Crosbie Smith, *Engineering Empires* (London: Palgrave Macmillan UK, 2005), 1.

4. Pretel and Camprubí, "Technological Encounters," 11.

5. Leida Fernández-Pieto, "Networks of American Experts in the Caribbean: The Harvard Botanic Station in Cuba (1898–1930)," in *Technology and Globalisation,* ed. David Pretel and Lino Camprubí, (London: Springer, 2018), 159–187.

6. Stephen Macekura, "The Point Four Program and US International Development Policy," *Political Science Quarterly* 128, no. 1 (2013): 132.

7. Macekura, "The Point Four," 131–133.

8. Philip M. Glick, *The Administration of Technical Assistance: Growth in the Americas* (Chicago: University of Chicago Press, 1957), 16–18.

9. Lloyd C. Gardner, *Economic Aspects of New Deal Diplomacy* (Madison: University of Wisconsin Press, 1964), 129–131.

10. Macekura, "Point Four," 127.

11. Macekura, "Point Four," 127.

12. Elizabeth A. Cobbs, *The Rich Neighbor Policy: Rockefeller and Kaiser in Brazil* (New Haven, CT: Yale University Press, 1992), 102.

13. Darlene Rivas, *Missionary Capitalist: Nelson Rockefeller in Venezuela* (Chapel Hill: University of North Carolina Press, 2002), chap. 3.

14. Macekura, "Point Four," 148.

15. Macekura, "Point Four," 143.

16. Illinois Institute of Technology (IIT), News Releases, October 13, 1942.

17. Leonard Sidney Reich, *The Making of American Industrial Research: Science and Business at GE and Bell, 1876–1926* (Cambridge, UK: Cambridge University Press, 1985), 3.

18. Armour Research Foundation/IITRI, https://www.informs.org/Explore/History-of-O.R.-Excellence/Non-Academic-Institutions/Armour-Research-Foundation-IITRI (viewed 11/11/2020); IIT News Releases, November 15, 1953.

19. IIT, News Releases, October 13, 1942.

20. IIT, News Releases, October 16, 1953.

21. Myron J, Rand, *Industrial Research Laboratories of the United States* (Washington, DC: National Research Council and National Academy of Sciences, 1950), 28; IIT, News Releases, January 13, 1950.

22. IIT, News Releases, February 22, 1943; August 28, 1943.

23. IIT, News Releases, October 13, 1942.

24. IIT, News Releases, November 13, 1944.

25. IIT, News Releases, August 17, 1945.

26. IIT, News Releases, October 13, 1942.

27. IIT, News Releases, September–December 1954, 165.

28. IIT, News Releases, November 6, 1953; December 11, 1953.

29. IIT, News Releases, November 17, 1958.

30. Vagtborg was born in Copenhagen, Denmark, in 1906. He obtained his Bachelor of Science degree in 1926 at the University of Illinois and his master's degree from Armour Institute of Technology. He worked in several companies related to the design and construction of sanitary engineering and municipal and industrial plants. He was professor of municipal and sanitary engineering at Armour Institute of Technology from 1931 to 1937 and director of the ARF from 1937 to 1944, when he became president of the Midwest Research Institute. IIT News Releases, March 2, 1941; November 13, 1944.

31. IIT, News Releases, March 2, 1941.

32. IIT, News Releases, March 2, 1941.

33. IIT, Press Releases, May 25, 1941.

34. IIT, Press Releases, May 25, 1941.

35. IIT, Press Release, November 19, 1941.

36. IIT, Press Release, May 25, 1941; November 19, 1941.

37. IIT, Press Release, May 25, 1941.

38. IIT, Press Release, November 19, 1941.

39. Prebisch was an influential economist known for his contributions to structuralist economics, which formed the basis of the economic dependency theory. He became the executive director of the Economic Commission for Latin America (ECLA) in 1950.

40. The CPI was created when the conservative government of Ramón Castillo attempted to introduce proposals that would favor industry. In 1941, the government funded the CPI to foster technological activities and carry out policies that improved trade.

41. Claudio Bellini, "Industrial Exports and Peronist Economic Policies in Post-War Argentina," *Journal of Latin American Studies* 44, no. 2 (2012): 294.

42. Armour Research Foundation, *Technological and Economic Survey of Argentine Industries with Industrial Research Recommendations* (Ann Arbor, MI: Edward Brothers Inc, 1947), xi.

43. Leslie Bethel, ed., *Historia de América Latina. El Cono Sur desde 1930* (Barcelona: Cambridge University Press and Editorial Crítica, 2002), 4.

44. Bethel, *Historia de América Latina*, 41–42.

45. Godwin was born in Washington, DC. He received a bachelor's degree from San Diego State College and master's and doctoral degrees from the State University of Iowa. Fluent in Spanish, he joined the ARF in 1938 as director of coal research, then head of the chemical engineering department, and in September 1941, was appointed assistant director before serving as director of Inter-American Research Service in New York for some years. From May 1949 to July 1950, he served as director of the ARF International Division and later remained related to the ARF as a consultant. IIT News Releases, September 8, 1941; August 17, 1942; July 31, 1950.

46. IIT News Releases, October 24, 1943.

47. IIT News Releases, August 17, 1942.

48. IIT News Releases, September 24, 1942.

49. Armour Research Foundation, *Technological and Economic Survey of Argentine Industries*. The report was published in Spanish by the CPI as separate reports authored by John Shellenberg, John Hopkins, and Francis Godwin.

50. Bellini, "Industrial Exports and Peronist Economic Policies in Post-War Argentina," 297–298.

51. IIT News Releases, February 13, 1943.

52. IIT News Releases, May 8, 1943.

53. Bethel, *Historia de América Latina*, 46–55.

54. "ARF Research Extends Beyond National Borders," *Technology News*, March 10, 1950, 4.

55. "ARF Research Extends Beyond National Borders," 4.

56. Susan M. Gauss, *Made in Mexico. Regions, Nation, and the State in the Rise of Mexican Industrialism 1920s–1940s* (University Park: The Pennsylvania State University Press, 2010), 115.

57. IIT News Releases, May 20, 1944. Holland was an industrial research adviser and adviser to the ARF who had formerly worked with the NRC and formed part of the group who traveled through South America in 1941. IIT News Releases, October 6, 1944.

58. All quotations from IIT News Releases, October 6, 1944.

59. IIT News Releases, October 6, 1944.

60. "ARF Research Extends Beyond National Borders," 4.

61. IIT News Releases, December 10, 1944; December 17, 1944.

62. IIT News Releases, August 3, 1945; August 17, 1945.

63. IIT News Releases, September 27, 1945; October 11, 1945.

64. Aurora Gómez-Galvarriato, "La Construcción Del Milagro Mexicano: El Instituto Mexicano de Investigaciones Tecnológicas, El Banco de México y La Armour Research Foundation," *Historia Mexicana* 69, no. 3 (2020): 1260–1263.

65. IIT News Releases, August 3, 1945.

66. Gómez-Galvarriato, "La Construcción del Milagro Mexicano," 1264–1267.

67. Armour Research Foundation, *Technological Audit of Selected Mexican Industries with Industrial Research Recommendations* (Ann Arbor, MI: Edward Brothers Inc., 1946), 459.

68. "ARF Research Extends Beyond National Borders," 4.

69. IIT News Releases, May 13, 1945.

70. IIT News Releases, June 27, 1949.

71. "ARF Research Extends beyond National Borders," 4.

72. IIT News Releases, March 16, 1948; August 5, 1949.

73. "Scientists Develop Dehydrated Tortilla Flour to Eliminate Future Mexican Shortages," *ARF NEWS*, February 1951, 3 and 5.

74. Guillermo Salorio, Jorge García, and Mario Valverde, "Substitución de los procedimientos primitivos en la elaboración de maíz nixtamalizado, artículo de consumo básico en la población mexicana," en Banco de México, Departamento de Investigaciones Industriales, *La Tecnología en el Desarrollo de la Industria Mexicana* (Mexico City: Banco de México, 1963), 89. Nixtamal is a key tortilla ingredient, a preparation of corn

and lime cooked in hot water. For the importance of the development of nixtamalized corn flour in Mexico, see Aurora Gómez-Galvarriato, *El Pan Nuestro: Una Historia de la Tortilla de Maíz* (Mexico City: El Colegio de México, 2024).

75. "Contributions to Mexican Economy Listed by International Division," *ARF News*, September 1949, 2.

76. Leedy was born in 1910 in Fremont, Ohio. He received his AB degree in physics at North Central College, Naperville, Illinois, in 1933 and his master's and doctorate degrees in physics at the University of Illinois in 1935 and 1938, respectively. In 1944, he was chairman of physics research at ARF, where he was engaged in various fields of acoustic research, and appointed acting director of the ARF in March 1948, director in April 1948, and vice president in 1950. IIT News Releases, March 12, 1948.

77. IIT News Releases, October 16, 1953.

78. IIT News Releases, July 6, 1950.

79. "A native Chicagoan, Collier earned his degree in civil engineering from Illinois Tech in 1937. He interrupted his college studies in 1934 for two years in the Army on a special construction and surveying assignment in the Panama Canal Zone," *ARF News,* April 1951, 7; IIT News Releases, June 14, 1951.

80. IIT News Releases, October 17, 1949.

81. IIT News Releases, May 13, 1945.

82. IIT News Releases, May 13, 1945.

83. IIT News Releases, September 25, 1957.

84. "ARF Research Extends Beyond National Borders," 4.

85. IIT News Releases, August 8, 1949.

86. IIT News Releases, June 3, 1949; IIT, University Archives and Special Collections, Heald Papers, President 1920–1953, Collection No. 024.04.01, Box, No. 54, "Annual Report of The Director to The President and Board of Trustees of Armour Research Foundation of Illinois Institute of Technology for the Year Ending August 31, 1949," October 10, 1949, 14.

87. Benjamin Hardy, a former employee of Nelson Rockefeller in the IAA who sent a proposal to Truman aides Clark Clifford and George Elsey on US development foreign policy to help them draft Truman's address, must have been acquainted with ARF projects. See Macekura, "The Point Four Program," 138–139.

88. Macekura, "The Point Four Program," 143–144, 149, 153–158.

89. IIT News Releases, October 17, 1949.

90. IIT News, March 15, 1949. Armstrong obtained his bachelor's degree at Michigan College of Mining Technology. He received his master's degree from the State University of Iowa and had been a member of ARF staff since August 1941 in the chemistry

department. He was associated with ARF's program in Mexico in 1945. In 1950, he returned to Chicago and became assistant manager of the International Division. In June 1951, he joined the International Bank for Reconstruction and Development Engineering Department in Washington, DC. IIT News Releases, March 15, 1949; July 31, 1950; "Walter Armstrong Manages Mexican Operations," *ARF News*, April 1949, 2.

91. IIT News Releases, January 13, 1950.

92. IIT News Releases, April 18, 1949.

93. "Latin-American Consuls Club Tours ARF," *ARF News*, July 1949, 7.

94. IIT News Releases, June 3, 1949.

95. IIT News Releases, October 17, 1949.

96. IIT News Releases, July 31, 1950.

97. IIT News Releases, June 29, 1951.

98. IIT News, June 14, 1951. Gollong was born in Berlin, Germany, and went to the United States in 1925. He graduated from the University of Cincinnati in 1941 and majored in mechanical engineering and business administration. He worked as a principal associate in the management consulting firm Griffernhagen and Associates; as a research engineer for the Celotex corporation, Chicago; and as a project engineer for Armstrong Coal Company, Lancaster, Pennsylvania. IIT News Releases, June 11, 1954; IIT News Releases, June 11, 1954.

99. IIT News Releases, January 11, 1952.

100. Schnur installed and operated a foundry in the Netherlands West Indies from 1931 to 1942 for Standard Oil Company of New Jersey and a foundry in Nicaragua from 1946–1946, training native crews in plant operation and molding, core making, and pattern making. He served in China as a gray iron consultant for the US Donald M. Nelson Mission in 1944–1945, responsible for training unskilled Chinese in improved foundry practices, then worked for the McLean Iron Works in Palatka, Florida. In February 1952, he was appointed foundry technologist for the ARF. IIT News Releases, February 27, 1952.

101. IIT News Releases, August 25, 1952.

102. IIT News Releases, September–December, 1954.

103. IIT News Releases, October 23, 1952.

104. IIT News Releases, September–December, 1954.

105. IIT News Releases, September–December, 1954.

106. IIT News Releases, April 25, 1960.

107. IIT News Releases, August 4, 1952.

108. Magis graduated from the University of Pittsburg in 1926 with a BS in metallurgical engineering and was an assistant open-hearth superintendent for Jones and

Laughlin Steel Co., Pittsburgh, Pennsylvania. During World War II, he was a special consultant to Henry Disston and Sons in Philadelphia and the American Rolling Mills Co., in Middleton, Ohio, working with armor plate, gun, and shell metals. He set up a plant for the Boiardi Steel Corp. in Milton, Pennsylvania, and put a steel plant into operation in Caracas, Venezuela, supervising its construction, designing equipment, and training its staff, and then became a consulting engineer. In September 1952, he was named a steel technologist at ARF. IIT News Releases, September 23, 1952.

109. IIT News Releases, September 30, 1952.

110. IIT News Releases, October 14, 1953.

111. IIT News Releases, October 5, 1953.

112. IIT News Releases, December 1, 1949.

113. "Foundation Agrees to Study Technical Needs of El Salvador," *ARF News*, January 1950, 3.

114. Armour Research Foundation, "Preliminary Technological Audit for the Industrial Development of El Salvador," Project No. 1–804.1, Chicago, May 16, 1950, unpublished document.

115. IIT News Releases, April 6, 1950.

116. IIT News Releases, April 6, 1950; June 19, 1951; "El Salvador Project Pays Dividends as Country Acts on 7 Proposals," *ARF News*, August 1951, 3.

117. IIT News Releases, October 14, 1953.

118. IIT News Releases, October 5, 1953.

119. IIT News Releases, December 17, 1953.

120. Gómez-Galvarriato, "La Construcción Del Milagro Mexicano: El Instituto Mexicano de Investigaciones Tecnológicas, El Banco de México y La Armour Research Foundation," 1283. It is unclear what specific date the office of the International Division of the ARF in Mexico City stopped operations.

121. IIT News Releases, March 11, 1954.

122. IIT News Releases, May 14, 1956.

123. IIT News Releases, June 6, 1956.

124. IIT News Releases, May 7, 1957.

125. IIT News Releases, July 15, 1959.

126. IIT News Releases, January 4, 1954.

127. IIT News Releases, September 10, 1958.

128. IIT News Releases, July 15, 1959; February 13, 1959.

129. Jaime Ayala Ramírez, "The Colombian Institute for Technological Research at Bogotá." UNIDO, DO3946, 1970, 12–16.

130. IIT News Releases, August 15, 1957.

131. IIT News Releases, June 26, 1959.

132. IIT News Releases, March 3, 1954.

133. IIT News Releases, September 23, 1953; December 2, 1953.

134. IIT News Releases, October 14, 1953.

135. IIT News Releases, December 2, 1953.

136. IIT News Releases, January 14, 1954, 63.

137. IIT News Releases, April 27, 1960.

138. IIT News Releases, October 14, 1953.

139. IIT News Releases, March 14, 1952.

140. IIT News Releases, November 14, 1957.

141. IIT News Releases, June 26, 1959.

142. IIT News Releases, November 14, 1957.

143. IIT News Releases, February 13, 1959.

144. Gabriel Kolko and Joyce Kolko, *The Limits of Power: The World and the United States Foreign Policy, 1945–1954* (New York: Harper & Row, 1972), 623.

CHAPTER 10

1. "FirstLook: Aswan, Egypt, May 14, 1964—AramcoWorld," accessed March 20, 2023, https://www.aramcoworld.com/Articles/May-2016/FirstLook-Aswan-Egypt; "(Khrushchev and Nasser at Aswan Dam Ceremony), *British Pathé*," accessed March 20, 2023, https://www.britishpathe.com/asset/193045/; "Khrushchev and Nasser Join In Diverting the Nile at Aswan; Leaders Set Off Explosion to Complete First Stage of Large Dam Project," *The New York Times*, May 15, 1964, see Archives, https://www.nytimes.com/1964/05/15/archives/khrushchev-and-nasser-join-in-diverting-the-nile-at-a-swan-leaders.html; also Alia Mossallam, "'We Are the Ones Who Made This Dam "High"!' A Builders' History of the Aswan High Dam," *Water History* 6, no. 4 (December 1, 2014): 297–314.

2. "Khrushchev and Nasser Join in Diverting the Nile at Aswan."

3. Quoted in Johnson, "Technology's Cutting Edge," 691.

4. Loren R. Graham, "The Formation of Soviet Research Institutes: A Combination of Revolutionary Innovation and International Borrowing," *Social Studies of Science* 5, no. 3 (1975): 303–329.

5. Kudayarova, *Unintended Community*; Dewitt, "Main Issues Concerning the Soviet Scientific, Engineering, and Educational Challenge," 26.

6. Abdel-Aziz I. Kashef, "Technical and Ecological Impacts of the High Aswan Dam," *Journal of Hydrology* 53, no. 1–2 (1981): 73–84. For a broader discussion of engineering projects in Egypt at the time, see Timothy Mitchell, *Rule of Experts: Egypt, Techno-Politics, Modernity* (Berkeley: University of California Press, 2002).

7. Silvia Borzutzky and David Berger, "Dammed If You Do, Dammed If You Don't: The Eisenhower Administration and the Aswan Dam," *The Middle East Journal* 64, no. 1 (2010): 84–102. The World Bank would play a significant role in forcing the Sudanese government in Khartoum to sign the technically focused Nile River Agreement with Egypt in 1959.

8. Seymour Rosen, "The USSR and International Education: A Brief Overview," *The Phi Delta Kappan* 51, no. 5 (1970): 247–250; Constantin Katsakioris, "Soviet Lessons for Arab Modernization: Soviet Educational Aid towards Arab Countries after 1956," *Journal of Modern European History* 8, no. 1 (2010): 95.

9. Katsakioris, "Soviet Lessons for Arab Modernization." Today, the US Bureau of Reclamation and American engineers are leading studies and early construction of the "Grand Ethiopian Renaissance dam" in Sudan and considering its impact on Egypt's hydrological needs and infrastructure; see Hisham Eldardiry and Faisal Hossain, "A Blueprint for Adapting High Aswan Dam Operation in Egypt to Challenges of Filling and Operation of the Grand Ethiopian Renaissance Dam," *Journal of Hydrology (Amsterdam)* 598 (2021): 125708.

10. Mossallam, "'We Are the Ones Who Made This Dam "High"!' A Builders' History of the Aswan High Dam."

11. Mimi Sheller and John Urry, "The New Mobilities Paradigm," *Environment and Planning. A* 38, no. 2 (2006): 207–226, https://doi.org/10.1068/a37268.

12. The critical nature of interpersonal interaction has been extensively highlighted in the historical literature on technology transfer. See, for example, Edward Beatty, "Approaches to Technology Transfer in History and the Case of Nineteenth-Century Mexico," *Comparative Technology Transfer and Society* 1, no. 2 (2003): 167–197; Bruce Edsall Seely, "Historical Patterns in the Scholarship of Technology Transfer," *Comparative Technology Transfer and Society* 1, no. 1 (2003): 7–48; Kristine Bruland, *Technology Transfer and Scandinavian Industrialisation* (New York: Berg, 1991).

13. Köll, "The Making of the Civil Engineer in China," 63–66; see also Kirby, "Engineering China," 146–148.

14. See also Aparajith Ramnath, *The Birth of an Indian Profession: Engineers, Industry, and the State, 1900–47* (London: Oxford University Press, 2017).

15. On congresses as sites of international interaction and exchange, see Bruce Sinclair, "The Power of Ceremony: Creating an International Engineering Community," History *of Technology* 21 (2000): 203–212.

16. Tuffnell, this volume.

17. See citations in the Introductory chapter, especially note 4.

18. Ramnath, *Birth*, 63, 101–102.

19. Meiksins, "Engineers in the United States," 64–65; Per-Olof Gronberg, *The Peregrine Profession: Transnational Mobility of Nordic Engineers and Architects, 1880–1930* (Leiden: Brill, 2019).

20. Robert O. Collins, "In Search of the Nile Waters, 1900–2000," *The Nile: Histories, Cultures, Myths*, ed. by Ḥagai Erlikh and I. Gershoni (London: Lynne Rienner, 2000), 259.

21. See the chapters by Ramnath and Köll, this volume, as well as Ramnath, *Birth*, and Köll, "Making."

22. Swedish Engineers' Society of Chicago, *A History of the Swedish Engineers' Society of Chicago, 1908–1948*. (Chicago[?]: s. n., 1948).

23. Ramnath, *Birth*, chap. 2.

24. See also Kirby, "Engineering China," 148–150.

25. Aurora Gómez-Galvarriato, "La Construcción Del Milagro Mexicano: El Instituto Mexicano de Investigaciones Tecnológicas, El Banco de México y La Armour Research Foundation," *Historia Mexicana* 69, no. 3 (2020): 1247–1309.

26. See, for example, Terri Kim, "Shifting Patterns of Transnational Academic Mobility: A Comparative and Historical Approach," *Comparative Education* 45, no. 3 (2009): 387–403.

27. Wright, "Can a Nation Learn?" 300, 303, noting that one study of the careers of ten leading machinists found that they worked for an average of 5.5 employers in 4.2 industrial centers.

28. As had German research and technical training centers generally; see Kim, "Shifting Patterns of Transnational Academic Mobility."

29. Solares and Beatty, "Engineering Database." See the beta version at https://github.com/solaresig/Blueprint-for-Modernity.

30. For example, see Bassett, *The Technological Indian*; also Gabriela Recio, "Engineers, Innovation and Globalization in Mexico: The MIT Education, 1880–1920," unpublished manuscript, 2020; Gronberg, *The Peregrine Profession: Transnational Mobility of Nordic Engineers and Architects, 1880–1930*, 43, 53.

31. The international footprint of Freiberg climbed steadily from about 10 percent in the early nineteenth century to over 50 percent by the 1880s. MIT and CSM climbed from about 10 percent in the mid-1880s to about 40 percent by 1910. Calculated from Solares and Beatty, "Engineering Database."

32. See Köll and Ramnath chapters, this volume, as well as their earlier publications, cited above.

33. See Köll's chapter, this volume.

34. See also Ramnath, *Birth*, 61–62.

35. Solares and Beatty, "Engineering Database."

36. For example, the *Transactions of the AIME* and, independent of the AIME, the aforementioned *Engineering and Mining Journal* out of New York, or the *Mining and Scientific Press* out of San Francisco. See Sinclair, "The Power of Ceremony," and Stephen Tuffnell, "Engineering Gold Rushes: Engineers and the Mechanics of Global Connectivity," in *A Global History of Gold Rushes*, ed. Benjamin Mountford and Stephen Tuffnell (Berkeley: University of California Press, 2018), 229–251.

37. Solares and Beatty, "Engineering Networks, Knowledge, and Technology Diffusion in Mining, 1870–1930," unpublished manuscript.

38. T. A. Rickard, *Interviews with Mining Engineers* (San Francisco: Mining and Scientific Press, 1922), with thanks to Ashlyn Poppe for research assistance.

39. Solares and Beatty, "Engineering Database."

40. Ranestad, "The Mining Sectors in Chile and Norway, ca. 1870–1940," 159.

41. See an example of this exercise in https://engineeringhistoryproject.org/.

42. On national innovation systems, see Nelson, *National Innovation Systems*; on learning, see Gavin Wright, "Can a Nation Learn? American Technology as a Network Phenomenon," in *Learning by Doing in Markets, Firms, and Countries*, ed. Naomi R. Lamoreaux, Daniel M. G. Raff, and Peter Temin (Chicago: University of Chicago Press, 1999), 295–332.

43. Solares and Beatty, "Engineering Database."

44. Rosenberg, *A World Connecting, 1870–1945*, 919.

SELECTED BIBLIOGRAPHY

All bibliographic and archival sources are detailed in the endnotes for each chapter. The introductory chapter to this volume aims, among other objectives, to provide a near-comprehensive survey of the literature on the history of engineering during our time period, focused on both national and global contexts. We hope this might prove useful to those entering the field as well as to seasoned researchers. Finally, we asked each chapter author to suggest five secondary sources that might serve as particularly useful readings on their topic. The collated list follows here:

Andersen, Casper. *British Engineers and Africa, 1875–1914*. London: Routledge, 2015.

Bassett, Ross Knox. *The Technological Indian*. Cambridge, MA: Harvard University Press, 2016.

Beatty, Edward. *Technology and the Search for Progress in Modern Mexico*. Oakland: University of California Press, 2015.

Bertilorenzi, Marco, Anne-Françoise Garçon, and Jean-Philippe Passaqui, eds. *Entre Technique et Gestion. Une Histoire Des Ingénieurs Civils Des Mines*. Paris: Presses des Mines, 2016.

Chakrabarty, Dipesh. *Provincializing Europe: Postcolonial Thought and Historical Difference-New Edition*. Princeton, NJ: Princeton University Press, 2008.

Eckes, Alfred E. *The United States and the Global Struggle for Minerals*. Austin: University of Texas Press, 1979.

Fox, Robert, and Anna Guagnini. *Education, Technology and Industrial Performance in Europe, 1850–1939*. Cambridge, UK: Cambridge University Press, 1993.

Garçon, and Bruno Belhoste, eds. *Les Ingénieurs Des Mines: Cultures, Pouvoirs Pratiques*. Paris: Comité pour l'histoire économique et financière de la France, 2012.

Gispen, Kees. *New Profession, Old Order: Engineers and German Society, 1815–1914*. Cambridge, UK: Cambridge University Press, 1989.

Gooday, Graeme J. N., and Morris F. Low. "Technology Transfer and Cultural Exchange: Western Scientists and Engineers Encounter Late Tokugawa and Meiji Japan." *Osiris* 13, no. 1 (1998): 99–128.

Kerr, Ian J. *Building the Railways of the Raj, 1850–1900*. New Delhi: Oxford University Press, 1995.

Kirby, William C. "Engineering China: Birth of the Developmental State, 1928–1937." In *Becoming Chinese*, edited by Wen-Hsin Yeh, 137–160. Berkeley: University of California Press, 2020.

Klingensmith, Daniel. *"One Valley and a Thousand": Dams, Nationalism, and Development*. Oxford: Oxford University Press, 2007.

Köll, Elisabeth. "The Making of the Civil Engineer in China: Knowledge Transfer, Institution Building, and the Rise of a Profession." In *Knowledge Acts in Modern China: Ideas, Institutions, and Identities*, edited by Robert J. Culp, U. Eddy, and Wen-Hsin Yeh, 148–173. Berkeley: University of California Press, 2016.

Lammel, Annamária, and Esther Katz. *Aires y Lluvias. Antropología Del Clima En México*. Mexico City: Centro de estudios mexicanos y centroamericanos, 2008.

Layton, Edwin T. *The Revolt of the Engineers: Social Responsibility and the American Engineering Profession*. Baltimore, MD: Johns Hopkins University Press, 1986.

Layton, Edwin T., Jr. "American Ideologies of Science and Engineering." *Technology and Culture* 17, no. 4 (1976): 688–701.

Limbaugh, Ronald H. *Tungsten in Peace and War, 1918–1946*. Reno: University of Nevada Press, 2010.

Lucier, Paul. *Scientists and Swindlers: Consulting on Coal and Oil in America, 1820–1890*. Baltimore, MD: Johns Hopkins University Press, 2008.

Marsden, Ben, and Crosbie Smith. *Engineering Empires: A Cultural History of Technology in Nineteenth-Century Britain*. London: Palgrave Macmillan UK, 2005.

McGowan, Abigail. *Crafting the Nation in Colonial India*. New York: Springer, 2009.

McGrath, Sylvia Wallace. *Charles Kenneth Leith: Scientific Adviser*. Madison: University of Wisconsin Press, 1971.

Noble, David F. *America by Design: Science, Technology, and the Rise of Corporate Capitalism*. New York: Oxford University Press, 1979.

Oldenziel, Ruth. *Making Technology Masculine: Men, Women and Modern Machines in America, 1870–1945*. Amsterdam: Amsterdam University Press, 1999.

Parsons, William Barclay. *An American Engineer in China*. Cambridge, UK: Cambridge University Press, 2012.

Pastrana, Manuel E. *El Servicio Meteorológico de La República Mexicana*. Mexico City: Imprenta de la Secretaría de Fomento, 1906.

Phimister, I. R. "The Reconstruction of the Southern Rhodesian Gold Mining Industry, 1903–10." *The Economic History Review* 29, no. 3 (1976): 465–81.

Picon, Antoine. "Engineers and Engineering History: Problems and Perspectives." *History and Technology* 20, no. 4 (December 2004): 421–436.

Pretel, David, and Lino Camprubí. *Technology and Globalisation: Networks of Experts in World History*. London: Springer, 2018.

Priest, Tyler. *Global Gambits: Big Steel and the US Quest for Manganese*. Westport, CT: Praeger, 2003.

Rand, Myron J. *Industrial Research Laboratories of the United States Including Consulting Research Laboratories*. Washington, DC: NAS/NRC, 1950.

Reich, Leonard S. *The Making of American Industrial Research: Science and Business at GE and Bell, 1876–1926*. Studies in Economic History and Policy. Cambridge, UK: Cambridge University Press, 1985.

Seow, Victor. *Carbon Technocracy: Energy Regimes in Modern East Asia*. Studies of the Weatherhead East Asian Institute, Columbia University. Chicago: University of Chicago Press, 2022.

Sheller, Mimi, and John Urry. "The New Mobilities Paradigm." *Environment and Planning. A* 38, no. 2 (2006): 207–226.

Shenhav, Yehouda A. *Manufacturing Rationality: The Engineering Foundations of the Managerial Revolution*. Oxford: Oxford University Press, 2002.

Thépot, André. *Les Ingénieurs Des Mines Du XIXe Siècle, Histoire d'un Corps Technique d'État, Tome I: 1810–1914*. Paris: Éditions Eska, 1998.

Turrell, Robert Vicat, and Jean-Jacques Van Helten. "The Rothschilds, the Exploration Company and Mining Finance." *Business History* 28, no. 2 (1986): 181–205.

Van Onselen, Charles. *The Cowboy Capitalist: John Hays Hammond, the American West, and the Jameson Raid in South Africa*. Charlottesville: University of Virginia Press, 2018.

Vincenti, Walter G. *What Engineers Know and How They Know It*. Baltimore, MD: Johns Hopkins University Press, 1990.

Wang, Hsien-chun. *Western Technology and China's Industrial Development: Steamship Building in Nineteenth-Century China, 1828–1895*. New York: Palgrave/Macmillan, 2022.

Wolfe, Mikael D. "The Climate of Conflict: Politico-Environmental Press Coverage and the Eruption of the Mexican Revolution, 1907–1911." *Hispanic American Historical Review* 99, no. 3 (2019): 467–499.

Wu, Shellen Xiao. *Empires of Coal: Fueling China's Entry into the Modern World Order, 1860–1920*. Stanford, CA: Stanford University Press, 2015.

CONTRIBUTORS

Marco Bertilorenzi is Associate Professor in Economic History at the University of Padova. He is the author of *The International Aluminium Cartel. The Business and Politics of a Cooperative Industrial Institution (1886–1978)* (Routledge, 2015) and coauthor of *Le Traité et la Notice. Relire Fayol avec Fayol* (Presses des Mines, 2020).

Edward (Ted) Beatty is Professor of History and Global Affairs at the University of Notre Dame. He is author of *Institutions and Investment: The Political Basis of Industrialization in Mexico before 1911* (Stanford University Press, 2001) and *Technology and the Search for Progress in Modern Mexico* (University of California Press, 2015), which won the Friedrich Katz prize for best book on Latin America in 2015 from the American Historical Association.

Aurora Gómez-Galvarriato is Professor at the Centro de Estudios Históricos, El Colegio de México, and served as Director of the Mexican National Archive from 2009 to 2013. She is the author of *Industry and Revolution: Social and Economic Change in the Orizaba Valley, Mexico* (Harvard University Press, 2013) and *El Pan Nuestro. Una historia de la tortilla de maíz* (El Colegio de México, 2024). She has published numerous articles and edited volumes in economic, business, and social history.

Mark Hendrickson is Associate Professor of history at the University of California, San Diego. He is a historian of the United States and author of *American Labor and Economic Citizenship: New Capitalism from WWI to the Great Depression* (Cambridge University Press, 2013) and the recipient of a Mellon New Directions Fellowship for his work on American mining engineers and geologists abroad, circa 1880–1930. He is preparing a book on the global history of antimony.

Douglas R. Jones holds a PhD in History from the University of Illinois Urbana-Champaign, where he researches the globally entangled histories of North America and southern Africa. He is currently transforming his dissertation, "Ownership of the

Earth Forever: Engineering Nature, Race, and Labor in North America and Southern Africa, 1860–1922," into a book and teaches American Studies at the University of Toronto.

Elisabeth Köll is Professor of History and Chair of the department at the University of Notre Dame. A historian of modern China, she has published *From Cotton Mill to Business Empire: The Emergence of Regional Enterprises in Modern China* (Harvard University Press, 2003) and *Railroads and the Transformation of China* (Harvard University Press, 2019) and is currently working on a study of informal finance in China.

Aparajith Ramnath is Associate Professor in the School of Arts and Sciences at Ahmedabad University. With a PhD from Imperial College London and an MSc from Oxford, he works on the history of science, technology, and business in India. He is author of *The Birth of an Indian Profession: Engineers, Industry, and the State, 1900–47* (Oxford University Press, 2017) and *Engineering a Nation: The Life and Career of M. Visvesvaraya (1861–1962)* (Penguin/Viking, 2024).

Israel G. Solares is Associate Researcher (Professor) at the Department of Mathematical Modeling of Social Systems, Institute for the Research in Applied Mathematics and Systems, at the Mexican National University (UNAM). He is the author of *Underground Leviathan: Corporate Sovereignty and Mining in the Americas* (University of Nevada Press, 2024).

Stephen Tuffnell is Associate Professor of Modern US history at the University of Oxford and Senior Research Fellow at the Rothermere American Institute. He researches the global histories of American emigration, gold mining, and US imperialism in Africa in the long nineteenth century. He is the author of *Made in Britain: Nation and Emigration in Nineteenth-Century America* (University of California, 2020) and coeditor, with Benjamin Mountford, of *A Global History of Gold Rushes* (University of California, 2018).

Mikael Wolfe is Associate Professor of History at Stanford University. He is author of *Watering the Revolution: An Environmental and Technological History of Agrarian Reform in Mexico* (Duke University Press, 2017), which won the Elinor K. Melville Prize for Latin American Environmental History, as well as articles on Mexican and environmental history. His second book, *Climates of Revolution: How Extreme Weather Shaped the Cuban Development*, will be published by Oxford University Press.

INDEX

Page numbers in italics refer to figures and tables.

Publisher contact:
The MIT Press
Massachusetts Institute of Technology
77 Massachusetts Avenue, Cambridge, MA 02139
mitpress.mit.edu

EU Authorised Representative:
Easy Access System Europe, Mustamäe tee 50,
10621 Tallinn, Estonia
gpsr.requests@easproject.com

Printed by Integrated Books International,
United States of America